© 2018 Massachusetts Institute of Technology

All rights reserved. No part of this book may be reproduced in any form by any electronic or mechanical means (including photocopying, recording, or information storage and retrieval) without permission in writing from the publisher.

This book was set in LaTeX by the authors. Printed and bound in the United States of America.

Library of Congress Cataloging-in-Publication Data

Names: Page, Rex L., author. | Gamboa, Ruben, author.
Title: Essential logic for computer science / Rex Page and Ruben Gamboa.
Description: Cambridge, MA ; London, England : The MIT Press, [2018] | Includes bibliographical references and index.
Identifiers: LCCN 2018016822 | ISBN 9780262039185 (hardcover : alk. paper)
Subjects: LCSH: Computer logic.
Classification: LCC QA76.9.L63 P34 2018 | DDC 005.101/5113—dc23 LC record available at https://lccn.loc.gov/2018016822

10 9 8 7 6 5 4 3 2 1

Essential Logic for Computer Science

Rex Page and Ruben Gamboa

The MIT Press
Cambridge, Massachusetts
London, England

Essential Logic for Computer Science

To Lucy Saarni, artist and philosopher
— Rex

and

To Mona Gamboa and the kids, who enrich life and laugh, sometimes, at logic jokes
— Ruben

Contents

Preface		xiii

I LOGIC AND EQUATIONS

1 Computer Systems: Simple Principles Lead to Complex Behavior   3
- 1.1 Hardware and Software   3
- 1.2 Structure of a Program   5
- 1.3 Deep Blue and Inductive Definitions   9
- Exercises   12

2 Boolean Formulas and Equations   15
- 2.1 Reasoning with Equations   15
- Exercises   18
- 2.2 Boolean Equations   19
- Exercises   26
- 2.3 Boolean Formulas   27
- Exercises   32
- 2.4 Digital Circuits   33
- Exercises   36
- 2.5 Deduction   37
- Exercises   49
- 2.6 Predicates and Quantifiers   51
- Exercises   54
- 2.7 Reasoning with Quantified Predicates   55
- Exercises   62
- 2.8 Boolean Models   63
- Exercises   68

	2.9	More General Models with Predicates and Quantifiers	68
3	**Software Testing and Prefix Notation**		71
		Exercises	76
4	**Mathematical Induction**		79
	4.1	Lists as Mathematical Objects	79
		Exercises	84
	4.2	Mathematical Induction	85
		Exercises	91
	4.3	Defun: Defining Operators in ACL2	92
	4.4	Concatenation, Prefixes, and Suffixes	93
		Exercises	99
5	**Mechanized Logic**		101
	5.1	ACL2 Theorems and Proofs	102
	5.2	Using Books of Proven Theorems	103
		Exercises	104
	5.3	Theorems with Constraints	105
		Exercises	107
	5.4	Helping Mechanized Logic Find Its Way	107
		Exercises	111
	5.5	Proof Automation and Things That Can't Be Done	112
		Exercises	119
II	**COMPUTER ARITHMETIC**		
6	**Binary Numerals**		123
	6.1	Numbers and Numerals	123
		Exercises	128
	6.2	Numbers from Numerals	129
		Exercises	133
	6.3	Binary Numerals	134
		Exercises	135
7	**Adders**		139
	7.1	Adding Numerals	139
		Exercises	140
	7.2	Circuits for Adding One-Bit Binary Numerals	140

	7.3	Circuit for Adding Two-Bit Binary Numerals	143
		Exercises	145
	7.4	Adding w-Bit Binary Numerals	145
		Exercises	148
	7.5	Numerals for Negative Numbers	150
		Exercises	153
8	**Multipliers and Bignum Arithmetic**		**157**
	8.1	Bignum Adder	158
		Exercises	161
	8.2	Shift-and-Add Multiplier	161
		Exercises	165

III ALGORITHMS

9	**Multiplexers and Demultiplexers**		**169**
	9.1	Multiplexer	169
		Exercises	172
	9.2	Demultiplexer	173
		Exercises	175
10	**Sorting**		**177**
	10.1	Insertion-Sort	178
		Exercises	180
	10.2	Order-Preserving Merge	182
		Exercises	183
	10.3	Merge-Sort	184
		Exercises	185
	10.4	Analysis of Sorting Algorithms	186
		10.4.1 Counting Computation Steps	186
		Exercises	188
		10.4.2 Computation Steps in Demultiplex	189
		Exercises	190
		10.4.3 Computation Steps in Merge	191
		Exercises	192
		10.4.4 Computation Steps in Merge-Sort	192
		Exercises	194
		10.4.5 Computation Steps in Insertion-Sort	196

		Exercises	199
11	**Search Trees**	201
	11.1	Finding Things	201
	11.2	The AVL Solution	203
	11.3	Representing Search Trees	206
	11.4	Ordered Search Trees	207
		Exercises	208
	11.5	Balanced Search Trees	208
		Exercises	210
	11.6	Inserting a New Item in a Search Tree	210
		Exercises	212
	11.7	Insertion, Case by Case	212
		Exercises	217
	11.8	Double Rotations	218
		Exercises	222
	11.9	Fast Insertion	223
		Exercises	225
12	**Hash Tables**	227
	12.1	Lists and Arrays	227
	12.2	Hash Operators	229
		Exercises	234
	12.3	Some Applications	236
IV	**COMPUTATION IN PRACTICE**	
13	**Sharding with Facebook**	243
	13.1	The Technical Challenge	243
	13.2	Stopgap Remedies	245
		13.2.1 Caching	245
		13.2.2 Sharding	246
	13.3	The Cassandra Solution	247
	13.4	Summary	249
14	**Parallel Computation with MapReduce**	251
	14.1	Vertical and Horizontal Scaling	251
	14.2	The MapReduce Strategy	252
	14.3	Data Mining with MapReduce	256

		14.4 Summary	261
15		**Generating Art with Computers**	**263**
		15.1 Representing Images in a Computer	263
		15.2 Generating Images Randomly	266
		15.3 Generating Purposeful Images	270
		Index	273

Preface

Computers are logic in action. Literally. Computer components are realizations of formulas in logic, and when activated by Boolean signals, those components compute the value of the formula that they actualize. Software too is an exhibit in logic. A software component is a specification in a formal language with underpinnings in logic, and some software components are, literally, algebraic formulas. Big formulas, but formulas nevertheless.

  Therefore, people studying computer science benefit from studying logic, and most computer science students are exposed to logic in their education. Often this exposure comes in the form of a few lectures and a problem set or two in a discrete math course. Applications of logic that students see usually have more to do with traditional mathematics than with computer science. Even when the course is "discrete math for computer science," the computer science part often has more to do with writing programs to solve problems in traditional mathematics than it has to do with concepts in computer science. We think computer science students will benefit from a substantially more extensive and rigorous exposure to logic and to seeing many applications of logic in their chosen field of study. All the examples in this text arise from issues in computer science.

  This book focuses directly on central themes in computer science. It frames the discussion in terms of logic, and it applies logic to problems in the domain of computer science. Hardware components, software components, testing and verification, and analysis of algorithms are some examples. Instead of illustrating mathematical induction by proving that a formula represents the sum of a sequence of numbers, we begin with an inductive proof of an important property of a software component that concatenates lists, and we proceed to verify properties of many other software and hardware components. It's the same old mathematical induction but presented in the context of topics that interest students of computer science. The logic of induction is at the forefront, unobscured by tricks in numeric algebra that many exercises on induction require in discussions grounded in topics of particular interest to mathematicians.

  We hope that readers will be inclined to devote substantial effort, on the order of what it takes to absorb a few dozen fifty-minute lectures at the college level, to understand some important problems in computer science and to pursue solutions to many of those problems through formal reasoning. Formalism is a watchword in this presentation, even to the point of using the mechanized logic of a partially automated proof engine, ACL2, that checks proofs to the last detail and can sometimes bridge, on its own, gaps that traditional mathematical proofs, even rigorous ones, leave open.

The text employs three formal notations: traditional algebraic formulas of propositional and predicate logic (and numeric algebra, occasionally), digital circuit diagrams, and ACL2, which is syntactically like Lisp, the programming language, but it is embedded in a mechanized logic that assists in producing formal proofs in first-order logic. ACL2 serves as a mathematical notation, and all of the material can be understood in terms of traditional, paper-and-pencil reasoning without putting formal models through their paces on a computer system. For readers who want to see formalization in action, the text presents examples using Proof Pad, a lightweight ACL2 environment. ACL2 experts use emacs or the ACL2 Sedan as their interface, and readers of this text can use those tools if they prefer, but the text illustrates the process in the Proof Pad framework, which in our experience puts a low burden on novices. In any case, Proof Pad will be adequate to support study with this text.

We chose ACL2 as the proof engine for this work because in our judgment it provides a more accessible introduction to mechanized formalism than any other available tool. We do not anticipate that any reader will become an accomplished ACL2 user, much less an ACL2 expert. We bring ACL2 into the discussion to show how logic, including mechanized logic, can benefit practicing software and hardware engineers. Readers who want to realize those benefits in large-scale projects will need to learn a lot more about ACL2 or another mechanized logic than they will glean from this introduction. In former years, we presented logic in the classroom without a mechanized logic, but we found through experience that most students are more comfortable and motivated when formal methods are backed up by tools that check proofs and help push through details.

Logic is the central topic of the text but not the only topic. Readers interested in the broad outlines of computer science will find material useful in that pursuit. The text can provide a basis for a serious introduction to computer science concepts, both for computer science students and for students in other fields who want to know what computer science is about. Earlier versions of the text have been used many times by the authors and other instructors as the primary text in two types of courses: logic for computer science and introduction to computer science for both computer science students and students in other disciplines. It has also been used as a supplementary text in discrete math courses for computer science students. The text has served well in all three realms.

There are no prerequisites beyond college prep, high school math. Even less, really. High school algebra is helpful, but no geometry, trigonometry, or calculus is needed. Programming experience is not a prerequisite either, and the equation-based approach that informs this presentation can help level the playing field between people with programming experience and those without it. Students whose goal is to certify that they already know the material are surprised to find that they don't, whereas those who arrive with less background tend to invest the necessary effort from the outset.

The required learning is far from easy. Successful students will have to do a lot of hard thinking to work their way through a few dozen exercises, and they will surely need to do a few dozen exercises to grasp the concepts. Reading alone won't be enough. Exercises in the text (over 180 in all) afford students with opportunities for problem solving. Fortunately, the work pays off, both in terms of immediate satisfaction and in the long term, if the testimonies of former students are a reliable measure. We hope readers will take some pleasure in working their way through the book, and that they will find what they learn edifying as they go on to other projects.

Prerequisite Knowledge and Chapter Dependencies. Equations provide a foundation for rigorous, formal methods throughout the presentation. Readers will need to understand equations at a level typically acquired in high school algebra, but they will need no other prerequisite knowledge. Instructors, on the other hand, will find a solid footing in equation-based programming advantageous. Chapters on the practice of computation are more descriptive than rigorous. These topics can be woven into the sequence at any point and can usefully slow the pace of introduction of challenging concepts in equations and reasoning. The following diagram elucidates some possible paths through the material.

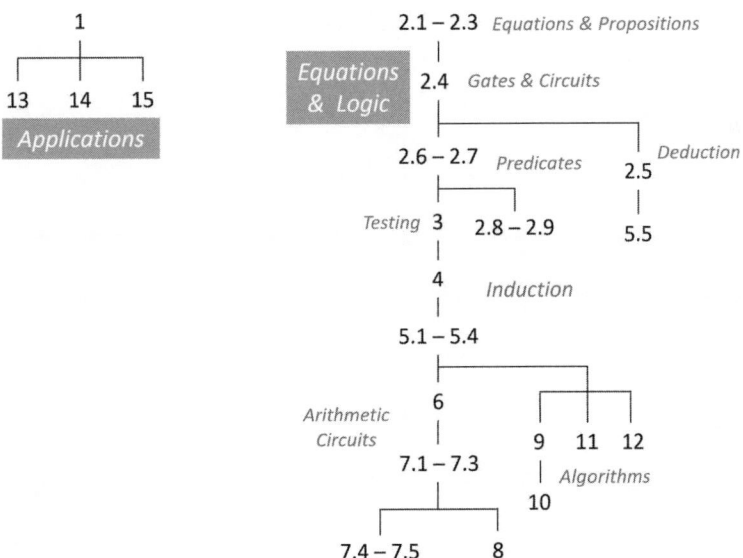

Acknowledgments. The authors would like to thank Caleb Eggensperger for developing the Proof Pad environment that has eased many students through early experiences with mechanized logic. Carl Eastland, Dale Vaillancourt, and Matthias Felleisen built an ACL2 environment that one of the authors relied on for many course offerings and which included DoubleCheck, the predicate-based, automated testing facility based on ideas in QuickCheck, the granddaddy of such tools, invented by John Hughes and Koen Claessen, which was later incorporated into Proof Pad. We thank them for their pioneering work. Qi Cheng used prototype versions of the text in applied logic courses and suggested improvements that made it a better book. The authors also want to credit the students, numbering more than a thousand, who applied themselves to early versions of the text and provided their feedback. Thank you, every one.

Rex Page and Ruben Gamboa
January, 2018

I Logic and Equations

1 Computer Systems: Simple Principles, Complex Behavior

1.1 Hardware and Software

Computer systems, both hardware and software, are some of the most complicated artifacts that people have ever created. Yet, computer systems are applications of principles of logic that philosophers have been developing for thousands of years. The logic that arose from this long effort, together with some engineering artifacts that emerged from its framework, will be the primary topics of this book.

The hardware component of a computer system is made up of physical pieces of equipment such as monitors, keyboards, printers, web cameras, and USB drives. Hardware also includes components inside the computer enclosure, such as chips, cables, hard drives, and circuit boards. The properties of hardware devices are largely fixed when the system is constructed. For example, the capacity of a hard drive is determined when it is manufactured. A three-terabyte drive stores three terabytes of data, not more. A particular cable may be constructed to carry twenty-five different signals simultaneously. Need to send twenty-six signals at the same time? Get a different cable, or more of them.

The hardware that makes up a computer system is not much different than the electronics inside a television set or a navigational device, but computer hardware can do something that most consumer electronic devices cannot do. It can respond to the software component of the computer system to exhibit an unlimited range of behaviors. The software component is a collection of computer programs, and programs can be added to the collection or removed from it at any time. So, a computer system is a multipurpose device in the extreme, capable of doing things that its designers never thought about.

You have no doubt heard of various computer chips. Maybe your laptop is powered by an Intel i7 or an Apple A10. Software for these chips consists of a list of instructions that, when carried out one by one, in sequence, modify the state of the computer system to make it perform the function the software designer had in mind. However, software designers rarely put together programs using the instruction set of the computer chip that performs the computation. Instead, they use a programming language in which the program tends to be shorter and in some ways easier to understand than it would be if it were a list of chip instructions. Another piece of software translates the program into instructions that the chip is wired to carry out.

There are hundreds of programming languages that software designers can choose from. In most cases, software expressed in these languages consists of a list of instructions to be performed, one by one, in the manner of chip instructions, but each instruction in the programming language gives rise to a sequence of at least a few and sometimes many commands in the instruction set of the underlying computer chip.

However, software as a list of instructions is not the only alternative. Another alternative uses drawings of system configurations to build software. For example, scientists use a programming language called LabView to design software to control laboratory equipment. Engineers use it to design digital systems, and college students use it in projects to design electronic devices. Computer programs in LabView look like engineering drawings, not like lists of instructions. Another visual language, Scratch, created by the MIT Media Lab, is used by children to construct software for games and other educational projects. LabView and Scratch programs don't look anything like chip instructions, but they have the same capability for specifying computations.

A related alternative is to specify software as a collection of equations in which the left-hand side of the equation names a computation to be carried out and the right-hand side reduces the computation to more basic components, each of which produces part of the result. The partial results are then combined to deliver the whole. An advantage of the equation-based approach is that a form of reasoning similar to that used in ordinary, high school algebra can be used to derive guarantees that the software produces results consistent with design requirements. Another advantage is that the equations themselves are often inspired directly by design requirements, so that if the software designer can think of tests that the software should pass, the software itself can emerge from specifications of the tests.

Instruction lists, visual programming languages, and equation-based software are fundamentally different ways to describe computer programs, but they are fully equivalent in terms of capabilities. Since the primary focus of this text is on logical reasoning about hardware and software components of computer systems, we will use the equation-based approach, but specification of the hardware and software artifacts that we study is not limited to that approach.

It is software that gives a computer system its power and flexibility. An iPhone, for example, has a screen that can display millions of pixels, but it is the software that determines whether the iPhone displays an album cover or a weather update. Software makes hardware useful by extending its range of behavior. Although iPhone audio hardware may be able to produce only a single tone at a time, the software controlling it can direct it to produce a sequence of tones that sound like Beethoven's Ninth.

1.2 Structure of a Program

> **Box 1.1**
> Models of Computation
>
> Logicians and mathematicians have been studying models of computation since before computers were invented. This was done in part to answer a deep mathematical question: What parts of mathematics can, in principle, be fully automated? Is it possible to build a machine that can prove all mathematical truths?
>
> Many different models of computation were developed to study this question: Turing machines, lambda calculus, partial recursive functions, unrestricted grammars, Post production rules, and random-access machines, to name a few. Historically, Turing machines have provided the canonical foundation for computation, but the random-access model conforms more closely to modern computers, and the lambda calculus model is used in both computer science theory and practice. The equation-based model of computation that is the focus of this book falls into the lambda calculus bailiwick.
>
> It is truly remarkable that all of these models of computation are equivalent. That is, any computation that can be described in one of the models can also be described in any of the others. A conjecture known as the Church–Turing thesis is based on this equivalence. It says that all realizable computations can be carried out by a Turing machine or by any equivalent model of computation, such as the lambda calculus model. Much of computer science theory revolves around this hypothesis.
>
> Another remarkable fact is that some problems cannot be solved by computer hardware or software. Alan Turing, one of the first computer scientists, was the first to describe uncomputable problems, and soon afterwards the logician Kurt Gödel showed that all mathematical systems that are at least as powerful as arithmetic are incomplete. No formal system can be used to prove all mathematical truths. The theorems of Turing and Gödel prove that it is not possible to build a machine that can verify all mathematical truths, even in principle.

You can think of the hardware as the parts of a computer that you can see and the software as information that tells the computer what to do. However, the distinction between hardware and software is not as clear cut as this suggests. Many hardware components encode software directly and control other pieces of the system, and many techniques used in the design of hardware were originated to build software. Major elements of hardware designs look like software and can use the same language and notation. A major theme of this book is that both hardware and software are realizations of formal logic. In this sense, computer systems are logic in action.

1.2 Structure of a Program

The distinction between hardware and software leaves many questions to the imagination:

1. How can software control hardware?
   Example: Instruct an audio device to emit a sound.

2. How can software detect the status of hardware?
   Example: Determine whether or not a switch is pressed.
3. What kinds of instructions can software give to hardware?
   Examples: Add numbers. Replace one formula by another. Select a formula.

A model of computation (box 1.1, page 5) is a way for software to control hardware, so there are as many answers to the first of these questions as there are models of computation, and there are a lot of models. Logicians have made sure of that. Since there are many equivalent models, the requirements of a project can guide the choice of models. Generally, a programming language based on any model of computation has intrinsic arithmetic and logic operators (adders, multipliers, inverters, and so on) and provides the ability to combine intrinsic operators and operators defined elsewhere in the software to define new operators.

Box 1.2
Operators, Operands, Functions, Parameters, Arguments

We will use the terms "operator" and "function" interchangeably. Some treatments use these terms to mean different things, but we find it more convenient in this presentation to think of them as the same thing. So, when we say function we mean operator and vice versa. What we're talking about when we use either term is a transformation that delivers results when supplied with input. We refer to the results delivered by an operator as its *value*, and we refer to the input supplied to the operator as its *operands*. Sometimes we use the term "argument" or "parameter" instead of "operand," usually in connection with the term "function," but the terms carry the same meaning in any case.

To summarize, an operator (aka, function) is supplied with operands (aka, parameters or arguments) and delivers a value. A formula like $x + y$ denotes the value delivered by the addition operator when supplied with the operands x and y. More generally, an operator f, when supplied with operands x and y, produces a value, usually denoted algebraically as $f(x, y)$. If f were the arithmetic addition operator (+), then $f(x, y)$ would stand for the value $x + y$, $f(2, 2)$ would denote 4, $f(3, 7)$ would stand for 10, and $f(2x, 5)$ would mean $2x + 5$.

Once we take the familiar operators and the ability to define new ones as basic elements of a computation model, we can talk about what it is possible for software to do. Software can affect hardware by the values that an operator delivers. For example, a computer program can tell an iPhone what to display on its screen by delivering a matrix of pixels. Each entry in the matrix can be a number that represents a color: 16,777,215 for red or 65,280 for green, for example. Similarly, the hardware can inform the software of the status of a component by invoking an operator and supplying the status as an operand. For example, touching the screen of an iPhone might trigger a software operator and supply the operator with the coordinates of the pixel selected by the touch. The operator could then control the device's response. Other gestures, such as tapping or scrolling, would trigger different operators.

To be more specific, let's consider a program to control a device that plays the game of rock-paper-scissors against a human opponent. The device has three buttons, allowing the human player to select rock, paper, or scissors. It also has a display unit. When the human player presses a button, the display unit shows the device's choice (rock, paper, or scissors) and shows the winner of the round. The program will compute the device's choice by invoking an operator called emily.[1] This operator will deliver rock or paper or scissors as its value. We could use numbers like 0, 1, and 2 as shorthand for these values, but most programming languages include many kinds of information, not just numbers, so we are going to stick with the longer names to make it easier for us to keep track of what things mean. Another part of the software will compare the human player's selection with that of the device and determine the winner of the round.

To make the game fair, the program will use separate operators for the device's choice and for examining the choice of the human player to determine the winner. That way, the device won't know the human player's choice before making its own. The operator emily will deliver the device's choice for a round of the game. It needs some kind of information because an operator with no operands always delivers the same value, and that would make for a very boring game. Any player will have seen the choices of the other player for previous rounds, so it seems fair to use some of that information in making a choice for the next round. The player for the device, which is the operator emily, will receive as its operand the choice that the human player selected in the immediately previous round. The operator emily (and this is the heart of the device's "prowess" at the game, if prowess is the right term, and it probably isn't) will, in turn, choose a play according to the following scheme. The operand u is the human player's choice for the previous round.

$$\text{emily}(u) = \begin{cases} \text{rock} & \text{if } u = \text{scissors} \\ \text{paper} & \text{if } u = \text{rock} \\ \text{scissors} & \text{otherwise} \end{cases}$$

The other operator, which we will call score, has two operands: (d, u). The first operand, d, indicates the device's choice for the round. It will be rock, paper, or scissors. The second operand, u, indicates the human player's choice for the round.

The score operator delivers a value with two components: (w, u). The first component, w, indicates the winner of the round. It will be device, human, or none (in case both players make the same choice in a round). The second component, u, indicates the human player's choice for the round (rock, paper, or scissors). The score operator reports

---

[1] This operator is named after a daughter of one of the authors who, when she was young, played the rock-paper-scissors game just like the program described here.

the human player's choice so that in the next round the software can tell the operator emily the choice that the human player made in the previous round, and emily can use that information in deciding on a choice for the next round.

The following table specifies the two-component value, (w, u), that the score operator delivers, given the two operands indicating the player's choices, (d, u):

$$\text{score}(d, u) = \begin{cases} (\text{none}, u) & \text{if } d = u \\ (\text{device}, u) & \text{if } (d, u) = (\text{rock}, \text{scissors}) \\ (\text{human}, u) & \text{if } (d, u) = (\text{rock}, \text{paper}) \\ (\text{device}, u) & \text{if } (d, u) = (\text{paper}, \text{rock}) \\ (\text{human}, u) & \text{if } (d, u) = (\text{paper}, \text{scissors}) \\ (\text{device}, u) & \text{if } (d, u) = (\text{scissors}, \text{paper}) \\ (\text{human}, u) & \text{if } (d, u) = (\text{scissors}, \text{rock}) \end{cases}$$

More sophisticated software could make the device play a better game of rock-paper-scissors. For example, the score operator might keep track of the number of times that the human player has selected scissors or rock or paper in previous rounds or it might keep track of all three. The software controlling the device's choice of plays could then take more information into account and play a better game. That is, the operator emily could be upgraded. Such is the flexibility of software.

The way we have described the operators emily and score illustrates an important point about our computational model. A program in our model consists of a collection of equations defining mathematical functions that start each computation from scratch based only on their input. They cannot "remember" anything from previous computations. This will come as a surprise to programmers accustomed to other computational models such as those on which programming languages like Java or C++ are based. In those models, programs use variables to record values and can update recorded values later in the computation. In our equation-based model, programs consist of collections of mathematical functions (operators) that base their results entirely on their operands. Operators do not make use of recorded values beyond the data supplied in operands and unchanging elements in the equations as they stand. This makes it possible to understand equation-based programs in terms of classical logic and traditional algebraic formulas.

To get back to our discussion of the game, after the first round the display component of the device would show the first element w_1 of the pair (w_1, u_1) delivered by the formula score(emily(none), b_1), where b_1 stands for rock, paper, or scissors, depending on which button the human player pressed. For the next round, the outcome would be (w_2, u_2) = score(emily(u_1), b_2), where b_2 is the button pressed by the human player for the new round, and the display would show the value of w_2. A game of five rounds would correspond to a sequence of equations, each showing the outcome of a particular round. The display hardware would show the winner of each round, and the software would supply the button press of the previous round, which is delivered by the score operator as the second element of a pair to the emily operator for the next round.

$$(w_1, u_1) = \text{score}(\text{emily}(\text{none}), b_1)$$
$$(w_2, u_2) = \text{score}(\text{emily}(u_1), b_2)$$
$$(w_3, u_3) = \text{score}(\text{emily}(u_2), b_3)$$
$$(w_4, u_4) = \text{score}(\text{emily}(u_3), b_4)$$
$$(w_5, u_5) = \text{score}(\text{emily}(u_4), b_5)$$

A more complicated version of the software could define an operator *play* that would generate a sequence of *n* plays, where *n* stands for an operand supplied to *play*. In such a scenario, there would need to be some way to supply an operand *n* to the operator *play*, where *n* might be a fixed number, like five or ten, or it might be a number selected by having the human player press the buttons in a special sequence. The point is that the hardware provides a fixed set of components such as displays and buttons, and the software specifies a way to use those components.

1.3 Deep Blue and Inductive Definitions

The equation-based model of computation has the same capabilities for specifying computations as any other known model. Knowing the great diversity of things that computers can do, it seems reasonable to expect them to be much more complicated than, say, the rock-paper-scissors device, but in fundamental terms, they have the same basic structure. Complex behavior from modest beginnings, a bargain if there ever was one.

Consider Deep Blue, the computer that beat world champion Gary Kasparov in a chess match on May 11, 1997. It was the first time a computer had performed so well in a game that challenges the mental capacities of human players. The Deep Blue chess-playing software can be specified as an operator with an operand that is an 8×8 matrix showing the positions of the pieces on the board after a move by the human player (Kasparov, in the 1997 game). The operator would deliver a result showing either the positions on the board after its move or a special white-flag token used to resign the game.

In principle, a chess-playing computer program can be an operator in software with the following characteristics. Given a matrix describing the board at a particular point in a game, the operator determines all possible legal moves. If there are no legal moves, the operator delivers the white flag, signaling that Deep Blue has resigned. If there is a legal move that results in checkmate, the operator delivers the 8×8 matrix specifying the new board position after that move. Otherwise, for each legal move, the operator considers each of the possible moves in response by the human opponent. Each one of those moves leads to a new board that can be examined by the chess-playing operator to choose a move based on a calculation that estimates the value of board positions.

An operator that follows this strategy can be defined in a circular way, circular in the sense that the operator invokes itself with new operands.[2]

Circular definitions of this kind are common in mathematics. We call them inductive definitions.[3] The trick that makes inductive definitions useful in mathematics is that at the point of circularity, the invocation of the operator being defined has operands that are closer to those in a noncircular part of the definition than to the original operands.

Inductive definitions will play a central role throughout the book and will be discussed in great detail, but we want to provide an introduction of sorts at this point to get the ball rolling. A natural number is a whole number that is not negative: $0, 1, 2, 3, \ldots$. The sum of the five reciprocals of the natural numbers 1 through 5 can be written as an algebraic formula, as in the following equation:

$$h5 = \frac{1}{1} + \frac{1}{2} + \frac{1}{3} + \frac{1}{4} + \frac{1}{5}$$

This is a noninductive definition of $h5$, and $h5$ has no operands, so it simply stands for a number, namely, $\frac{137}{60}$, which is about 2.3.

We could, of course, have written a formula for the sum of the first ten reciprocals or a sum with any number of terms. Sums of this form comprise a mathematical entity known as the harmonic series. Specifically, they are the partial sums of the series, and we can define an operator h that computes such a sum with any specified number n of terms.

$$h(n) = \frac{1}{1} + \frac{1}{2} + \frac{1}{3} + \ldots \frac{1}{n}$$

As a step in that direction, observe that for any natural number n, the sum $h(n+1)$ of the first $n+1$ reciprocals is the same as the sum of the first n reciprocals, that is, $h(n)$, plus $\frac{1}{n+1}$. Suppose we specify that $h(0)$ stands for zero.[4] Then, the following equations express the specification for $h(0)$ and the observation about the relationship between $h(n)$ and $h(n+1)$:

$$h(n+1) = h(n) + \frac{1}{n+1} \quad \{h1\}$$
$$h(0) = 0 \quad \{h0\}$$

It will be a common practice in this text to attach names to equations to make it easy to discuss them. In this case, we have used the names $\{h1\}$ and $\{h0\}$. The equation $\{h1\}$

---

[2] To *invoke* an operator is to apply it to its operands to make a computation. An *invocation* is a formula that invokes an operator.

[3] Operators with inductive definitions are sometimes called "recursive" functions. We try to avoid that term because it is often associated with specialized ways to carry out the computation, but we may slip up from time to time and call them recursive functions. Besides, you should know the term so you'll know what people are talking about if they mention it.

[4] Our informal definition is that $h(n)$ stands for the sum of the first n reciprocals. When n is zero, there would be no terms in the sum, so the standard convention that the sum of an empty set of numbers is zero is consistent with the equation $h(0) = 0$.

1.3 Deep Blue and Inductive Definitions

is circular because both sides of the equation refer to a value delivered by the operator h. However, the operand n in the circular reference (that is, the formula $h(n)$ on the right-hand side of the {h1} equation) is closer to zero than the operand $(n + 1)$ on the left-hand side of the equation. Furthermore, in equation {h0}, the operand on the left-hand side is zero and the equation is not circular. So, the operand on the right-hand side of the circular equation, {h1}, is closer to the operand in the part of the definition that is not circular, {h0}, than it is to the operand on the left-hand side of the circular equation.

It turns out, and we will study this in great detail later, that a collection of circular equations with these two characteristics (a reduced operand on the right-hand side in circular equations and a specific operand on the left-hand side in noncircular equations) provides a useful definition of an operator. Therefore, we can say that equations {h1} and {h0} define the operator h. The equations comprise an inductive definition.

The definition is mathematically rigorous and also computational. To see how this works, observe that, according to equation {h1}, $h(5) = h(4) + \frac{1}{5}$. Furthermore, {h1} also says that $h(4) = h(3) + \frac{1}{4}$. Combining these equations algebraically leads to the equation $h(5) = h(3) + \frac{1}{4} + \frac{1}{5}$. Continuing the analysis in this way, using equation {h1} at each step, produces the following sequence of equations:

$$h(5) = h(4) + \frac{1}{5}$$
$$h(5) = h(3) + \frac{1}{4} + \frac{1}{5}$$
$$h(5) = h(2) + \frac{1}{3} + \frac{1}{4} + \frac{1}{5}$$
$$h(5) = h(1) + \frac{1}{2} + \frac{1}{3} + \frac{1}{4} + \frac{1}{5}$$
$$h(5) = h(0) + \frac{1}{1} + \frac{1}{2} + \frac{1}{3} + \frac{1}{4} + \frac{1}{5}$$

Equation {h0} says $h(0) = 0$, so the last equation is equivalent to the following one:

$$h(5) = 0 + \frac{1}{1} + \frac{1}{2} + \frac{1}{3} + \frac{1}{4} + \frac{1}{5}$$

This equation needs no analysis, just a little arithmetic to compute $h(5) = \frac{137}{60}$. In this way, the equations {h1} and {h0} provide not just a definition of the operator h but also a scheme for computing $h(n)$ for any natural number n. Furthermore, other properties of the operator h, such as the fact that when n is a natural number, the formula $h(n + 1)$ must in all cases stand for a strictly positive number and that its value must be a ratio of whole numbers, can be derived from {h1} and {h0}. The fact that for any number x, no matter how large, there is a natural number n such that $h(n) > x$ is another property of the harmonic series that can be derived from equations {h0} and {h1}. The fact that $h(n)$ grows larger at about the same rate as $log(n)$ is also a consequence of {h1} and {h0}. All the properties of the harmonic series are a consequence of those two equations. Simple beginnings. Big results.

It turns out that every computable function has a definition consisting of some equations specifying properties of the function and that all the properties of the function derive from those equations. This notion provides the basis for most of the reasoning discussed in this text. The implications of inductive equations are broader than at first they seem.

To get back to the chess-playing computer, the problem with our naïve approach is that there are too many moves to consider. Although the function can, *in principle*, select the best move given a particular arrangement of pieces on the chess board, *in practice* the computation would take too much time. The sun in our solar system would be long dead before the computer could decide on its first move.

Deep Blue did play in more or less this way, but it didn't look at board positions all the way to the end of the game. Most of the time, it looked six to eight moves ahead. Even that is a very big computation but feasible for Deep Blue because it had thousands of processing units and could consider many moves at the same time. That gave it the ability to analyze about 200 million board positions per second. The combination of massive computational power and limited look-ahead made it possible for Deep Blue to outplay the most accomplished human chess player of the time. Now, two decades after the big match, chess-playing software on laptops gives accomplished players a run for their money. They use chess-playing software as part of their practice regimen. They learn things from the way the computer plays. Simple beginnings. Astonishing outcomes.

Deep Blue is a combination of a complicated computer and a computer program with a long, complex definition. The Deep Blue software conformed to a conventional computation model, not the equation-based model we used to describe it, but the program could have been designed using equations in a manner similar to the way equations provided a definition of the operator h but a lot more of them, of course. Not so simple, but founded on a few simple principles.

Exercises

1. Can you use equations {h0} and {h1} to compute $h(-1)$? Or $h(\frac{1}{2})$? Why not?

2. Define an operator *factor(k, n)*: true if k is a factor of n, false otherwise.
   Note: Natural numbers only. And k is a factor n if there is a number m such that $n = km$.
   Note: $mod(k, n)$ = remainder in $k \div n$: $mod(17, 5) = 2$, $mod(9, 2) = 1$, $mod(12, 4) = 0$.
   Hint: One equation is enough. Use *mod*.

3. Define an operator $lf(n)$ = largest factor of n other than n: $lf(30) = 15$, $lf(15) = 5$.
   Note: Use $lft(n, k)$ = largest factor of n up to k: $lft(30, 7) = 6$, $lft(120, 10) = 10$.
   Note: Use $\lfloor n \div k \rfloor = n \div k$ rounded down to an integer.

4. Define an operator to compute $p(n)$, which is true if n is a prime number and false otherwise. Refer to the operator lf from exercise 3 in your definition.
   Note: A natural number $p \geq 2$ is a prime number if it has no factors other than p and 1.

Exercises

5. Define an operator to compute $rp(n)$, the sum of the reciprocals of all the prime numbers that are less than or equal to the natural number n. Use the equations that define the operator h (page 10) as a model for your definition of $rp(n)$.[5]

    Hint: Use three equations, one for numbers n that are prime numbers, another for nonzero numbers that are not prime, and a third for the case when $n = 0$. It will be helpful to refer to the operator p from exercise 4.

---

[5] The harmonic series, $h(n)$, grows without bound as n becomes large but very slowly, at about the same rate of growth as $log(n)$. The number $rp(n)$ grows even more slowly, of course, but it too grows without bound. However, the sum of the squares of the reciprocals is bounded by $\pi^2/6$, which is about 1.64. Leonhard Euler proved these facts over 200 years ago, during a kind of Cambrian explosion of mathematics initiated a hundred years earlier by Isaac Newton and Gottfried Leibniz with their invention of the infinitesimal calculus. In their development of calculus, Newton and Leibniz conjured up mathematical entities known as infinitesimal numbers to justify the new mathematics that they introduced in the 1600s, but it wasn't until the 1960s that Abraham Robinson invented a formal logic for reasoning about infinitesimal numbers. The proof engine ACL2r, which is an extension of ACL2, the mechanized logic used extensively in this book, employs Robinson's nonstandard analysis to support partially automated, formal reasoning about functions with numeric operands of infinite precision.

2 Boolean Formulas and Equations

2.1 Reasoning with Equations

Symbolic logic, like other parts of mathematics, starts from a small collection of axioms and employs rules of inference to find additional propositions consistent with those axioms. This chapter will define a grammar of logic formulas, postulate a few equations stating that certain formulas carry the same meaning as others, and derive new equations using substitution of equals for equals as the rule of inference.

You will probably find this familiar from your experience with numeric algebra, but the discourse here will attend carefully to details, and this formality may extend beyond what you are accustomed to. What it buys is mechanization. That is, logic formulas and reasoning about them will amount to mechanized computation, and this will make it possible for computers to check that our reasoning follows all the rules, without exception. This justifies more confidence in conclusions than would otherwise be possible.

We will be doing all of this in the domain of symbolic logic, which includes operations like "logical-or" and "logical negation," rather than arithmetic operations, such as addition and multiplication. We will be doing Boolean algebra rather than numeric algebra, but the underlying rule of inference, namely, substitution of equals for equals, applies equally well to both Boolean and numeric formulas. To illustrate the level of formality that we are shooting for, let's see how it works with a problem in the familiar domain of numeric algebra.

You are surely familiar with the equation $(-1) \times (-1) = 1$, but you may not know that it is a consequence of some basic facts about arithmetic. That is, the fact that multiplying two negative numbers produces a positive one is not independent of other facts about numbers, and it is not an arbitrary decision either. Instead, it is an inference one can draw from an acceptance of other familiar equations. We will derive the equation $(-1) \times (-1) = 1$ from equations that you have accepted without question for a long time.

The equations in figure 2.1 (page 16) express some standard rules of numeric computation. In those equations, the letters stand in place of numbers. They can also stand in place of other formulas. So, the variable x stands for a grammatically correct formula, which

$$x + 0 = x \qquad \{+ \text{ identity}\}$$
$$(-x) + x = 0 \qquad \{+ \text{ complement}\}$$
$$x \times 1 = x \qquad \{\times \text{ identity}\}$$
$$x \times 0 = 0 \qquad \{\times \text{ null}\}$$
$$x + y = y + x \qquad \{+ \text{ commutative}\}$$
$$x \times y = y \times x \qquad \{\times \text{ commutative}\}$$
$$x + (y + z) = (x + y) + z \qquad \{+ \text{ associative}\}$$
$$x \times (y \times z) = (x \times y) \times z \qquad \{\times \text{ associative}\}$$
$$x \times (y + z) = (x \times y) + (x \times z) \qquad \{\text{distributive law}\}$$

Figure 2.1
Equations of numeric algebra.

could be something simple, such as 2, or it could be a more complicated formula, such as $3 \times (y + 1)$. (Some treatments call x a metavariable in this context.)

We refer to letters used in this way as variables, even though within a particular equation they stand for a fixed number or a particular formula. The formula associated with a variable, although unspecified, is the same for every occurrence of the variable in the equation. If x stands for $3 \times (y + 1)$ at one point in the equation, then everywhere else x occurs in the equation, it stands for that same formula, $3 \times (y + 1)$. This is the usual custom in algebra.

Box 2.1
Hold on to Your Seat

Mathematical formulas communicate information using a formal grammar imbued with specific meanings. The grammar determines which phrases are well formed and which are not. For example, $x + 3 \times (y + z)$ conforms to a grammar of numeric formulas. It is grammatically correct, and it stands for a particular and carefully prescribed calculation. The nonformula $x + 3 \times (y+) \times z$ does not conform to that grammar and therefore carries no meaning. Reasoning about computer hardware and software calls for a high degree of formality to ensure a level of consistency that is essential to its usefulness.

Rigorous formality takes many readers by surprise. It takes some getting used to. Things may seem overly simple in the beginning. Then, suddenly, you may find yourself thrashing around in deep water. Take a deep breath and slowly work through the material. It provides a basis for everything to follow. The ideas and methods call for careful study and frequent review. When things start to go off track, slow down, back up a little, and try again. Gradually, the pieces will fall into place, but you can expect to run into some bumps. *Hold on to your seat.*

If we accept the equations of figure 2.1, we can apply one of them to transform the formula $(-1) \times (-1)$ to a new formula that stands for the same number. Then, we can apply

2.1 Reasoning with Equations

another equation to transform that formula to a new one, and so on. We look for a way to apply the accepted equations one by one, so that in the end we arrive at the formula 1. At every step, we know that the new formula stands for the same number as the old one, so in the end we know that $(-1) \times (-1) = 1$.

Figure 2.2 (page 18) displays this sort of equation-by-equation derivation of the formula 1 from the formula $(-1) \times (-1)$. To understand figure 2.2, you must remember that each variable can denote any grammatically correct formula. For example, in the equation $x + 0 = x$, which is called the {+ identity} equation, the variable x could stand for a number, such as 3, or it could stand for a more complicated formula, such as $(1 + 3)$. It could even stand for a formula with variables in it, such as $(a + (b \times c))$ or $(((-1) \times (x + 3)) + (x + y))$.

Another crucial point is that each step cites exactly one equation from figure 2.1 (page 16) to justify the transformation from the formula in the previous step. We are so accustomed to calculating with numeric formulas that we often combine many basic steps into one. When we reason formally, we are careful to do just one step at a time. We justify each step by citing an equation from a list of known equations. In our proof of $(-1) \times (-1) = 1$, we will justify steps by citing equations from figure 2.1 and from no other source. We will not skip steps. Keep that in mind as you go through the proof, line by line.

The first step in the proof (figure 2.2, page 18) uses a version of the {+ identity} equation in which the variable x stands for the formula $((-1) \times (-1))$. In this context, the {+ identity} equation leads to the new formula $((-1) \times (-1)) + 0$.

The second step in the proof reads the {+ complement} equation backwards (equations go both ways) and in a form where the variable x stands for the number 1. When x is 1, the {+ complement} equation is $(-1) + 1 = 0$. Reading the equation backwards, we can substitute $((-1) + 1)$ for 0, and that leads to the formula $((-1) \times (-1)) + ((-1) + 1)$. So, we know now that $(-1) \times (-1) = ((-1) \times (-1)) + ((-1) + 1)$.

Doesn't really seem like progress, does it? But, we press on anyway, one step at a time. The transformations, step by step, finally confirm that the two formulas $(-1) \times (-1)$ and 1 represent the same number.

Pay particular attention to the last three lines of the proof. Most people tend to jump from the formula $0 + 1$ to the formula 1 in one step. That jump requires knowing the equation $0 + 1 = 1$. However, that equation is not among those listed in figure 2.1. We want to do the proof without citing any equations other than those in figure 2.1, so we need two steps to get from $(0 + 1)$ to 1, and those are the last two steps in the proof.

One of the things we hope you will glean from this derivation is that the equation $(-1) \times (-1) = 1$ does not depend on vague, philosophical assertions like "two negatives make a positive." Instead, the equation $(-1) \times (-1) = 1$ is a consequence of some basic arithmetic

$$
\begin{aligned}
& (-1) \times (-1) \\
= \ & ((-1) \times (-1)) + 0 & & \{+ \text{ identity}\} \\
= \ & ((-1) \times (-1)) + ((-1) + 1) & & \{+ \text{ complement}\} \\
= \ & (((-1) \times (-1)) + (-1)) + 1 & & \{+ \text{ associative}\} \\
= \ & (((-1) \times (-1)) + ((-1) \times 1)) + 1 & & \{\times \text{ identity}\} \\
= \ & ((-1) \times ((-1) + 1)) + 1 & & \{\text{distributive law}\} \\
= \ & ((-1) \times 0) + 1 & & \{+ \text{ complement}\} \\
= \ & 0 + 1 & & \{\times \text{ null}\} \\
= \ & 1 + 0 & & \{+ \text{ commutative}\} \\
= \ & 1 & & \{+ \text{ identity}\}
\end{aligned}
$$

Figure 2.2
Why $(-1) \times (-1) = 1$.

equations. If you accept the basic equations and the idea of substituting equals for equals,[6] you must, as a rational consequence, accept the equation $(-1) \times (-1) = 1$.

Using this same kind of reasoning, we will derive new Boolean equations from a few basic ones postulated as axioms. A Boolean *axiom* is a Boolean equation that we assume to be true, without proof. We will also learn that digital circuits are physical manifestations of logic formulas, and we will be able to parlay that idea to derive behavioral properties of computer components.

Likewise, because a computer program is, literally, a formula, we will be able to derive properties of software directly from the programs themselves. This makes it possible for us to be entirely certain about some of the behavioral characteristics of software and of the digital circuits that comprise hardware components. Our certainty stems from the mechanistic formalism that we insist on from the beginning, which can be checked to the last detail with automated computation.

Exercises

1. Use the equations of figure 2.1 (page 16), together with the additional equation $(1 + 1) = 2$, to derive the equation $(x + x) = (2 \times x)$.

2. Derive the following equation using the equations of figure 2.1 (page 16):

$$((-1) \times x) + x = 0 \qquad \{\times \text{ negation}\}$$

---

[6] Substituting equals for equals applies the first of Euclid's common notions: Things which are equal to the same thing are also equal to one another. People have used equations in logical reasoning for a long time.

2.2 Boolean Equations

$$x \lor False = x \quad \{\lor \text{ identity}\}$$
$$x \lor True = True \quad \{\lor \text{ null}\}$$
$$x \lor y = y \lor x \quad \{\lor \text{ commutative}\}$$
$$x \lor (y \lor z) = (x \lor y) \lor z \quad \{\lor \text{ associative}\}$$
$$x \lor (y \land z) = (x \lor y) \land (x \lor z) \quad \{\lor \text{ distributive}\}$$
$$x \to y = (\neg x) \lor y \quad \{\text{implication}\}$$
$$\neg(x \lor y) = (\neg x) \land (\neg y) \quad \{\lor \text{ DeMorgan}\}$$
$$x \lor x = x \quad \{\lor \text{ idempotent}\}$$
$$x \to x = True \quad \{\text{self-implication}\}$$
$$\neg(\neg x) = x \quad \{\text{double negation}\}$$

Figure 2.3
Boolean axioms (basic equations).

3. Derive the equation $((x + (((-1) \times (x + y)) + z)) + y) = z$ using the equations of figure 2.1 (page 16) and, if you like, the $\{\times \text{ negation}\}$ equation from exercise 2.

2.2 Boolean Equations

Let's start with the Boolean equations in figure 2.3. These equations, which we will call the Boolean axioms, are the starting point for our system of reasoning. They form the basis from which we will derive many other equations. If these axiomatic equations are new to you and seem strange, try to view them as ordinary algebraic equations but with a different collection of operators. A formula in numeric algebra has operations like addition (+) and multiplication (×). Boolean formulas employ logic operations: logical-and (∧), logical-or (∨), logical-negation (¬), and implication (→). Furthermore, Boolean formulas stand for logic values ($True$, $False$), rather than for numbers ($\ldots -2, -1, 0, 1, 2 \ldots$).

When we derive a new equation by citing an equation that was itself derived rather than by citing an axiomatic equation from figure 2.3, we refer to the cited equation as a *theorem* to distinguish it from an axiom. The new equation is also a theorem, and it can be cited in derivations of other new equations. We call such a derivation of a new equation a *proof* of the equation.

The first equation in the theorem $\{\lor \text{ truth table}\}$ (figure 2.4, page 20) is a special case of the $\{\lor \text{ identity}\}$ axiom (figure 2.3), and the proof of that equation simply amounts to making that observation. That is, the proof just rewrites the $\{\lor \text{ identity}\}$ axiom with $False$ in place of x. The proof of the second equation is equally short, but it cites a different axiom. For practice, try to prove the other two equations in the $\{\lor \text{ truth table}\}$ theorem by citing axioms in a similar way.

We are serious about that. Did you prove the other two equations? No? Well ... go back and do it, then. Without participation, there is no learning. ... *We'll wait here.*

Theorem 2.1 (*{∨ truth table}*)

- *False ∨ False = False*
- *False ∨ True = True*
- *True ∨ False = True*
- *True ∨ True = True*

Proof.

      False ∨ False
= *False*   {∨ identity}   —replace x in the axiom with *False*

      False ∨ True
= *True*   {∨ null}   —replace x in the axiom with *False*

   ...for practice, prove the other two equations yourself...

Q.E.D.

Figure 2.4
Proof of theorem {∨ truth table}.

Finished now? Good for you. You cited the {∨ identity} axiom in your proof of the third equation in the theorem and the {∨ null} axiom in your proof of the fourth equation, right? We knew you could do it.

Box 2.2
Truth Tables

A *truth table* for a formula is a list of equations stating the values that the formula represents. There is one equation in the truth table for each possible combination of values for the variables in the formula. If there is only one variable in the formula, there will be two equations in its truth table, one for the case when the variable has the value *True* and one for the case when it has the value *False*. If there are two variables in the formula, there will be four equations in the truth table because for each choice of value for the first variable, there are two choices for the other. Three variables lead to eight equations. The number of equations in the truth table doubles with each additional variable in the formula.

A truth table for a logic operator is the truth table for the formula that has variables in place of the operands. For example, the truth table for the logical-or operator (∨) is the truth table for the formula ($x \lor y$). That formula has two variables, so the truth table has four equations.

Derivations are usually more than one step, of course. The {∨ complement} theorem (figure 2.5, page 21) has a two-step proof, citing the {implication} axiom and the {self-

2.2 Boolean Equations

Theorem 2.2 ({∨ complement}) $(\neg x) \vee x = True$

Proof.

$$\begin{aligned}
& (\neg x) \vee x \\
= \quad & x \rightarrow x \quad \{\text{implication}\} \\
= \quad & True \quad \{\text{self-implication}\}
\end{aligned}$$

Q.E.D.

Figure 2.5
Proof of theorem {∨ complement}.

implication} axiom. The {∨ complement} theorem is often called the "law of the excluded middle" because it says that any logic formula, together with its negation, covers all of the possibilities. A formula in logic is either true or false. There is no middle ground.

All of the logic operators have truth tables, and we can derive the equations in those truth tables from the axioms. Figure 2.6 (page 22) displays the truth table for the negation operator ($\neg$). The figure includes a four-step proof of the first equation in the table. To beef up your comprehension of the ideas, construct your own proof of the second equation in the theorem.

An important facet of these proofs is that they are entirely syntactic. That is, they apply axioms by matching the grammar of a formula f (or a subformula of f) in the proof with a formula g from one side of an equation in the axioms. The matching associates the variables in g with corresponding subformulas of f. Then, the formula h on the other side of the axiomatic equation is rewritten, replacing each variable in h with the subformula of f that the matching process established. The rewritten version of h becomes the new, derived formula. We know that the derived formula stands for the same value as the original formula because the axiom asserts this equivalence, and we assume the axioms are right.

Let's prove another truth-table theorem, partly to practice reasoning with equations but also to discuss a common point of confusion about logic. The implication operator ($\rightarrow$) is a cornerstone of logic in real-world problems, but it is common to get tripped up when it comes to reasoning with implication.

The {$\rightarrow$ truth table} theorem (figure 2.7, page 24) provides the truth table for the implication operator. An important aspect of the proof is that it cites not only axioms from figure 2.3 (page 19) but also equations from the {$\neg$ truth table} theorem. This is the way mathematics goes. Once we have derived a new equation from the axioms, we can cite the new equation to derive still more equations.

Theorem 2.3 ({¬ truth table})

- ¬*True* = *False*
- ¬*False* = *True*

Proof.

    ¬*True*
=   ¬(*False* → *False*)   {self-implication}
=   ¬((¬*False*) ∨ *False*)   {implication}   —replace x and y in axiom with *False*
=   ¬(¬*False*)   {∨ identity}   —replace x in the axiom with ¬*False*
=   *False*   {double negation}   —replace x in axiom with *False*

    ¬*False*
=   ... you fill in the details here ...
=   *True*

                                                               Q.E.D.

Figure 2.6
Proof of theorem {¬ truth table}.

2.2 Boolean Equations 23

> **Box 2.3**
> Truth Tables and Feasibility
>
> Reasoning from equations is one way to prove that two formulas stand for the same value. Another way is to build truth tables for both formulas. If the tables match, value for value, you can conclude that the formulas are equal. When there are only a few variables and therefore only a few combinations, the truth tables can be compared quickly and accurately. Unfortunately, this approach quickly goes south when there are many variables. With two variables, as in the truth table for logical-or, there are four combinations of values (two choices for each variable, *True* or *False*, so two times two combinations in all). With three variables, there are eight (2^3) combinations, which makes the truth-table method tedious but not infeasible.
>
> With ten variables, there are 1,024 (2^{10}) combinations, and with twenty, over a million. Too much for people to handle but easy for computers. However, a formula specifying a computing component, hardware or software, has hundreds of variables. We want to reason about computing components, and there is no hope of doing that with truth tables. With a hundred variables, there are 2^{100} combinations, and that number is so large that no computer could finish checking all the cases before the sun runs out of fuel. Truth tables are infeasible in this realm.
>
> Proofs like those that make up the core of this book are not the only effective way to work with large formulas. Hardware and software designers sometimes use SMT solvers (satisfiability modulo theories), BDD tools (binary decision diagrams), or other similar methods to verify properties of circuits and computer programs, but such tools don't apply to all formulas. Reasoning based on grammatical form makes it feasible to deal with any logic formula, regardless of the number of variables, because the formulas can be split into parts small enough to manage, and those parts can be reintegrated, based on their grammatical relationships, to produce a full analysis. Feasible doesn't mean easy, though. It takes a lot of effort, but it can pay off.

In day-to-day life outside the domain of symbolic logic, the usual interpretation of the logical implication $x \rightarrow y$ is to conclude that y is true whenever x is true. However, the implication says nothing about y when x is not true. In particular, it does not say that y is not true when x is not true. Theorem $\{\rightarrow$ truth table$\}$ shows that the formula *False* $\rightarrow y$ has the value *True* when y is *True* and also when y is *False*. In other words, the truth of the formula $x \rightarrow y$ in the case where the operand on the left of the arrow (which is known as the *hypothesis* of the implication), x, is *False* provides no information about the right-hand operand, y (which is known as the *conclusion* of the implication).

A common mistake in everyday life is to infer from the truth of the implication $x \rightarrow y$ that the implication $(\neg x) \rightarrow (\neg y)$ is also true. Sometimes this leads to bad results, even

Theorem 2.4 ({→ truth table})

- $False \to False = True$
- $False \to True = True$
- $True \to False = False$
- $True \to True = True$

Proof.

$$
\begin{array}{lll}
& False \to False & \\
= & (\neg False) \vee False & \{\text{implication}\} \quad \text{—put } False \text{ for } x \text{ and for } y \text{ in axiom} \\
= & \neg False & \{\vee \text{ identity}\} \\
= & True & \{\neg \text{ truth table}\}
\end{array}
$$

...for practice, prove the other equations yourself...

Q.E.D.

Figure 2.7
Proof of theorem {→ truth table}.

in everyday life.[7] In symbolic logic, it is worse than that. Such a conclusion puts an inconsistency into the mathematical system, which renders the system useless.

Over half of the Boolean axioms in figure 2.3 (page 19) have names associated with the logical-or ($\vee$) operation. One of them, the {$\vee$ DeMorgan} equation, establishes a connection between logical-or and logical-and. It converts the negation of a logical-or to the logical-and of two negations: $\neg(x \vee y) = (\neg x) \wedge (\neg y)$. We can use this connection to prove some logical-and equations that are similar to the logical-or axioms. An example is the null law for logical-and (figure 2.8, page 25).

This regime of theorem after theorem, proof after proof, is tiresome, isn't it? Nevertheless, let's push through one more. Then you can work some out on your own before going on to another topic. It's going to be one proof after another, all the way down the line.

Some equations simplify the target formula when used in one direction but make the target formula more complicated when used in the other direction. For example, applying the null law for logical-or ({$\vee$ null}: $x \vee True = True$) from left to right simplifies a logical-or formula to $True$. When the equation is applied in the other direction, however, it transforms the simple formula $True$ into something more complicated ($x \vee True$). When you apply the equation from right to left, the variable x on the left-hand side stands for any

---

[7] If turtles are reported in the park, you can conclude that there are reptiles in the park, but if no turtles are reported, you cannot conclude that there are no reptiles. There might be rattlesnakes. In the notation of logic, $turtle \to reptile$ is true but $(\neg turtle) \to (\neg reptile)$ isn't.

2.2 Boolean Equations

Theorem 2.5 ({∧ null}) $x \wedge False = False$

Proof.

	$x \wedge False$		
=	$x \wedge (\neg True)$	$\{\neg$ truth table$\}$	
=	$(\neg(\neg x)) \wedge (\neg True)$	$\{$double negation$\}$	
=	$\neg((\neg x) \vee True)$	$\{\vee$ DeMorgan$\}$	—*put $(\neg x)$ for x, True for y in axiom*
=	$\neg True$	$\{\vee$ null$\}$	
=	$False$	$\{\neg$ truth table$\}$	

Q.E.D.

Figure 2.8
Proof of theorem {∧ null}.

formula you want to make up (as long as it's grammatically correct). It can have hundreds of variables and thousands of operations. This may seem perverse, but if that's what it takes to complete a proof, so be it.

The null law for logical-and ({∧ null}: $x \wedge False = False$) is similarly asymmetric. It goes from complicated to simple in one direction and from simple to complicated in the other. A particularly interesting and important asymmetric equation is the absorption law (figure 2.9, page 26). It has two variables and two operations on one side, but only one variable and no operations on the other.

Box 2.4
Abstraction

Citing proven theorems to prove new ones is similar to an idea known as abstraction in engineering design. Instead of citing an old theorem to prove a new one, we could copy the proof of the old theorem into the new proof. However, that would make the proof longer, harder to understand, and more likely to contain errors.

Computer programs are built from components that are themselves also computer programs. As components become available, they are used to build more complex components. Sometimes, a component has almost the right form to be used in a new program but not quite. Maybe the existing component doubles a number, but the new usage needs to triple it. It is tempting to copy the old component, change $2 \times x$ to $3 \times x$, and paste the revised component into the program.

In our experience, copy-and-paste programming is a common source of errors in software, especially in programs maintained over time. When a maintainer finds an error in a component that was copied from elsewhere, there is nothing to direct the maintainer to fix the same error in the original component. A better choice, most of the time, is to make a new component with

Theorem 2.6 ({∧ absorption}) $(x \vee y) \wedge y = y$

Proof.

$$
\begin{aligned}
&\quad (x \vee y) \wedge y \\
&= (x \vee y) \wedge (y \vee False) &&\{\vee \text{ identity}\} \\
&= (y \vee x) \wedge (y \vee False) &&\{\vee \text{ commutative}\} \\
&= y \vee (x \wedge False) &&\{\vee \text{ distributive}\} \\
&= y \vee False &&\{\wedge \text{ null}\} \\
&= y &&\{\vee \text{ identity}\}
\end{aligned}
$$

Q.E.D.

Figure 2.9
Proof of theorem {∧ absorption}.

---

a variable, say m, in place of the 2. This is known as creating an abstraction of the component ("abstract" as opposed to "specific" or "concrete").

The new component can be used for both doubling and tripling simply by specifying 2 for m in one case and 3 for m in the other. If later in the project an error is discovered in the component, the error only needs to be fixed in one place, not two, or maybe ten or a hundred places, depending on how many engineers copied the original component to make a change. Abstraction is an important engineering method. Citing old theorems to prove new ones has similar advantages.

We hope the gauntlet of theorems and proofs so far helps you understand how to derive a new equation from equations you already know. The technique requires matching a formula to one side of a known equation, then replacing it by the corresponding formula on the other side of the equation. The "matching" process is a crucial step. It involves replacing the variables in the known equation by constituents of the formula you are trying to match. This is based in the mechanics of a formal grammar.

Unfortunately, it is surprisingly easy to have a lapse of concentration and make a mistake while trying to substitute equals for equals. Fortunately, it is easy for computers to verify correct matchings and report erroneous ones. A computer system that does this is known as a "mechanized logic." After you have enough practice to gain a good understanding of the process, we will begin to use a mechanized logic to make sure our reasoning is correct.

Exercises

1. Use the Boolean axioms and the theorems of this section to derive the truth table for the formula $(x \vee ((\neg y) \wedge (\neg z)))$.

Note: Since there are three variables in the formula, the truth table will have eight entries, which means you will need to prove eight equations. Each equation will have on the left-hand side a different combination of values (*True* or *False*) for the variables x, y, and z, and on the right-hand side each equation will have the value (*True* or *False*) of the formula for that combination.

2. Use the Boolean axioms (figure 2.3, page 19) and the {∧ absorption} theorem (figure 2.9, page 26) to derive the {∨ absorption} equation: $(x \wedge y) \vee y = y$.

3. Derive the equation $((x \vee y) \wedge (\neg(x \wedge y))) = (((\neg x) \wedge y) \vee (x \wedge (\neg y)))$ from the Boolean axioms. *Note*: These formulas define the exclusive-or operator.

2.3 Boolean Formulas

We have been doing proofs based on the grammatical elements of formulas, but we never tried to put together a precise definition of that grammar. We have been relying on your experience with numeric algebra. However, we really should have a precise definition of the grammar. We start with the most basic elements, then go on to more complicated ones.

The simplest Boolean formulas are the basic constants (*True* and *False*) and variables (x, y, ...). We normally use ordinary, lowercase letters, for variables, but sometimes variables are letters with subscripts, such as x_3, y_i, or z_n. This gives us sufficient variety for any formula, but we don't need to limit ourselves to lowercase Roman letters. We could use Greek letters or even make up recognizable squiggles, like Dr. Seuss.

So, if you write *True*, *False*, or a letter from the alphabet, you have composed a grammatically correct Boolean formula. This is the first rule of Boolean grammar. Formulas conforming to this rule have no substructure, so we call them *atomic* formulas.

Boolean operators make it possible to construct more complicated formulas. We refer to operators that require two operands as *binary operators* (∧, ∨, and →). These operators lead to the second rule of Boolean grammar: If a and b are grammatically correct Boolean formulas and ∘ is a binary operator (such as one of the symbols ∧, ∨, or →), then $(a \circ b)$ is also a grammatically correct Boolean formula.

For example, the first rule confirms that x and *True* are grammatically correct Boolean formulas. Since ∧ is a binary operator, $(x \wedge True)$ is a grammatically correct Boolean formula by the second rule of grammar. Furthermore, since → is a binary operator and y is a grammatically correct Boolean formula (by the first rule), $((x \wedge True) \rightarrow y)$ must be a grammatically correct Boolean formula (by the second rule).

The third rule of Boolean grammar shows how to incorporate the negation operator into formulas. If x is a grammatically correct formula, then so is $(\neg x)$.

The three rules of grammar are sufficient to cover a full range of grammatically correct Boolean formulas, and they lead to an infinite variety of grammatically correct formulas. However, there is a fine point to discuss about parentheses. Parentheses are important because they make it easy to define the grammar and to explain the meaning of a formula. The formulas covered by the three rules are fully parenthesized, including a top level of

v	{atomic}	
$(a \circ b)$	{bin-op}	*all grammatically correct formulas*
$(\neg a)$	{negation}	*must match one of these templates*
(a)	{group}	

<u>*requirements on symbols*</u>
- v is a variable or *True* or *False*
  (a variable is a letter or a letter with a subscript)
- a and b are grammatically correct Boolean formulas
- $\circ$ is a binary operator

Figure 2.10
Rules of grammar for Boolean formulas.

parentheses enclosing the entire formula when an operator is involved. Top-level parentheses are often omitted in informal presentations, and we have usually omitted them.

For example, we have been writing formulas like $x \vee y$, without the top-level parentheses that the grammar requires. To conform to the grammar, we would have to write $(x \vee y)$, with the parentheses. Because we have omitted top-level parentheses, requiring them probably comes as a surprise. But allowing nonatomic formulas without top-level parentheses requires additional rules of grammar, and we think that the added value of omitting parentheses fails to compensate for the extra complexity.

Here is a more complex formula with incorrect grammar: $x \wedge y \vee z$. This formula is missing two levels of parentheses. Even worse, there are two options for the inner parentheses. Does $x \wedge y \vee z$ mean $((x \wedge y) \vee z)$ or $(x \wedge (y \vee z))$? There are ways to deal with formulas that omit parentheses, but to avoid confusion, we are not going to allow such formulas. The same problem occurs with formulas in numeric algebra. We know that $x \times y + z$ means $((x \times y) + z)$ and not $(x \times (y + z))$ because we know the convention that gives multiplicative operators a higher precedence than additive operators. But that takes some getting used to, and we want to minimize the possibility of misinterpretation, especially because Boolean formulas may be new to you.

We will sometimes be informal enough to omit the top level of parentheses around the whole formula, but we will not omit interior parentheses. The grammar does allow redundant parentheses, however. For example, the formula $(x \vee ((x \wedge y)))$ is grammatically correct and has the same meaning as the formula $(x \vee (x \wedge y))$. The first formula has redundant parentheses but the second one doesn't. Allowing redundant parentheses requires a fourth rule of grammar, which says that (a) is a grammatically correct formula if a is.

With the four rules of figure 2.10, we can determine whether or not any given sequence of symbols is a grammatically correct Boolean formula. The definition of the grammar

2.3 Boolean Formulas

$(x \lor False) = x$ {∨ identity}
$(x \lor True) = True$ {∨ null}
$(x \lor y) = (y \lor x)$ {∨ commutative}
$(x \lor (y \lor z)) = ((x \lor y) \lor z)$ {∨ associative}
$(x \lor (y \land z)) = ((x \lor y) \land (x \lor z))$ {∨ distributive}
$(x \to y) = ((\neg x) \lor y)$ {implication}
$(\neg(x \lor y)) = ((\neg x) \land (\neg y))$ {∨ DeMorgan}
$(x \lor x) = x$ {∨ idempotent}
$(x \to x) = True$ {self-implication}
$(\neg(\neg x)) = x$ {double negation}
$((x)) = (x)$ {redundant grouping}
$(v) = v$ {atomic release}

requirements on symbols

- x, y, and z are grammatically correct Boolean formulas
- v is a variable or $True$ or $False$
  (a variable is a letter or a letter with a subscript)

Figure 2.11
Axioms of Boolean algebra.

$$(x \rightarrow False) = (\neg x) \qquad \{\neg \text{ as } \rightarrow\}$$
$$(\neg(x \wedge y)) = ((\neg x) \vee (\neg y)) \qquad \{\wedge \text{ DeMorgan}\}$$
$$(x \vee (\neg x)) = True \qquad \{\vee \text{ complement}\}$$
$$(x \wedge (\neg x)) = False \qquad \{\wedge \text{ complement}\}$$
$$(\neg True) = False \qquad \{\neg True\}$$
$$(\neg False) = True \qquad \{\neg False\}$$
$$(True \rightarrow x) = x \qquad \{\rightarrow \text{ identity}\}$$
$$(x \wedge True) = x \qquad \{\wedge \text{ identity}\}$$
$$(x \wedge y) = (y \wedge x) \qquad \{\wedge \text{ commutative}\}$$
$$(x \wedge (y \wedge z)) = ((x \wedge y) \wedge z) \qquad \{\wedge \text{ associative}\}$$
$$(x \wedge (y \vee z)) = ((x \wedge y) \vee (x \wedge z)) \qquad \{\wedge \text{ distributive}\}$$
$$(x \wedge x) = x \qquad \{\wedge \text{ idempotent}\}$$
$$(x \rightarrow y) = ((\neg y) \rightarrow (\neg x)) \qquad \{\text{contrapositive}\}$$
$$(x \rightarrow (y \rightarrow z)) = ((x \wedge y) \rightarrow z) \qquad \{\text{currying}\}$$
$$((x \wedge y) \vee y) = y \qquad \{\vee \text{ absorption}\}$$
$$((x \rightarrow y) \wedge (x \rightarrow z)) = (x \rightarrow (y \wedge z)) \qquad \{\wedge \text{ implication}\}$$
$$((x \rightarrow y) \wedge (x \rightarrow (\neg y))) = (\neg x) \qquad \{\text{absurdity}\}$$
$$(x \rightarrow (\neg x)) = (\neg x) \qquad \{\text{contradiction}\}$$

Figure 2.12
Some Boolean theorems.

is circular, but in a useful way that shows how to build more complicated formulas from simpler ones. To verify that a formula is grammatically correct, find the rule of grammar that matches it, then verify that each part of the formula that matches with a variable in the rule of grammar is also grammatically correct. Atomic formulas have no substructure, so they require no further analysis when checking for grammatical correctness.

For example, consider the formula $((x \vee (\neg y)) \wedge (x \rightarrow z))$. It matches with the {bin-op} rule. The variables in the rule match with elements of the formula in the following way:

symbol from {bin-op} rule	matching element in $((x \vee (\neg y)) \wedge (x \rightarrow z))$
a	$(x \vee (\neg y))$
o	$\wedge$
b	$(x \rightarrow z)$

The only other symbols in the rule are the top-level parentheses, and these match identically with the outer parentheses in the target formula. Therefore, the target formula is grammatically correct if the formulas $(x \vee (\neg y))$ and $(x \rightarrow z)$ are grammatically correct. We use the same approach to verify the grammatical correctness of those formulas.

2.3 Boolean Formulas

The first one, $(x \lor (\neg y))$, again matches with the {bin-op} rule, and the following table shows how the analysis continues in this case:

symbol from {bin-op} rule	matching element in $(x \lor (\neg y))$
a	x
$\circ$	$\lor$
b	$(\neg y)$

This reduces the verification of the grammatical correctness of $(x \lor (\neg y))$ to the verification of the two formulas x and $(\neg y)$. Since x matches with the {atomic} rule, it must be grammatically correct. The $(\neg y)$ element matches with the {negation} rule, with y from the formula matching a in the rule. So, $(\neg y)$ is grammatically correct if y is, and y is grammatically correct because it matches with the {atomic} rule. Altogether, this verifies that $(x \lor (\neg y))$ is grammatically correct.

The second element of the original formula, $(x \to z)$, is easier to verify. It matches the {bin-op} rule with x corresponding to a in the rule, y corresponding to b, and $\to$ corresponding to $\circ$ in the rule. Since x and z match the {atomic} rule, they are grammatically correct. This completes the verification that $((x \lor (\neg y)) \land (x \to z))$ is grammatically correct.

Let's look at another example: $(x \lor (\land y))$. This sequence of symbols matches with the {bin-op} rule, with x corresponding to a in the rule, $\lor$ corresponding to $\circ$, and $(\land y)$ corresponding to b. So, the formula is grammatically correct if x and $(\land y)$ are. However, there is no rule that matches $(\land y)$. The only place the symbol $\land$ could match a rule in the table is in the {bin-op} rule. In the {bin-op} rule, there must be a formula between the opening parenthesis and the operator. Since there is nothing between the opening parenthesis and the $\land$ operator in the target formula, it cannot be grammatically correct.

That covers the grammar of Boolean formulas. What about meaning? Every grammatically correct Boolean formula denotes, when the values of its variables are specified, either the value *True* or the value *False*. Each of the binary operators, given specific operands (*True* or *False*), delivers a specific result (*True* or *False*). Truth-table theorems (as in figure 2.4, page 20) derive the values that the operators deliver when supplied with *True*/*False* operands. We can derive the meaning of any grammatically correct formula in that same way, down to *True* or *False* if the formula doesn't have any variables in it.

However, to deal with parentheses in a completely mechanized way, we need to add two equations to those of figure 2.3 (page 19). The axioms of figure 2.11 (page 29) provide all of the information needed to determine the value of any grammatically correct formula. In fact, the equations in the figure have even more general applicability. They provide all the information needed to verify not only whether a given formula has the same meaning as the formula *True* or the formula *False* but also to verify whether or not any two given grammatically correct formulas have the same meaning.

Exercises

1. Determine which of the following are Boolean formulas by the rules of grammar (figure 2.10, page 28):

$$((x \wedge y) \vee y)$$
$$((x \rightarrow y) \wedge (x \rightarrow (\neg y)))$$
$$((False \rightarrow (\neg y)) \neg (x \vee True))$$

2. Derive the truth tables (see page 20) of the formulas from the previous exercise.

3. Use the axioms of Boolean algebra (figure 2.11, page 29) to prove the equations in figure 2.12 (page 30).
   Note: After proving an equation, you may cite it in subsequent proofs.

4. Prove the following equation:
$$((x \rightarrow y) \wedge (y \rightarrow x)) = ((x \rightarrow y) \wedge ((\neg x) \rightarrow (\neg y)))$$

5. Prove the following equation:
$$((x \rightarrow y) \wedge (y \rightarrow x)) = (((\neg x) \vee y) \wedge (x \vee (\neg y)))$$

6. Prove the following equation:
$$((x \rightarrow y) \wedge (y \rightarrow x)) = (\neg((x \wedge (\neg y)) \vee ((\neg x) \wedge y)))$$

7. Prove the following equation:
$$((x \rightarrow y) \wedge (y \rightarrow x)) = (\neg((x \vee y) \wedge (\neg(x \wedge y))))$$

8. Prove the following equation:
$$((x \rightarrow y) \wedge (y \rightarrow x)) = ((x \wedge y) \vee ((\neg x) \wedge (\neg y)))$$

Box 2.5
Boolean Equivalence ($\leftrightarrow$)

The following equation defines the Boolean equivalence operator ($\leftrightarrow$):

$$(x \leftrightarrow y) = ((x \rightarrow y) \wedge (y \rightarrow x)) \quad \{\text{equivalence}\}$$

When an implication goes both ways ($x \leftrightarrow y$), both operands have the same value ($x = y$). This is the negation of the exclusive or (exercise 3, page 27), which is true when the operands have different values. Exercise 7 proves this relationship between Boolean equivalence and exclusive or. Altogether, exercises 4–8 show six formulas that deliver the value of the Boolean equivalence operator. In other words, the exercises provide six formulas that are all equivalent to Boolean equivalence.

2.4 Digital Circuits

Logic formulas provide a mathematical notation for operations in symbolic logic. These same operations can be materialized as electronic devices. The basic operators of logic represented in the form of electronic devices are called *logic gates*. There are logic gates for the logical-and, the logical-or, negation, and several other operators not yet discussed.

Box 2.6
Implication Gate Is Universal

One of the operators discussed at length, implication ($\rightarrow$), is not among the operators commonly fabricated as a logic gate. This does not restrict the kinds of operations that can be performed by digital logic because, as we know from the {implication} axiom of Boolean algebra (figure 2.11, page 29), $(x \rightarrow y) = ((\neg x) \vee y)$. So, anything we can do with the implication operator we can also do with negation and logical-or.

The lack of a conventional logic gate for the implication operator is ironic. George Boole himself, the inventor of Boolean algebra, called implication the queen of logic operators. It is one of only a few basic operators that are *functionally complete* (or *universal*) in the sense that, for any formula in logic, there is an equivalent formula with no operators other than implication. That is, given any grammatically correct logic formula, there is a formula using only implication operators (no logical-and, logical-or, negation, or any other operators) that has the same meaning. Logical-or and logical-and are not functionally complete, but their negations, nor and nand, are. Figure 2.14 (page 36) shows how to write logic formulas using only nand gates.

There may be a way to fabricate implication gates that allows extensive, three-dimensional stacking of circuits, which is infeasible with most gates. If that pans out, it could be possible to build faster circuits with more components in a smaller space. R. Stanley Williams discusses this idea in an interesting YouTube video on memristor chips. You can look it up.

A logic gate takes input signals that correspond to the operands of logic operators and delivers output signals that correspond to the values delivered by those operators. A logic gate with two inputs is a physical representation of a binary operator. The negation operator corresponds to a logic gate with one input.

Logic operands and values always stand for either *True* or *False*. Similarly, an input line of a logic gate can distinguish between only two distinct signals and deliver only two distinct signals on an output line. Conventionally, those signals are written as 1 (for *True*) and 0 (for *False*). Of course, logic gates are electronic devices, so 1 and 0 are just labels for the signals. Any two different symbols could be used to represent them in writing. The choice of 1 and 0 is more or less arbitrary.

There are many ways to handle the electronics, but we are going to leave the physics to the electrical engineers. A voltage at a point in a circuit might represent the signal 1 (*True*) and the lack of voltage, the signal 0 (*False*), but physical representations depend on the

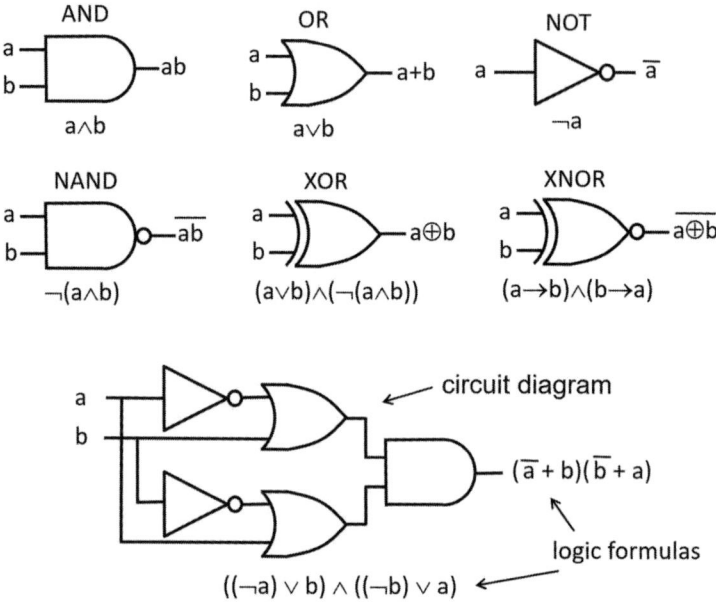

Figure 2.13
Digital circuits = logic formulas.

technology employed. We are going to focus on the logic and trust that the electronic hardware can faithfully deal with a signal representing *True* and a different one representing *False*.

Circuits can be depicted as wiring diagrams in which wires, represented by lines, carry signals between gates. Distinctive shapes in the diagrams represent different logic gates. Circuits can be represented as formulas as well as diagrams. The logic formulas we have been using would serve the purpose, but traditionally the algebraic notation for circuits takes a different form. In algebraic formulas for digital circuits, circuit designers usually represent the logical-and by the juxtaposition of the names being used for the input signals (in the same way that juxtaposition of variables is used to denote multiplication in numeric algebra). They represent logical-or by a plus sign (+) and negation by a bar over the formula. For example, the formula $\overline{ab}$ denotes the negation of the logical-and of the signals *a* and *b*.

Figure 2.13 presents symbols used for logic gates in circuit diagrams and annotates them with both the algebraic notation used by circuit designers and the logic formulas we have been using. The important fact to remember is that all three notations represent the same concepts in logic. Circuit diagrams, logic formulas, and the algebraic notation used by circuit designers are three different notations for exactly the same mathematical objects. In

2.4 Digital Circuits

this sense, digital circuits, and, therefore, computers, are materializations of logic formulas. Computers are logic in action.

The logic operators that we have been using ($\land$, $\lor$, $\neg$, $\rightarrow$) make it possible to write a formula that delivers precisely the values in the truth table for any given formula. The {implication} axiom (figure 2.11, page 29) expresses implication in terms of logical-or and negation, which means we lose no expressive power by discarding implication from the set of logic operations.

Surprisingly, the reverse is also true. That is, for any given input/output relationship that can be expressed in a formula using logical-and, logical-or, and negation, there is an equivalent logic formula using implication as the only operator. The new formula, using only the implication operator, produces the same results as the original formula. The {$\neg$ as $\rightarrow$} equation (figure 2.12, page 30) provides a start in this direction by showing how to express negation in terms of the implication operator. Furthermore, implication is not the only logic operator that is universal in this sense. Another one is the negation of logical-and, which is called *nand*. When fabricated using certain widespread technologies, nand gates can run faster and be more reliable than other gates, so many integrated circuits make frequent use of nand gates.

It is interesting to see how to put together digital circuits for basic operators ($\land$, $\lor$, and $\neg$) using only nand gates. Consider negation, for example. Negation has only one input signal and nand has two. Feeding the same signal into both inputs of a nand gate produces the behavior of the negation operator, as the following equation confirms:

$$(\neg a) = (\neg(a \land a)) \quad \{\neg \text{ as nand}\}$$

In this way, a nand gate can serve in place of a negation gate (also known as an *inverter*). There is a one-step proof of the equation, citing the {$\land$ idempotent} theorem (page 30).

A nand-only circuit for logical-and can be put together from two nand gates in sequence. The signal from the first nand gate is inverted by feeding it into both inputs of a second nand gate. Algebraically, this circuit corresponds to the following {$\land$ as nand} equation. It takes a two-step proof to verify the equation. The first step converts the outside nand to negation using the {$\neg$ as nand} equation, and the second step cites the {double negation} axiom from figure 2.11 (page 29).

$$(a \land b) = (\neg((\neg(a \land b)) \land (\neg(a \land b)))) \quad \{\land \text{ as nand}\}$$

Negation took one nand gate and logical-and took two. Logical-or can be implemented with three nand gates, as shown in the following equation, which can be verified using the {$\neg$ as nand} equation, DeMorgan's laws, and double negation:

$$(a \lor b) = (\neg((\neg(a \land a)) \land (\neg(b \land b)))) \quad \{\lor \text{ as nand}\}$$

Figure 2.14 (page 36) diagrams the digital circuits corresponding to the formulas that express logical-and, logical-or, and negation in terms of nand operations.

¬ circuit

a ─▷o─ ā

(¬a) = (¬(a ∧ a))

∧ circuit

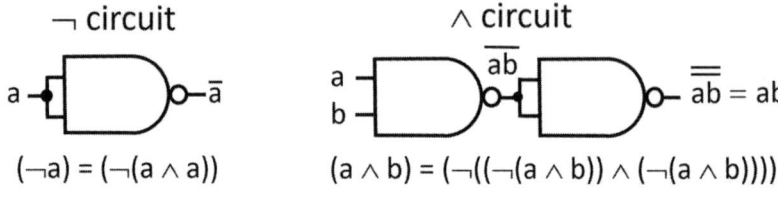

(a ∧ b) = (¬((¬(a ∧ b)) ∧ (¬(a ∧ b))))

∨ circuit

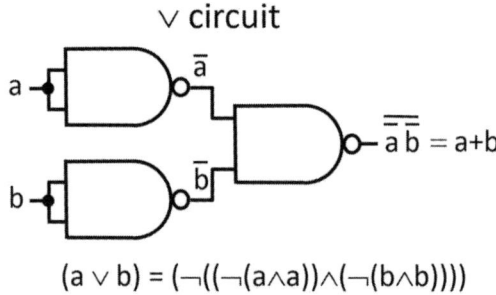

(a ∨ b) = (¬((¬(a∧a))∧(¬(b∧b))))

Figure 2.14
Nand is all you need.

Exercises

1. Using a negation-gate and an or-gate, draw a circuit diagram for an "implication circuit" that has the same input/output behavior as the implication operator.
   Hint: Follow the example of the {implication} axiom (figure 2.11, page 29). One of the inputs will need to be a constant rather than a variable.

2. For each of the following logic formulas, draw an equivalent circuit diagram. Since we don't have a symbol for an implication gate, you can either make up your own symbol or materialize logic gates where you need them using the circuit diagram from exercise 1.

$$((a \vee (b \wedge (\neg a))) \vee (\neg(a \vee b)))$$
$$(((\neg a) \wedge (\neg b)) \wedge (b \wedge (\neg c)))$$
$$(a \to (b \to c))$$
$$((a \wedge b) \to c)$$

3. Rewrite each of the formulas in the previous exercise in the algebraic notation used by electrical engineers: juxtaposition for ∧, + for ∨, and $\bar{a}$ for (¬a). Use the {implication} axiom to represent implications using negation and logical-or.

4. Draw circuit diagrams with behavior of the and-gate, the or-gate, and the negation-gate using implication operators only.

2.5 Deduction 37

Box 2.7
Struggling? Join the Club

Reasoning with Boolean equations requires a lot of intellectual effort. Almost everyone struggles when trying to master the concepts and apply them to solve problems. So, if you're struggling, you're normal. If you get discouraged to the point of despair, you're normal. It gets easier with every successful solution, but it never gets easy. What you are doing here is real mathematics, and it places the same kind of burden on the intellect as real engineering. Engineering is the application of principles of math and science to the design of useful things, so real engineering and real mathematics share a great deal of common ground.

If it does not appeal to you to struggle through many failed attempts and beyond them to a solution, only to go on to the next problem and start the process once again, you are going to find engineering to be an unpleasant activity. It's frustration, frustration, frustration ad nauseam, then a solution, then on to the next problem. Finding the solution through all of that fog brings a lot of satisfaction, and for engineers and mathematicians, that satisfaction makes it all worthwhile.

So, take heart. Keep trying. Hundreds of students have worked their way through the reading and exercises of this chapter and the ones that follow, and almost all of them have succeeded. To do so, they invested a great deal of energy in solving problems, reading, again and again, the examples, and applying the ideas. Sometimes it takes many hours, just to solve one problem. Don't give up.

2.5 Deduction

We have been reasoning with equations, which means we are reasoning in two directions at the same time. Equations go both ways. Deductive reasoning is one-directional. It derives a conclusion from hypotheses using one-directional rules of inference. A proof shows that the conclusion is true whenever the hypotheses are true, but it provides no information about the conclusion when the truth of one or more of the hypotheses is unknown.

In the following discussion of proof by deduction, theorems will be stated using a *turnstile* ($\vdash$) to separate the hypotheses from the conclusions. Hypotheses go on the left of the turnstile and the conclusion on the right. All of the hypotheses are formulas in logic, as is the conclusion. A turnstile asserts that there is a derivation of the conclusion from the hypotheses using the rules of inference. The commutativity law for logical-and, for example, can be stated as follows:

$$\text{Theorem } \{\wedge \text{ commutes}\}: a \wedge b \vdash b \wedge a$$

Later, we will prove the {∧ commutes} theorem using a formal apparatus for deductive reasoning known as *natural deduction*.[8] Theorems proved by natural deduction have zero or more logic formulas on the left of the turnstile and exactly one formula on the right. The formulas on the left are the *hypotheses* of the theorem and the one on the right is the *conclusion*. A proof begins with the formulas on the left, which are assumed to be true. At each step, the proof cites either a rule of inference or a previously proven theorem to derive a new formula. The rule or theorem ensures that when the hypotheses are true, so is the derived formula. Derived formulas can play the role of hypotheses in subsequent steps of the proof, and new formulas can be derived from them in the same manner. The derived formula at the end of the proof is the conclusion. A deductive proof of a theorem with hypothesis h and conclusion c verifies that the implication $(h \to c)$ is true.

$$h \vdash c \quad \text{ensures that} \quad (h \to c) = True$$

Of course, the truth of an implication formula doesn't say anything about the value of the left-hand operand of the implication operator. That value could be either $True$ or $False$. The implication formula just says that the only combination of values that can make $(h \to c)$ have the value $False$ (namely, $h = True$, $c = False$, as verified in theorem {→ truth table}, page 24) cannot occur. In the same way, a deductive proof of a theorem does not provide any information about the hypotheses. It only says that the conclusion will be true whenever all of the hypotheses are true.

Sometimes a theorem has several hypotheses. A proof by deductive reasoning of the theorem $h_1, h_2 \vdash c$, which has two hypotheses, ensures that $((h_1 \land h_2) \to c) = True$. A theorem with no hypotheses at all would have no formulas on the left-hand side of the turnstile: $\vdash c$. A proof of such a theorem would verify the equation $c = True$.

All of the axioms of Boolean algebra (figure 2.11, page 29) can be derived through deductive reasoning. Many presentations of classical logic begin with deductive reasoning, but we started with Boolean algebra because we will be using logic to reason about digital circuits and software specified in the form of equations. So, equations play a central role throughout the discussion.[9]

Figure 2.15 (page 39) provides schematics of the rules of inference of *natural deduction*. A citation of an inference rule derives a new formula from hypotheses (assumed true) or from formulas derived through prior reasoning in the proof. Each rule citation has three parts:

1. A proof above the line (or multiple proofs, depending on the rule),

---

[8] Natural deduction is a formal system of logic pioneered in the 1930s by the mathematician Gerhard Gentzen and refined in the 1960s by the logician Dag Prawitz.

[9] An accessible, more extensive discussion of natural deduction can be found in O'Donnell, Hall, and Page, *Discrete Mathematics Using a Computer* (2nd ed.), Springer, 2006.

2.5 Deduction

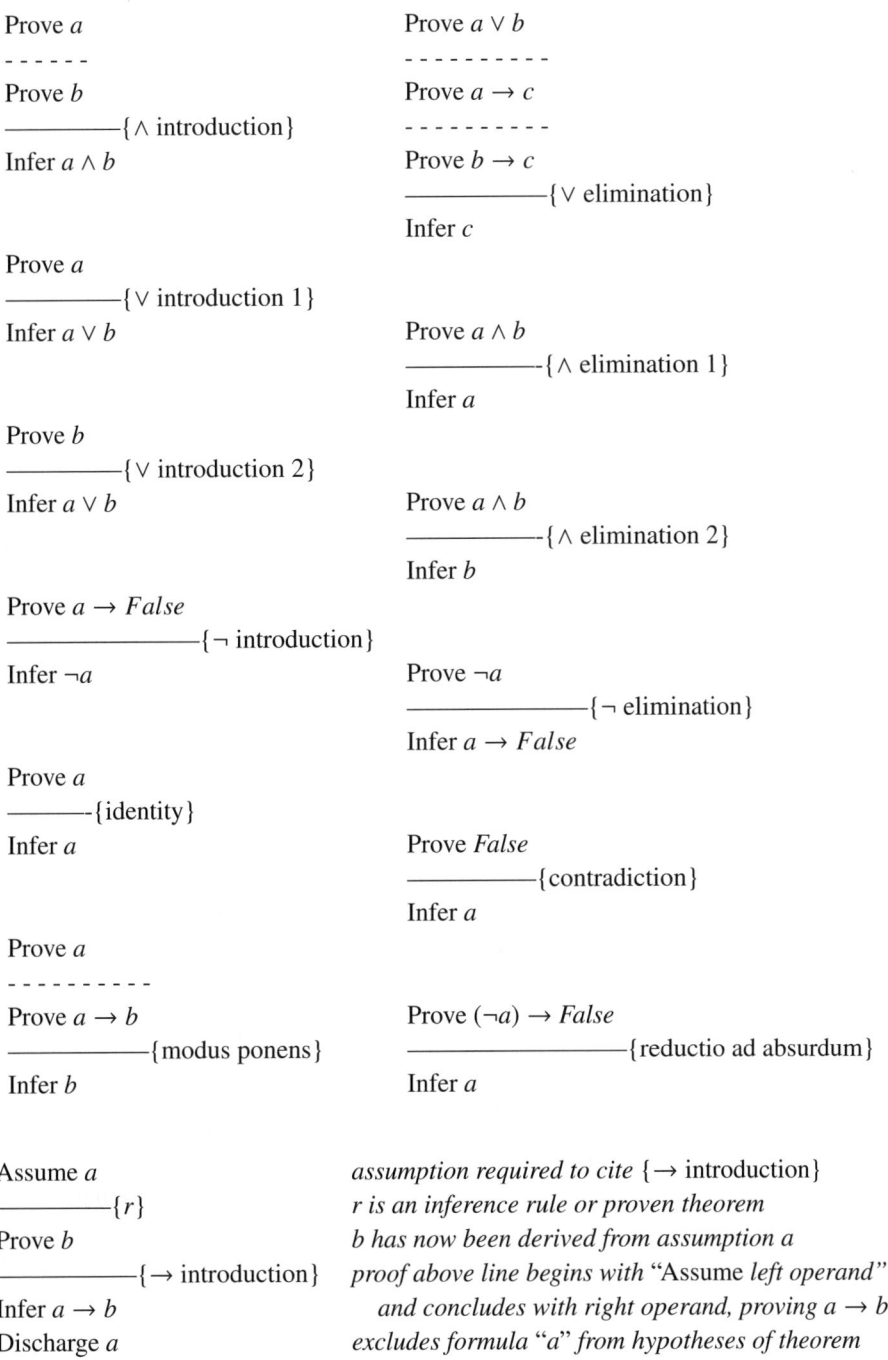

Figure 2.15
Rules of inference for natural deduction.

Theorem {Socrates was mortal}: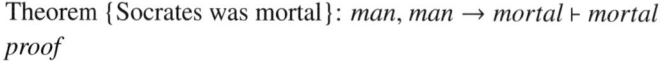
proof

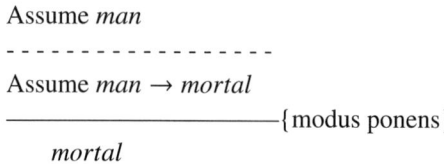

Figure 2.16
Theorem {Socrates was mortal}: citing modus ponens.

2. a line annotated with the name of the cited inference rule, and
3. exactly one logic formula below the line.

A *deductive proof* is a sequence of citations of inference rules in which the final citation has, below the line, the formula that is the conclusion of the theorem that the proof verifies. An inference rule can place specific constraints on the formula that is its conclusion and/or on the formulas that are the conclusions of the proofs that the rule requires above the line. For example, the $\{\wedge$ elimination 1$\}$ inference rule (figure 2.15, page 39) requires one proof above the line and constrains the conclusion of that proof to be a logical-and formula $(a \wedge b)$. The $\{\wedge$ introduction$\}$ rule, on the other hand, requires two proofs above the line and constrains its conclusion (below the line) to be a logical-and formula $(a \wedge b)$. Some rules place constraints both on the formulas above the line (which have been derived earlier in the proof) and on the formula below the line. For example, the $\{\neg$ introduction$\}$ rule requires the formula above the line to be an implication whose conclusion is the logical constant *False* $(a \rightarrow False)$ and constrains the formula below the line to be a negation formula $(\neg a)$. Furthermore, the hypothesis of the implication above the line must be the same formula that is negated below the line.

When a rule is cited in a deductive proof, the name of the rule is written just to the right of the line that separates the proofs that the rule requires above the line from the conclusion that the citation derives, which is the formula written below the line. When an inference rule requires multiple proofs above the line, dashed lines separate those proofs. Each of the proofs that the rule requires above the line is itself a proof. That is, it is also a sequence of citations ending in a conclusion formula.

The *scope* of a citation of an inference rule extends upward to the beginning of the first of the proofs above the line that the rule requires. The scopes of citations in a proof can overlap, and when they do, some proofs are nested inside others. In fact, the scope of the

2.5 Deduction

last citation of a rule in a proof always extends upward to the beginning of the proof, so the scopes of all of the other citations are nested within the scope of the last citation.[10]

Wherever an inference rule requires a proof above the line, an assumption can take the place of that proof. That is, an assumption can always stand in lieu of a proof. A logic formula that is marked as an assumption is a hypothesis of the theorem verified by the proof (unless that assumption is subsequently *discharged*, a special dispensation that will be discussed later). So, any proof may begin with a formula that is marked as an assumption. However, no proof can have a formula marked as an assumption after the first citation of a rule in the proof. Assumptions can appear only above the line of the first citation in the sequence of citations comprising the proof, but since citations (and therefore proofs) can be nested, an assumption need not be the first line in the entire proof. It may, instead, be the first line in a proof that is nested inside another proof.

The {modus ponens} rule (figure 2.15, page 39) is probably the most widely recognized rule of inference because of the well-known "Socrates was mortal" syllogism (figure 2.16, page 40). The rule says that if there is a proof concluding in the formula a and a proof concluding in the formula $(a \rightarrow b)$, those proofs, together with a citation of modus ponens, derive the conclusion b.[11]

Proofs by natural deduction follow a strictly prescribed format, and it is worth going over that format again in slightly different terms. The proof of a theorem that has n hypotheses will have n different formulas representing those hypotheses marked as assumptions at the beginning of one or more of the proofs required by the inference rules that the proof cites. That is, each hypothesis, at the point where it is introduced into the proof, is marked as an assumption. An assumption so marked stands in lieu of the proof required at that point by whatever inference rule is being cited. A particular formula in a proof may be marked as an assumption at more than one point in a proof, but no matter how many places it appears as an assumption, it is still just one hypothesis of the theorem being proved. A proof of a theorem with n hypotheses will have at least n formulas marked as assumptions. It will have more than n formulas marked as assumptions if two or more of the assumptions specify the same formula (or if a formula is discharged).

Assumptions must appear at the beginning of a proof, before the citation of any inference rule in the proof. Assumptions cannot pop up after the citation of an inference rule in a proof. Of course, since dashed lines indicate separate proofs, the assumption need not be

---

[10] Because of nested scopes, proofs by natural deduction are sometimes written with parentheses, like algebraic formulas, or displayed as "tree diagrams," with the conclusion of the theorem at the bottom and the citations spread out upwards in a branching structure that makes the overlapping (and nonoverlapping) of scopes easy to see. We have chosen a vertical format with implicit overlapping because this notation is more compact than a tree diagram and, in our judgment, more readable than a parenthesized proof formula.

[11] Deductive proofs are one-directional, and so is the theorem about Socrates. One can conclude mortality from two hypotheses about circumstances of life, but one cannot derive those circumstances from the mortality of Socrates. Rabbits, for example, are mortal, but they are not men.

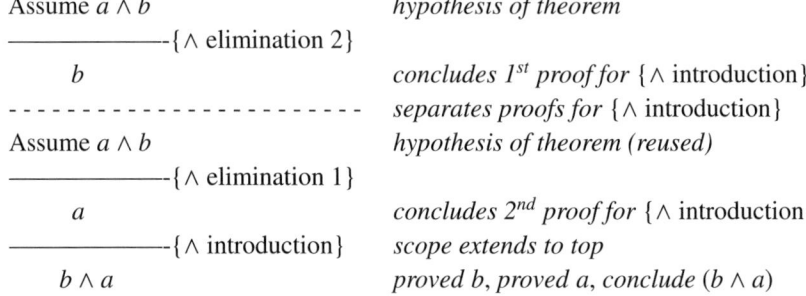

Figure 2.17
{∧ commutes}: citing three inference rules involving ∧.

at the beginning of the entire proof, but it could, instead, be at the beginning of a proof separated from another proof by a dashed line. The {Socrates was mortal} theorem (figure 2.16, page 40) has two hypotheses. One is marked as an assumption standing in lieu of the first proof required by a citation of {modus ponens} and the other is marked as an assumption standing in lieu of the second proof required by {modus ponens}.

Three of the inference rules of natural deduction (figure 2.15, page 39) involve the ∧ operator: {∧ introduction}, {∧ elimination 1}, and {∧ elimination 2}. Using these rules, we can construct a deductive proof of the commutativity law for ∧, and the proof will serve as a reasonably straightforward example to get started with natural deduction.

The proof in figure 2.17 cites the {∧ introduction} inference rule, a rule that requires two proofs above the line. The first of those proofs, in this example, has the hypothesis of the theorem above the line, marked as an assumption. It then cites the {∧ elimination 2} rule, which requires the right-hand operand (b) of the ∧ above the line to go under the line as the conclusion. Then comes the dashed line separating the two proofs required by {∧ introduction}. The second of the proofs comes next. It has the same form as the first proof except that it cites {∧ elimination 1} instead of {∧ elimination 2}. The {∧ elimination 1} rule brings the left-hand operand (a) of the ∧ down below the line. The {∧ introduction} rule requires the conclusion of the first proof above the line to become the left-hand operand of the ∧ operation that the rule introduces, and it requires the conclusion of the second proof to be the right-hand operand. The ∧ formula with those two operands is the conclusion below the line citing the {∧ introduction} inference rule. That final citation completes the proof of the {∧ commutes} theorem.

To recap, the proof of the {∧ commutes} theorem consisted of three proofs, in a sense, one being the whole proof, ending in the citation of the {∧ introduction} rule, and the other

2.5 Deduction

two being the proofs required by the citation of {∧ introduction}. Each of the three proofs in this example consisted of exactly one rule citation. Sometimes there are several rule citations in a proof, sometimes only one, and sometimes, when the proof is an assumption, none.

Box 2.8
Variables Stand for Formulas

As in Boolean algebra, variables in deductive reasoning stand for formulas (see page 15). Any grammatically correct formula can be plugged in for a variable (sometimes called a metavariable in this context), as long as that same formula replaces all instances of the variable. For example, theorem {∧ commutes} $(a \wedge b \vdash b \wedge a)$ has two variables, a and b. Since $(x \vee y)$ and $(y \rightarrow z)$ are formulas, the theorem justifies the following more specialized version:

$$(x \vee y) \wedge (y \rightarrow z) \vdash (y \rightarrow z) \wedge (x \vee y)$$

The theorem also justifies the following restatement of the theorem that uses the formula b in place of a and the formula a in place of b.

$$b \wedge a \vdash a \wedge b$$

Variables are used in this manner throughout the book. Nothing new here. We bring it up again because it is important to keep in mind when citing inference rules or theorems.

Two formulas are annotated as assumptions in the proof of the {∧ commutes} theorem. This suggests that the theorem has two hypotheses, but in this case, both assumptions are the same formula. A particular formula can be used as many times as necessary as an assumption in a proof, but it only counts as one hypothesis in the theorem. The number of hypotheses of the theorem is the number of distinct formulas annotated as assumptions in the proof minus the number of those assumptions that are discharged by citations of the {→ introduction} inference rule, which we will discuss shortly.

Given a proof, it is straightforward to extract the theorem it proves. On the left of the turnstile (⊢), put all of the different formulas annotated as assumptions in the proof except those that are discharged. After the turnstile, write the formula in the conclusion at the end of the proof.

Now we come to the issue of discharging formulas assumed in proofs. The inference rule {→ introduction} has some unique characteristics. It requires only one proof above the line, but that proof must begin with a formula (let's call it a) marked as an assumption. To repeat, the proof above the line in a citation of {→ introduction} that concludes with the formula $(a \rightarrow b)$ below the line must begin with "Assume a" and then continue from there

Theorem {self-implication}: ⊢ a → a   *Note: This theorem has no hypotheses.*
proof

    Assume a *assumption discharged later*
    ――――――{identity}
      a
    ――――――{→ introduction}
    $a \to a$ *assumed a, proved a, conclude $a \to a$*
    Discharge a *discharged by {→ introduction} citation, as promised*

Figure 2.18
{self-implication}: citing {identity} and {→ introduction} inference rules.

to derive the formula b, just above the line. The scope of the {→ introduction} citation extends upward to the required assumption.

Normally, any formula assumed at the beginning of a proof becomes a hypothesis of the theorem that was proved. However, a discharged assumption is not added to the hypotheses of the theorem. A citation of the {→ introduction} rule triggers a discharge of the assumed formula that the rule requires at the beginning of the proof above the line. (Without that assumption, the citation, and therefore the proof, is not valid.) The reason for the discharge is that the truth of an implication formula doesn't place any constraints on the value of its left-hand operand. The implication says that if the left-hand operand has the value *True*, then so does the right-hand operand, and the proof confirms that relationship. Since the citation of the {→ introduction} rule simply confirms that the implication formula below the line is true, the citation places no constraints on the value of the left-hand operand. The assumption only applies within the scope of the citation of the {→ introduction} inference rule and therefore does not become a hypothesis of the theorem being proved (unless it is assumed elsewhere in the proof and not discharged).

Figure 2.18 displays a proof of the {self-implication} theorem, which says that formulas of the form $(a \to a)$ always have the value *True*. The proof cites the {identity} rule, which is included among the inference rules to make it possible for proofs by natural deduction to stay strictly within the formalism required by the system. The {identity} rule says that a proof of formula a can be followed by a citation of the {identity} rule with that same formula, a, as its conclusion below the line.

The application of any rule must match the template in the specification of the rule, and the {identity} rule is sometimes needed to make it possible to match a template. That is what happens in the proof of self-implication (figure 2.18). The proof cites the {→ introduction} rule to derive the formula $(a \to a)$, and that citation requires a proof above the line that begins with the assumption of a and concludes with the formula a just above

2.5 Deduction

Theorem {∨ commutes}: $a \lor b \vdash b \lor a$
proof

Assume $(a \lor b)$	*hypothesis of theorem*
- -	*separates 1st and 2nd proofs for* {∨ elimination}
Assume a	*this assumption will be discharged*
───────{∨ introduction 2}	*allows arbitrary left-hand operand in conclusion*
$(b \lor a)$	
───────{→ introduction}	*assumed a, proved* $(b \lor a)$, *conclude* $a \to (b \lor a)$
$a \to (b \lor a)$	{∨ elimination} *requires this conclusion here*
Discharge a	*discharged by* {→ introduction} *citation, as promised*
- -	*separates 2nd and 3rd proofs for* {∨ elimination}
Assume b	*this assumption will be discharged*
───────{∨ introduction 1}	*allows arbitrary right-hand operand in conclusion*
$(b \lor a)$	
───────{→ introduction}	*assumed b, proved* $(b \lor a)$, *conclude* $b \to (b \lor a)$
$b \to (b \lor a)$	{∨ elimination} *requires this conclusion here*
Discharge b	*discharged by* {→ introduction} *citation, as promised*
───────{∨ elimination}	*scope extends up to* Assume $(a \lor b)$
$(b \lor a)$	*3 req'd proofs above, conclude* $(b \lor a)$

Figure 2.19
{∨ commutes}: citing {∨ elimination}.

the line. The {identity} rule makes it possible to satisfy this requirement. In the proof of self-implication, the citation of the {→ introduction} rule follows the derivation of *a* from *a* and triggers a discharge of the assumption of *a* at the beginning of the proof. There are no other assumed formulas in the proof, so the theorem proved has no hypotheses. That is, the proof confirms that the conclusion formula ($a \to a$) has the value *True*, regardless of what formula *a* stands for.

We now take on a theorem whose proof is more complex than those we have studied so far. The {∧ commutes} theorem proved earlier is similar to the {∨ commutes} theorem that we will discuss now, but the proofs are very different. Figure 2.19, which displays the proof of the {∨ commutes} theorem, cites all three of the inference rules involving the ∨ operator and affords an example of how the {∨ elimination} rule works.

The {∨ elimination} rule calls for three proofs above the line. The first of the three proofs must conclude in a formula that is a logical-or, $(a \lor b)$, where, of course, *a* and *b* can be any grammatically correct logic formulas. The second proof must conclude in an implication, $a \to c$. In this implication, *a* is the left-hand operand of the ∨ formula that concluded the first proof above the line, and c (which of course can be a formula rather than just a variable) is the conclusion under the line citing the {∨ elimination} rule. The third proof must conclude in an implication, $b \rightarrow c$, with the same right-hand operand as the implication that concludes the second proof but with a left-hand operand that is the same as the right-hand operand of the logical-or that concludes the first proof above the line. The {∨ elimination} rule is complicated but surprisingly easy to cite because the rule places so many constraints on its various parts.

The proof in figure 2.19 cites both of the "or introduction" rules: {∨ introduction 1} and {∨ introduction 2}. These rules allow the introduction of an arbitrary formula into the proof. That is, when you cite the rule, you can make up one of the formulas in the conclusion (namely, the right-hand operand of the logical-or in the case of {∨ introduction 1} and the left-hand operand in the case of {∨ introduction 2}). The formula you choose can be as complicated or as simple as you like, whatever is needed to make the proof work. In the proof at hand, the made-up formulas are simple (b in one case and a in the other case), but they are exactly what the proof needs.

In addition to citing all three inference rules involving the ∨ operator, the proof cites the {→ introduction} rule twice. Both of those citations require discharges, so there's lot of action in the proof. Figure 2.19 elucidates the details with commentary intended to help you work through the proof to understand how the citations fit together and comprise a proof of the {∨ commutes} theorem.

Deductive proofs are one-directional, so it's a little ironic that most of the theorems we've proved so far using natural deduction turn out to be bidirectional. The proofs went in only one direction, but the theorems were provable in the other direction, too.

The theorem that we turn to now, the implication chain rule (figure 2.20, page 47), only goes in one direction. It derives a conclusion from two hypotheses, but the two hypotheses cannot be derived from the conclusion. Again, commentary with the proof is intended to help you work your way through it. Pay particular attention to the discharge of the assumption that is introduced at the top of the proof.

In deductive proofs, previously proven theorems can be cited as if they were inference rules. Of course, the proof could always be carried out using inference rules alone by copying the proof of the cited theorem in place of its citation, but that leads to very long proofs, just as writing a computer program without defining and invoking procedures encapsulating common operations leads to very long programs. Long proofs, like long programs, tend to be unreliable, maybe because it's so difficult to analyze such a large mass of detail without getting confused. But even if they weren't unreliable, they would be an eyesore, not to mention difficult to fix if there were an error. That's why the ability to cite proven theorems in deductive proofs is important. It makes them shorter and easier to comprehend incrementally, one short proof at a time.

2.5 Deduction

Theorem $\{\to \text{chain}\}$: $(a \to b), (b \to c) \vdash a \to c$
proof

Assume a	*assumption discharged later*
----------------------	*separates proofs required by* {modus ponens}
Assume $a \to b$	*hypothesis of theorem*
———————{modus ponens}	{modus ponens} *citation ...*
b	*scope of citation extends up to* Assume a
----------------------	*separates proofs for second* {modus ponens}
Assume $b \to c$	*hypothesis of theorem*
———————{modus ponens}	*second citation of* {modus ponens} *...*
c	*scope overlaps first citation*
———————$\{\to \text{introduction}\}$	
$(a \to c)$	*assumed a, proved c, conclude* $(a \to c)$
Discharge a	*discharged by citation of* $\{\to \text{introduction}\}$

Figure 2.20
Proving the implication chain rule.

Theorem {modus tollens}: $a \to b, \neg b \vdash \neg a$
proof

Assume $a \to b$	*hypothesis of theorem*
------------------	*separates proofs of hypotheses of* $\{\to \text{chain}\}$ *theorem*
Assume $\neg b$	*hypothesis of theorem*
————$\{\neg \text{elimination}\}$	
$b \to \text{False}$	
————$\{\to \text{chain}\}$	*citing theorem with 2 hypotheses ...*
$a \to \text{False}$	*scope extends to top of proof*
————$\{\neg \text{introduction}\}$	
$\neg a$	

Figure 2.21
{modus tollens}: citing a theorem to justify an inference.

Theorem {¬¬ forward}: (¬(¬a)) ⊢ a
proof

    Assume (¬(¬a))   *hypothesis of theorem*
    ————————-{¬ elimination}
    (¬a) → *False*
    ————————-{reductio ad absurdum}   *proved* (¬a) → *False*, *conclude a*
        a

Figure 2.22
{¬¬ forward}: citing reductio ad absurdum.

The citation of a theorem in a proof must be preceded by proofs of each of its hypotheses above the line, just as each inference rule citation must be preceded by a certain number of proofs above the line. As with inference rules that require multiple proofs above the line, we use a dashed line to separate the required proofs when citing a theorem that has more than one hypothesis and therefore requires more than one proof above the line. Figure 2.21 (page 47) displays a proof of the modus tollens theorem.[12] The proof cites the implication chain rule theorem. Since that theorem has two hypotheses, there are two proofs above the line where the theorem is cited. Those proofs conclude in the implication formulas that are the hypotheses of the implication chain rule. Finally, the citation of the {¬ introduction} rule completes the proof.

The {reductio ad absurdum} rule supports "proof by contradiction." It says that if you can prove that the formula (¬a) → *False* is true, you can conclude that the formula *a* is true. The proof in figure 2.22 cites the reductio ad absurdum rule to prove a theorem about double negation.

Citations in the example proofs so far have included all of the inference rules but one. The rule we haven't used yet is {contradiction}.[13] The proof of the {disjunctive syllogism} theorem displayed in figure 2.23 (page 50) exhibits a citation of that rule. The theorem says that if a logical-or is known to be true and its left-hand operand is known to be false, then its right-hand operand must be true. The strategy of the proof employs the {∨ elimination} rule, which calls for three proofs above the line. The first of those proofs is simply an assumption of the logical-or formula that is a hypothesis of the theorem. The second proof derives *False* from the other hypothesis of the theorem and an assumption of the left-hand

---

[12] The inference rule {modus ponens} says that the conclusion of an implication can be derived from a proof of its hypothesis. The modus tollens theorem says that the hypothesis of an implication can be derived from a proof of the negation of its conclusion.

[13] It is ironic that proofs citing the {reductio ad absurdum} rule are called proofs by contradiction, while proofs citing the {contradiction} rule have no special name. Nevertheless, that is the custom, maybe because the {contradiction} rule, like the {identity} rule, is needed primarily to facilitate the formalities of natural deduction.

Exercises

operand of the logical-or. That assumption is discharged when the proof cites the {→ introduction} rule. The third proof is similar to the second proof but cites the {identity} rule at the point corresponding to the citation of the {contradiction} rule in the second proof.

Creating proofs by natural deduction is hard. It requires a lot of practice just to get a firm grasp of the ideas. The following exercises provide an opportunity to get some of that practice. As a rule of thumb, it often helps to start at the bottom of a proof by natural deduction. Write the conclusion formula at the bottom (it will have to go there anyway) and draw a line above it. Choose an inference rule that might be cited on that line, and think about how you might be able to cite the rule, possibly considering the hypotheses of the theorem you are trying to prove or other formulas that might be derivable from them. Working from the bottom of the proof in this way can be an effective strategy for finding the insights needed to create a proof.

Exercises

1. Use natural deduction to prove theorem {∧ complement}: $a, \neg a \vdash False$

2. Use natural deduction to prove the following theorem: $a, a \rightarrow b, b \rightarrow c \vdash c$

3. Derive the equation $((a \wedge ((a \rightarrow b) \wedge (b \rightarrow c))) \rightarrow c) = True$ using the axioms of Boolean algebra (figure 2.11, page 29).

4. Explain the connection between exercises 2 and 3.

5. Use natural deduction to prove the following theorem: $\vdash (a \wedge b) \rightarrow a$

6. Use natural deduction to prove theorem {nor commutes}: $\neg(a \vee b) \vdash \neg(b \vee a)$
   Note: The {∨ commutes} theorem may help, but not directly because $\neg(a \vee b)$ is a negation formula, not a logical-or formula. It has a logical-or as a subformula, but natural deduction requires matching the whole formula, not a subformula.

7. Use natural deduction to prove theorem {nand commutes}: $\neg(a \wedge b) \vdash \neg(b \wedge a)$
   Note: The {∧ commutes} theorem will not help in this proof for the same reason that {∨ commutes} does not help in exercise 6.

8. Use natural deduction to prove theorem {nor elimination 1}: $\neg(a \vee b) \vdash \neg a$

9. Use natural deduction to prove {DeMorgan ∨ forward}: $\neg(a \vee b) \vdash (\neg a) \wedge (\neg b)$

10. Use natural deduction to prove {DeMorgan ∨ backward}: $(\neg a) \wedge (\neg b) \vdash \neg(a \vee b)$

11. Explain the connection between exercises 9 and 10.
    Hint: Review box 2.5 (page 32).

12. Use natural deduction to prove theorem {∨ complement}: $\vdash a \vee (\neg a)$
    Hint: Use the {reductio ad absurdum} inference rule, cite the {nor elimination 1} theorem from exercise 8, and remember that a particular assumption can occur at multiple points in a proof.

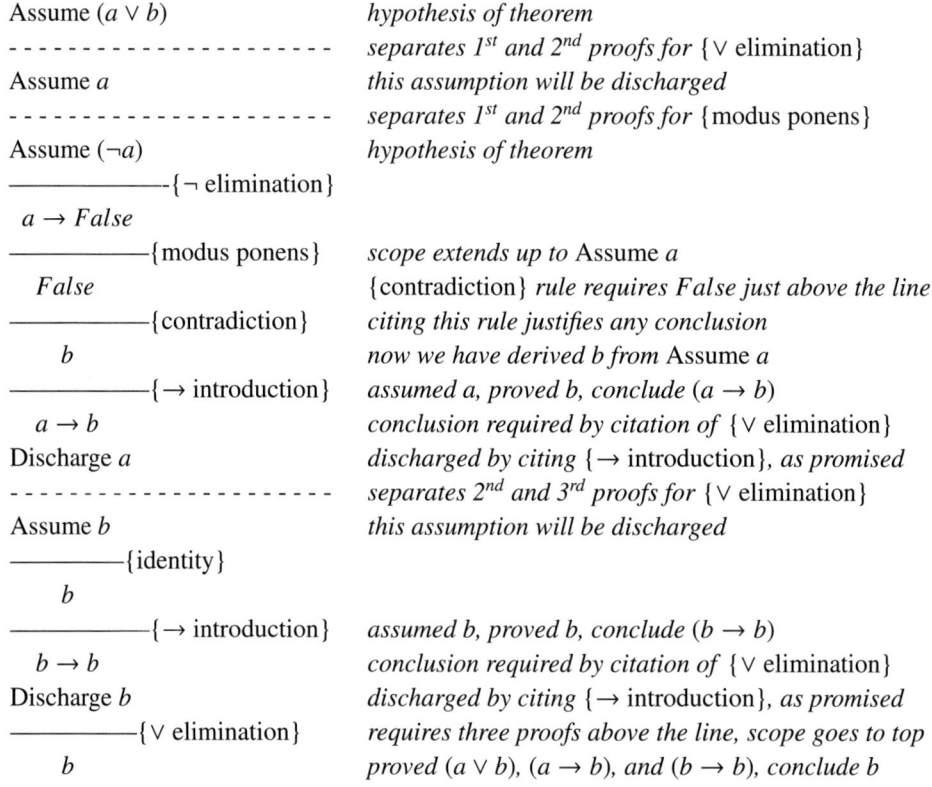

Figure 2.23
{disjunctive syllogism}: citing {contradiction}.

13. The proof of the {disjunctive syllogism} theorem in figure 2.23 (page 50) would be shorter if it cited the {self-implication} theorem (figure 2.18, page 44) to derive the formula ($b \rightarrow b$) instead of using the {identity} and {$\rightarrow$ introduction} inference rules to derive that formula. Change the proof to make it shorter in this way.

2.6 Predicates and Quantifiers

We have been using the term *proposition* to mean a formula that is either true or false. Any set[14] of propositions is called, when the set is taken as a whole, a *predicate*. We will require a predicate to be a collection of propositions indexed by a set known as the *universe of discourse*. If P is a predicate and x is an element from the universe of discourse, then $P(x)$ is the proposition selected from the predicate by the index x.[15]

If we write a formula that connects some of the propositions of the predicate P with logical-and, $(P(x_1) \wedge P(x_2) \wedge P(x_3) \wedge P(x_4))$, the formula has the value $True$ when all of the propositions in the formula $(P(x_1), P(x_2), P(x_3), P(x_4))$ are $True$. We would like to be able to write a formula for the logical-and of all the propositions in the predicate P. We could do this with an ordinary logical-and formula, but this gets bulky when there are a lot of elements in the universe of discourse, and it's impossible when the universe of discourse has an infinite number of elements.

The usual way to write the logical-and of all the propositions in a predicate makes use of a symbol that looks like an upside-down letter A and is known as the *universal quantifier* ($\forall$). The formula ($\forall x.P(x)$) stands for the logical-and of all the propositions in the predicate P.[16] The value of the formula is $False$ if there is an element x from the universe of discourse for which the proposition $P(x)$ is $False$. Otherwise, ($\forall x.P(x)$) is $True$.

---

[14] The term "set" has a checkered history in mathematics. It is tricky to define in a way that avoids contradictions like Russell's paradox, which you can read about in online articles or textbooks. Instead of dwelling on those issues, we are going to assume that, for any of the sets that we talk about, we have a way of figuring out whether any given item is an element of the set or not. Usually, our sets will be familiar ones, such as the set of natural numbers, which is the universe of discourse indexing the propositions in proofs by mathematical induction, or the set of lists that can be constructed by an ACL2 program. Occasionally, the universe of discourse will be the set of all programs that can be expressed in a given programming language. In that case, any interpreter for the language can determine whether or not a given item is in the set.

[15] You can think of the predicate as an operator that delivers the associated proposition as output when supplied with the index of the proposition as input, such as the ACL2 operator natp: (natp x) is true if x is a natural number and false otherwise. No matter whether you look at it as a set of propositions indexed by a universe of discourse or as an operator that delivers a true/false value given an element of the universe of discourse, the predicate is the same mathematical entity. The indexed-set approach is sometimes called an "extensional" view because it focuses on the externally observable characteristics of the predicate, whereas the operator perspective is called an "intensional" view because it involves the internal workings of a way to produce the true/false value of a proposition, given its index. Occasionally, a predicate will not correspond to a computation, and in that case the operator (intensional) view isn't valid because there will be no computation associated with the predicate. The extensional view is the way to think about predicates of that kind.

[16] The formula ($\forall x.P(x)$) reads "for all x, $P(x)$ is $True$."

For example, suppose n stands for a natural number and we use $E(n)$ as shorthand for the proposition "$2n$ is a nonnegative, even number." Then $\{E(0), E(1), E(2), \ldots\}$ is a set of propositions indexed by the natural numbers. For each natural number n, there is a corresponding proposition $E(n)$, so E is a predicate with the natural numbers as its universe of discourse.

If we write a formula that connects some of these propositions with $\wedge$, $(E(5) \wedge E(3) \wedge E(7) \wedge E(1))$, the formula has the value $True$ because all four of those propositions are $True$. In fact, all of the propositions in the predicate E are $True$. $E(n)$ is $True$ regardless of which natural number n stands for because any number of the form $2n$ is a nonnegative, even number when n is a natural number. Therefore, there is no element n in the universe of discourse of the predicate E for which the proposition $E(n)$ is $False$, which means that the value of the quantified formula $\forall n.E(n)$ is $True$.

A quantifier converts a set of propositions (that is, a predicate) into a single value, true or false. That is, a quantifier converts a predicate to a proposition. Syntactically, a quantified formula starts with a quantifier symbol followed by a variable, then a period, and finally a logic formula representing a proposition. The variable, which is known as the *bound variable* in the formula, stands for an element of the universe of discourse, and the quantification ranges over the entire universe of discourse. The universe of discourse is not specified directly in the formula, but the formula has no meaning unless the universe of discourse is known.

Box 2.9
Quantifier with Empty Universe

Let P be a predicate. The formula $\forall x.P(x)$ is false when there is at least one index x in the universe of discourse for which $P(x)$ is false. Otherwise, the $\forall$ quantification is true. If the universe of discourse is empty, there aren't any indexes at all, let alone one for which the predicate is false. Therefore, $\forall x.P(x)$ is true when the universe of discourse is empty.

Using a similar rationale, a $\exists$ quantification is false when the universe of discourse is empty because it can only be true if there is at least one element, x, in the universe of discourse for which $P(x)$ is true.

Any variable that is not bound is called a *free variable*. In the formula $(\forall x.P(x)) \vee Q(y)$, x is a bound variable and y is a free variable. This can get a bit tricky, but you have to keep it straight to understand how quantifiers work. An especially tricky case is the formula $(\forall x.P(x)) \vee Q(x)$. In this formula, x is a bound variable in the operand on the left-hand side of the $\vee$ but a free variable on the right-hand side.

The only other quantifier we will use is the *existential quantifier*. It forms the logical-or of all the propositions in a predicate and is represented by a symbol that looks like a

2.6 Predicates and Quantifiers

backwards letter E. The formula $(\exists x.P(x))$ has the value $True$ if there is an element x from the universe of discourse for which the proposition $P(x)$ is $True$.[17]

Consider the equation $((n + 7) = 12)$. Any equation is a proposition because it is either $True$ or $False$. Let's call this proposition $Q(n)$. We can take the view that Q is a predicate with the natural numbers as its universe of discourse because for each natural number n, $Q(n)$ stands for a proposition. You know from algebra that there is a natural number n for which equation $((n + 7) = 12)$ holds. That is, there is a value (namely, the number $n = 5$) in the universe of discourse for which the proposition $Q(n)$ is $True$. Therefore, according to the definition of the existential quantifier, the formula $(\exists n.Q(n))$ is $True$.

The formula $(\forall n.Q(n))$, however, is $False$ because there is a natural number n for which the proposition $Q(n)$ is $False$. In fact there are many of them, but the number of $False$ propositions in the predicate doesn't matter in a universal quantification. One is enough. By the definition of the universal quantifier, the formula $(\forall n.Q(n))$ is $False$ if there is even one element of the universe of discourse for which the proposition $Q(n)$ is $False$.

Box 2.10
Equal by Definition: $\equiv$

The three-line variation of the equals sign indicates that the term on the left stands for the formula on the right *by definition*.

$term \equiv \ldots some\ formula \ldots$	*definition of term*
$Q(n) \equiv ((n + 7) = 12)$	$Q(n)$ *stands for* $((n + 7) = 12)$

Predicates can have more than one index. For example, the predicate R, defined as follows, has two indexes:

$$R(m, n) \equiv ((n + 7) = m)$$

In this discussion, the universe of discourse for both indexes will be the natural numbers.[18] For each pair of natural numbers (n, m), $R(n, m)$ stands for a proposition (namely, the equation $((n + 7) = m)$, which is either $True$ or $False$). The formula $(\exists n.R(n, m))$ is a different proposition for each natural number m. That makes it a set of propositions indexed by the natural numbers, so it is a predicate with the natural numbers as its universe of discourse. To keep things straight, let's give this predicate a name.

---

[17] The formula $(\exists x.P(x))$ reads "there exists x such that $P(x)$ is $True$."

[18] A predicate with multiple indexes can have different universes of discourse for different indexes. One index could come from a set of numbers and the other from a set of words, for example, but both the first and second indexes of the particular predicate R discussed here are natural numbers.

$$S(m) \equiv (\exists n.R(n, m))$$

Let's convert this predicate to a proposition by quantifying it: $(\forall m.S(m))$. This is a proposition, so it is either *True* or *False*, but which is it? By the definition of the predicate R, $S(m)$ would be *True* if there were no natural numbers m for which the quantification $(\exists n.((n+7) = m))$ was *False*. Suppose m is the natural number zero. The proposition $S(0)$ says $(\exists n.((n + 7) = 0))$. There are no natural numbers n such that $((n + 7) = 0)$ because n would have to be negative, and all natural numbers are zero or bigger. Therefore, $S(0)$ is *False*, and that makes $(\forall m.S(m)) = False$.

By definition, $S(m)$ stands for the formula $(\exists n.R(n,m))$, so we can put that formula in place of $S(m)$ in $(\forall m.S(m))$. When we do this, the formula becomes $(\forall m.(\exists n.R(n, m)))$, in which an existential quantification is nested inside a universal quantification. It can go the other way too, and with any combination of quantifiers. All of the following formulas are propositions, and with your understanding of numbers, you can figure out which ones are *True* and which are *False*:

$$(\exists m.(\forall n.R(n, m)))$$
$$(\exists m.(\exists n.R(n, m)))$$
$$(\forall m.(\forall n.R(n, m)))$$

Nested quantifications like this are common when a predicate has multiple indexes.

Exercises

1. Work out the values of the following formulas, where $R(m, n) \equiv ((n + 7) = m)$:

    a) $(\exists m.(\forall n.R(n, m)))$

    b) $(\exists m.(\exists n.R(n, m)))$

    c) $(\forall m.(\forall n.R(n, m)))$

2. Mark the free variables in the following formulas and say how many bound variables each formula has:

    a) $(P(x) \lor (P(y) \to P(z)))$

    b) $(\forall x.(P(x) \land (\forall y.Q(y))))$

    c) $(P(x) \to (\exists y.Q(y)))$

    d) $(\exists x.(P(x) \land (\forall y.Q(y))))$

    e) $((\forall x.(P(x) \to Q(y))) \lor (\forall x.W(x)))$

    f) $(\forall x.(\forall z.R(x, y, z)))$

2.7 Reasoning with Quantified Predicates

Quantifiers provide a way to convert predicates to propositions, and you have some experience in reasoning about Boolean formulas constructed with propositions and operators. This section discusses some new methods and equations to make the same kind of reasoning possible with formulas containing quantifiers.

Let's start with a predicate P with two indexes. $P(x, y)$ will denote the proposition in the predicate P that is indexed by the pair (x, y), where x comes from the universe of discourse for the first index and y from the universe of discourse for the second index.

In our discussion, we will want to provide some examples of specific values in the universe of discourse. We could do this by making up some special symbols for those values, but to keep things uncomplicated at this point, let's say that natural numbers are the universe of discourse for both indexes. That will give us familiar symbols for particular indexes. $P(5, 2)$, $P(0, 6)$, and $P(3, 7)$ would be specific propositions in the predicate P. $P(x, y)$ would also be a proposition in predicate P, but it would not be a specific one unless we knew which natural numbers x and y stood for. Again, we choose natural numbers as the universe of discourse only to make it easy to designate specific elements in the domain. The points we make in the discussion about reasoning with quantified predicates would be the same for other choices of the universe of discourse.

Suppose we had already proved that the formula $(\forall x.(P(5, x)))$ is $True$. How can we use this predicate in another proof? One way is to observe that $(\forall x.(P(5, x)))$ means that all of the formulas $P(5, 0)$, $P(5, 1)$, $P(5, 2)$, ... are $True$, so we can assert in a proof that, for example, $P(5, 0) = True$. We could also assert that $P(5, 1)$ is $True$, as well as $P(5, 2)$, and so on. That is, once we have proved that a universally quantified formula has the value $True$, we can use that to justify a more specific theorem that eliminates the quantifier and replaces the bound variable by any specific value in the universe of discourse.

Not only that, but we could replace the variable by any formula representing a value in the universe of discourse. For example, we could assert that if x and y denote natural numbers, then $P(5, 2x + y + 4)$ is $True$.

Another formula to consider is the existential quantification $(\exists x.(P(5, x)))$. Suppose we know it has the value $True$, and we want to make use of that fact in a proof. The meaning of the formula $(\exists x.(P(5, x)))$ is that there is at least one x from the universe of discourse that makes the expression $True$, but it doesn't say which one. It could be that $P(5, 9)$ is $True$, or that $P(5, 3)$ is $True$, or any other proposition in P whose index is a pair with 5 as its first component. It could be exactly one of them, or just two or three, or it could even be all of them, but there must be at least one. That's all we know from a proof that $(\exists x.(P(5, x)))$ is $True$.

One way to make use of that fact is to use a notational convention to indicate that what looks like a variable (which could stand for any value in the universe of discourse or even a formula representing such a value) is not really a variable but is, instead, a specific value

Step 1. Rename bound variables
Step 2. Migrate quantifiers
Step 3. Eliminate quantifiers (leaving a proposition)
Step 4. Prove theorem about the proposition

Figure 2.24
A four-step strategy for reasoning with quantifiers.

in the universe of discourse. That is, the symbol is a constant, not a variable. One way to do this would be to designate a special symbol, capital C, perhaps, to use when we want to indicate that what looks like a variable is really a constant. If we need several different constants in our discussion, we could use different subscripts: C_x, C_y, C_{197}, C_ξ, and so on. Another approach is to use subscripts on ordinary variables (x_0, y_8, ...) to indicate that the symbols stand for constants, not variables. The important thing is (1) to say in the commentary of the proof that the new symbol stands for a specific value from the universe of discourse and is not a variable and (2) to make sure to use a different symbol for each different constant and to not use that symbol for any other purpose in the proof. In any case, what was a bound variable in an existential quantification becomes a free constant in the formula that represents a particular proposition in the predicate, and it is a proposition that has the value $True$.

So what about proofs? How can we prove a statement that uses quantifiers? One approach is to systematically remove the quantifiers so as to produce a formula that has no quantifiers. In other words, we are left with a formula without variables, such as $P(5, 3)$ or $P(5, x_0)$, where x_0 stands for a particular element in the universe of discourse whose value we don't know but which in any case stands for just one value and cannot be replaced by another variable or formula. Since there are no variables, this is really just a Boolean formula for an ordinary proposition, so it can be proved using the same methods we used with Boolean propositions.

This approach to reasoning about quantified formulas is a four-step process (figure 2.24). The last step is the already familiar area of reasoning with Boolean formulas that represent propositions, but the first three steps involve new ideas.

Renaming bound variables, which is the first step, is sometimes necessary to prevent a quantifier from accidentally referring to a different bound variable that happens to have the same name. For example, consider the formula $(\forall x.Odd(x)) \lor (\forall x.Even(x))$, where the universe of discourse for both quantifiers is the integers.[19] This formula has the value

---

[19] $Odd(x) \equiv (\exists y.(x = 2y + 1))$. $Even(x) \equiv (\exists y.(x = 2y))$. These formulas define the predicates Odd and $Even$. The integers are the universe of discourse, both for the predicates (variable x) and for the quantifications (bound variable y).

2.7 Reasoning with Quantified Predicates

False because there is an integer x for which $Odd(x)$ is false (the number 2, for example). That makes $(\forall x.Odd(x))$ false. Similarly, $(\forall x.Even(x))$ is false, and since both operands of $\vee$ in $(\forall x.Odd(x)) \vee (\forall x.Even(x))$ are false, the formula is false ({$\vee$ truth table}, figure 2.4, page 20). There are two bound variables in the formula. These are different variables, even though they both have the same name, x. The formula $(\forall x.(Odd(x) \vee Even(x)))$, on the other hand, has only one bound variable and is true. We have to be careful to keep these formulas straight. They look similar, but they have totally different meanings.

Another important aspect of quantified formulas is that the name of a bound variable has no effect on the value of the formula: $(\forall x.Even(x))$ has the same meaning as $(\forall y.Even(y))$. So, to avoid conflating two different bound variables, we start by choosing names for the bound variables so that each quantification uses a bound variable of a different name. For example, in the formula $(\forall x.Odd(x)) \vee (\forall x.Even(x))$, we could change the name of the bound variable in the second quantification: $(\forall x.Odd(x)) \vee (\forall y.Even(y))$. The meaning of the formula remains the same, but the change of name prevents confusion between the two different bound variables.

Suppose we have a quantified formula that contains many variables: $(\forall x.(P(\ldots x \ldots)))$. How do we rename the bound variable x in this formula? First, we insist on using a completely new name, one that does not appear elsewhere in the formula. This is for the same reason as before: we do not want to confuse the variable x with a different variable that happens to have the same name. Choosing a new name for x that is the same as the name of some other variable would defeat our purpose.

Second, we must ensure that when we change the name of the bound variable x to, say, y, we replace all occurrences of that bound variable x with y. However, we must be careful not to change any occurrence of a different variable that happens also to have the name x. For example, recall that there are two different but identically named bound variables in the formula $(\forall x.Odd(x)) \vee (\forall x.Even(x))$. We would want to change one of them to a new name, such as y. That would produce either the formula $(\forall x.Odd(x)) \vee (\forall y.Even(y))$ or the formula $(\forall y.Odd(y)) \vee (\forall x.Even(x))$.

For example, suppose we are trying to prove that if P is a predicate with two indexes, then the following formula has the value $True$:

$$((\exists y.(\forall x.P(x,y))) \rightarrow (\forall x.(\exists y.P(x,y)))) = True$$

The formula has four different bound variables but only two different names. We need to rename two of them to avoid conflicts. That is the first step in our four-step strategy. We apply the equation {R∃} from figure 2.25 (page 58) to justify renaming the bound variable y in $\exists y$ in the left-hand operand of the implication. Any name is okay as long as it is not already present in the formula. We choose v and get a new formula that, according to equation {R∃}, has the same value as the formula we started with.

$$((\exists y.(\forall x.P(x,y))) \rightarrow (\forall x.(\exists y.P(x,y)))) = ((\exists v.(\forall x.P(x,v))) \rightarrow (\forall x.(\exists y.P(x,y)))) \quad \{R\exists\}$$

$$((\forall x.P(\ldots x \ldots)) \wedge Q) = (\forall x.(P(\ldots x \ldots) \wedge Q)) \qquad \{\forall\wedge\}$$
$$((\exists x.P(\ldots x \ldots)) \wedge Q) = (\exists x.(P(\ldots x \ldots) \wedge Q)) \qquad \{\exists\wedge\}$$
$$((\forall x.P(\ldots x \ldots)) \vee Q) = (\forall x.(P(\ldots x \ldots) \vee Q)) \qquad \{\forall\vee\}$$
$$((\exists x.P(\ldots x \ldots)) \vee Q) = (\exists x.(P(\ldots x \ldots) \vee Q)) \qquad \{\exists\vee\}$$
$$((\forall x.P(\ldots x \ldots)) \to Q) = (\exists x.(P(\ldots x \ldots) \to Q)) \qquad \{\forall\to\}$$
$$((\exists x.P(\ldots x \ldots)) \to Q) = (\forall x.(P(\ldots x \ldots) \to Q)) \qquad \{\exists\to\}$$
$$(Q \to (\forall x.P(\ldots x \ldots))) = (\forall x.(Q \to P(\ldots x \ldots))) \qquad \{\to\forall\}$$
$$(Q \to (\exists x.P(\ldots x \ldots))) = (\exists x.(Q \to P(\ldots x \ldots))) \qquad \{\to\exists\}$$
$$(\neg(\forall x.P(\ldots x \ldots))) = (\exists x.(\neg P(\ldots x \ldots))) \qquad \{\neg\forall\}$$
$$(\neg(\exists x.P(\ldots x \ldots))) = (\forall x.(\neg P(\ldots x \ldots))) \qquad \{\neg\exists\}$$
$$(\forall x.P(\ldots x \ldots)) = (\forall y.P(\ldots y \ldots)) \qquad \{R\forall\}$$
$$(\exists x.P(\ldots x \ldots)) = (\exists y.P(\ldots y \ldots)) \qquad \{R\exists\}$$

x must not be a free variable in Q or in $P(\ldots y \ldots)$

y must not be a free variable in $P(\ldots x \ldots)$

Figure 2.25
Equations of quantifier reasoning.

Next, we rename the bound variable x in $\forall x$ on the left of the implication. We change x to u, again choosing a name that differs from all the other variable names in the formula.

$$((\exists v.(\forall x.P(x,v))) \to (\forall x.(\exists y.P(x,y)))) = (\exists v.(\forall u.P(u,v))) \to (\forall x.(\exists y.P(x,y))) \quad \{R\forall\}$$

We now have four variables in the formula: u, v, x, and y. Each quantifier is associated with a bound variable of a different name, so we won't mix up the roles of the variables. The next step is to migrate the quantifiers to the front of the formula. Then the implication will be inside the scope of all the quantifiers: $(\forall x.(\forall v.(\exists u.(\exists y.(\cdots \to \ldots)))))$. The equations in figure 2.25 provide the basis for quantifier migration. Applying the equations in a proof calls for a close look at parentheses to be sure that the cited equation from the figure matches the syntax of the formula that is under scrutiny. Migrating quantifiers is a tedious process and requires careful attention to detail.

Before continuing with the migration example, let's discuss the equations of figure 2.25, which show how quantifiers interact with certain logic operators. The figure lists some constraints on the names of variables. In the equations with binary operators ($\wedge$, $\vee$, $\to$), one of the operands is a proposition, Q. On the left-hand side of the equation, Q stands outside the quantification, but on the right-hand side it resides inside the quantification. The bound variable, x, in the quantification must not occur as a free variable in the formula for Q. If it did, the quantifier would *capture* the free variable on the right-hand side of the equation, where Q moves inside the scope of the quantifier. That capture would change the

2.7 Reasoning with Quantified Predicates

$$\begin{aligned}
& (\forall x.P(\ldots x \ldots)) \to Q \\
= \; & (\neg(\forall x.P(\ldots x \ldots))) \vee Q & \{\text{implication}\} \\
= \; & (\exists x.(\neg P(\ldots x \ldots))) \vee Q & \{\neg \forall\} \\
= \; & \exists x.((\neg P(\ldots x \ldots)) \vee Q) & \{\exists \vee\} \\
= \; & \exists x.(P(\ldots x \ldots) \to Q) & \{\text{implication}\}
\end{aligned}$$

Figure 2.26
Proof of equation $\{\forall \to\}$.

free variable x to a bound variable and change the meaning of the formula. That explains the note in figure 2.25 (page 58) prohibiting the occurrence of x as a free variable in Q.

A similar constraint applies to the renaming equations ($\{R\forall\}$ and $\{R\exists\}$) for the same reason: to avoid the capture of a variable by a quantifier. The constraints on variable names do not affect the applicability of the equations because the bound variables can always be renamed to avoid conflicts. Renaming a bound variable to a name that does not occur elsewhere in the formula never changes the meaning of the formula.

The rules $\{\neg\forall\}$ and $\{\neg\exists\}$ are quantifier analogs of the DeMorgan equations (figures 2.11 and 2.12, pages 29 and 30). We will sketch a proof of the $\{\neg\forall\}$ equation but leave the details and proofs of equations ($\{\neg\exists\}$, $\{\forall\wedge\}$, $\{\exists\wedge\}$, $\{\forall\vee\}$, and $\{\exists\vee\}$) as exercises.

Suppose that $\neg(\forall x.P(\ldots x \ldots)) = True$. Then, by the {double negation} equation and the truth table of the negation operator (figure 2.6, page 22), $(\forall x.P(\ldots x \ldots)) = False$. Therefore, according to the definition of universal quantification ($\forall$, page 51), there is some value x_0 in the universe of discourse for which $P(\ldots x_0, \ldots) = False$, which implies that $\neg P(\ldots x_0, \ldots) = True$. From the definition of existential quantification ($\exists$, page 52) we conclude that $(\exists x.(\neg P(\ldots x \ldots))) = True$.

Four of the equations in figure 2.25 (page 58) show how quantifiers interact with implication: $\{\forall\to\}$, $\{\exists\to\}$, $\{\to\forall\}$, $\{\to\exists\}$. In these equations, the implication is outside of the quantification on the left-hand side of the equation but inside the quantification on the right-hand side. In two of the equations, $\{\forall\to\}$ and $\{\exists\to\}$, the quantified formula is the left-hand operand of an implication ($\to$), and in those equations, the quantifiers flip when the implication moves inside the quantification: $\forall$ flips to $\exists$ and $\exists$ flips to $\forall$. The equations $\{\neg\forall\}$, $\{\neg\exists\}$, and {implication} team up to justify the flipping quantifiers. Figure 2.26 provides the details for equation $\{\forall\to\}$. The proof is similar for equation $\{\exists\to\}$. The other two equations, $\{\to\forall\}$ and $\{\to\exists\}$, have proofs similar to this one, but because the quantified formula is the right-hand operand of the implication, the quantifiers don't flip.

Let's continue the migration of quantifiers that we started earlier. We were working on a formula with quantifiers on both sides of an implication operator ($\to$), and we renamed some of the variables to avoid conflicts.

$$((\exists y.(\forall x.P(x,y))) \to (\forall x.(\exists y.P(x,y)))) = ((\exists v.(\forall u.P(u,v))) \to (\forall x.(\exists y.P(x,y))))$$

$$\begin{aligned}
&((\exists y.(\forall x.P(x,y)))) \rightarrow (\forall x.(\exists y.P(x,y)))) \\
= \ &((\exists v.(\forall x.P(x,v))) \rightarrow (\forall x.(\exists y.P(x,y)))) &&\{R\exists\} \\
= \ &((\exists v.(\forall u.P(u,v))) \rightarrow (\forall x.(\exists y.P(x,y)))) &&\{R\forall\} \\
= \ &(\forall x.((\exists v.(\forall u.P(u,v))) \rightarrow (\exists y.P(x,y)))) &&\{\rightarrow\forall\} \\
= \ &(\forall x.(\forall v.((\forall u.P(u,v)) \rightarrow (\exists y.P(x,y))))) &&\{\exists\rightarrow\} \\
= \ &(\forall x.(\forall v.(\exists u.(P(u,v) \rightarrow (\exists y.P(x,y)))))) &&\{\forall\rightarrow\} \\
= \ &(\forall x.(\forall v.(\exists u.(\exists y.(P(u,v) \rightarrow P(x,y)))))) &&\{\rightarrow\exists\}
\end{aligned}$$

Figure 2.27
Example: migrating quantifiers.

We are now working with the formula on the right-hand side of the equation, which has a different name for each bound variable. Presently, the implication resides outside of the scope of all four quantifiers. We want to migrate the quantifiers to the beginning of the formula and force the implication inside the scope of the quantifiers. We will migrate all four quantifiers, and we need to start somewhere. Later we will say something about how to choose a starting point, but for the moment we'll just pick one. Let's migrate the $\forall x$ quantification. Applying the $\{\rightarrow\forall\}$ equation (figure 2.25, page 58) leads to the following formula:

$$((\ldots) \rightarrow (\forall x.(\ldots))) = (\forall x.((\exists v.(\forall u.P(u,v))) \rightarrow (\exists y.P(x,y)))) \quad \{\rightarrow\forall\}$$

Again, we have a choice of quantifications to migrate. This time we choose $\exists v$ and come to the following formula by way of equation $\{\exists\rightarrow\}$:

$$(\ldots((\exists v.(\ldots)) \rightarrow (\ldots))\ldots) = (\forall x.(\forall v.((\forall u.P(u,v)) \rightarrow (\exists y.P(x,y))))) \quad \{\exists\rightarrow\}$$

At this point, the implication remains outside the scope of two quantifiers: $\forall u$ and $\exists y$. We migrate $\forall u$ first.

$$(\ldots((\forall u.(\ldots)) \rightarrow (\ldots))\ldots) = (\forall x.(\forall v.(\exists u.(P(u,v) \rightarrow (\exists y.P(x,y)))))) \quad \{\forall\rightarrow\}$$

Finally, we migrate $\exists y$.

$$(\ldots((\ldots) \rightarrow (\exists y.(\ldots)))\ldots) = (\forall x.(\forall v.(\exists u.(\exists y.(P(u,v) \rightarrow P(x,y)))))) \quad \{\rightarrow\exists\}$$

That completes the migration of quantifiers, which is the second step in our four-step strategy for reasoning about quantified formulas. Figure 2.27 puts the renaming and quantifier migration steps all in one place to make it convenient for you to check each step to be sure you see how they make use of the equations in figure 2.25 (page 58).

We now have a formula with all the quantifiers at the beginning. Step three is quantifier removal, and we start with the first quantifier, $\forall x$. A universal quantification is true if the quantified formula is always true regardless of what value from the universe of discourse is chosen for the bound variable. We want to prove that the formula is true, so if we remove

2.7 Reasoning with Quantified Predicates

$\forall x$, we will need to prove that the formula that remains is true, regardless of which value from the universe of discourse x denotes. We discard the $\forall$ quantifier and replace its bound variable by the symbol x_0. Then, we set out to prove that regardless of which value x_0 denotes, the formula has the value $True$.

We do the same for the second quantifier, $\forall v$, replacing its bound variable by the symbol v_0. Note that the symbols x_0 and v_0 are different from every other symbol in the formula $(P(u,v) \rightarrow P(x,y))$. That is crucial, as we pointed out earlier. Without the first two quantifiers, we are left with the following formula:

$$(\exists u.(\exists y.(P(u, v_0) \rightarrow P(x_0, y))))$$

If we can prove that it is true regardless of what values x_0 and v_0 stand for, we will have established that the original formula has the value $True$. By definition, an existential quantification is true if there is a value from the universe of discourse that makes the quantified formula true when it takes the place of the bound variable. So, we want to find a value for u and a value for y for which $(P(u, v_0) \rightarrow P(x_0, y)) = True$. The existential quantifications ($\exists$) are not particular about what these values are. We just need to find some that work.

Since u is the first index of the predicate P, and so is x_0, both u and x_0 come from the same universe of discourse. Therefore, we can choose $u = x_0$ if we want to. Let's do that. And let's choose $y = v_0$. We can do that because v_0 is in the universe of discourse for y. With these choices, we remove the existential quantifiers and get the following formula:

$$P(x_0, v_0) \rightarrow P(x_0, v_0)$$

That formula is an ordinary proposition in Boolean algebra. We can cite the {self-implication} axiom (figure 2.11, page 29) to conclude that $(P(x_0, v_0) \rightarrow P(x_0, v_0)) = True$, which proves the theorem.

This example illustrates a rationale for migrating $\forall x$, then $\exists v$, and finally $\forall u$ and $\exists y$. We were motivated by the desire to wait as long as possible before having to choose values for bound variables. That means that whenever we have a choice, we want to migrate a quantifier that will become a universal quantifier ($\forall$) before we migrate one that will become an existential quantifier ($\exists$). For example, when we were migrating quantifiers in the following formula, we considered migrating $\exists v$, and we also considered migrating $\forall x$:

$$((\exists v.(\forall u.P(u,v))) \rightarrow (\forall x.(\exists y.P(x,y))))$$

Both migrations led to a $\forall$ quantifier in the front, so they were equally advantageous. We chose to migrate the $\forall x$, but let's try migrating the $\exists v$ instead (citing $\{\exists\rightarrow\}$) to see how it goes. We get the following formula:

$$(\forall v.((\forall u.(P(u,v))) \rightarrow (\forall x.(\exists y.(P(x,y))))))$$

Now, there is a choice of migrating the $\forall u$ or $\forall x$. This time the choice matters. If we migrate $\forall x$ first (citing $\{\rightarrow\forall\}$), we get $\forall x$ at the front, but if we migrate $\forall u$ first (citing $\{\forall\rightarrow\}$), we get $\exists u$ at the front. We prefer to have universal quantifiers at the beginning of the formula because eliminating an existential quantifier forces us to choose a particular value from the universe of discourse for which the quantified formula is true. There might be only one such value, and we would be stuck with it. We want to leave the choice open if we can because that could make it possible to choose a beneficial value at a later stage. Therefore, we're better off migrating $\forall x$ first (citing $\{\rightarrow\forall\}$). That leaves us with the following formula:

$$(\forall v.((\forall x.((\forall u.P(u,v)) \rightarrow (\exists y.P(x,y))))))$$

To see the advantage of having universal quantifiers at the front of the formula, consider the formula $(\forall x.((\exists y.(x < y))))$, with pairs (x, y) of integers as the universe of discourse. We can remove the quantifier $(\forall x)$, choosing x_0 to represent an integer constant. This leaves the formula $(\exists y.(x_0 < y))$. Now it is easy to choose an integer for the variable y that makes the formula true. For example, $y = x_0 + 1$ leaves us with the formula $x_0 < x_0 + 1$. This exhibits an integer y for which $x_0 < y$, which means that $(\exists y.(x_0 < y)) = True$. Of course, there are many other choices for y that would also work, but by the definition of the existential quantifier $(\exists)$, we only need one.

On the other hand, suppose we were trying to prove the following statement, which is *actually false*: $(\exists y.((\forall x.(x < y))))$. The $\exists y$ quantifier has to be removed first, and we would like to choose an advantageous value for y. However, we don't get a choice with $\exists y$ because the truth of an existential quantification $(\exists)$ only guarantees that there is at least one value that works. So, there may be only one value available, and even if there were many, we could not pick one that would exceed all possible values of x. The upshot is that, whenever possible, we want to migrate quantifiers in such a way that universal quantifiers $(\forall)$ precede existential quantifiers $(\exists)$ to keep from getting trapped by early choices that don't work out in the end.

The four-step strategy is only one of many ways to prove theorems about formulas with quantifiers, and it happens to be an approach that makes it possible to use the mechanized logic of ACL2 to reason about quantified formulas. Natural deduction can also support reasoning about quantified formulas, but it calls for additional rules of inference to incorporate quantifiers into the scheme. The equations of figure 2.25 (page 58) extend the equations for propositions of figure 2.11 (page 29) into the realm of predicates and quantification.

Exercises

1. Prove $(\forall x.P(x)) \rightarrow (\exists x.P(x)) = True$ if the universe of discourse is not empty.

2. What is the value of $(\forall x.P(x)) \rightarrow (\exists x.P(x))$ if the universe of discourse is empty?

2.8 Boolean Models

3. What goes wrong in trying to prove $(\exists x.P(x)) \rightarrow (\forall x.P(x)) = True$? Explain why this is different from exercise 1.

4. Prove $(((\forall x.(P(x) \rightarrow Q(x))) \wedge (\forall x.(Q(x) \rightarrow R(x)))) \rightarrow (\forall x.(P(x) \rightarrow R(x)))) = True$.

5. Prove equation $\{\neg\exists\}$. *Hint*: Cite equations $\{\neg\forall\}$ and $\{$double negation$\}$.

6. Prove equation $\{\exists\wedge\}$.

7. Derive the equations $\{\forall\vee\}$ and $\{\exists\vee\}$ from axioms and proven equations.
   Hint: Cite $\{\exists\wedge\}$ and the DeMorgan equations.

8. Prove equations $\{\exists\rightarrow\}$, $\{\rightarrow\forall\}$, and $\{\rightarrow\exists\}$. *Hint*: Emulate the proof of $\{\forall\rightarrow\}$.

2.8 Boolean Models

A Boolean variable x stands for a proposition such as "Socrates is a man." In the domain of logic, this represents a true/false value, and Boolean formulas and equations provide a way to carry out a consistent analysis of relationships between propositions. However, stating the proposition in the form "Socrates is a man" suggests that we expect the proposition to have meaning in a real-world domain. If the variable x stands for the statement "Socrates is a man," we expect x to be *True* if the statement is consistent with a fact in the real world but *False* if it is inconsistent with observations. Even if we don't know the facts, we expect x to be either *True* or *False*, not both and not something else, according to the actual state of underlying conditions unknown to us.

In other words, we intend the proposition x to mean something in the real world, and we use logic to analyze relationships between this proposition and other assertions about the real world stated in the form of true/false propositions. In logic mode, we apply the mathematical axioms and rules of inference, and in the end we interpret our conclusions, which are statements in logic, as statements about the real world. This makes sense only so far as the propositions we started with accurately model the real-world domain that we want them to represent. In the domains of software and digital circuits, the correspondence is extremely reliable. It is rare for a digital circuit to go haywire or an interpreter of a programming language to go awry.

Accurate modeling of other domains by propositions in logic can be problematic, but we want to expand the horizon a bit with a discussion of the game of *Nim*. There are many variants of this game, and we will consider a simple one. The game starts with a pile of ten stones. Two players (Alice and Bob in this discussion) take turns removing one, two, or three stones from the pile. The player who picks up the last stone loses. The chart in figure 2.28 (page 64) summarizes a game that Alice won.

We will try to simulate the play of the game of Nim using propositional logic. We will use Boolean variables to represent the pile of stones. For example, we could use the variable x to mean "there are 10 stones in the pile" and y to mean "there are 9 stones in the pile." If we

Move	Alice	Bob	Stones
0			10
1	Remove 2		8
2		Remove 3	5
3	Remove 1		4
4		Remove 2	2
5	Remove 1		1
6		Remove 1	0

Figure 2.28
A game of Nim.

continue with this scheme, we will have a lot of names with no easily recalled connection to their meanings.

Another approach is to let $x10$ mean "there are 10 stones in the pile," $x9$ that "there are 9 stones in the pile," and so on down to $x0$ for an empty board. The collection of these 11 Boolean variables can describe any pile, but they could just as well describe an impossible situation. For example, if both $x1$ and $x5$ were $True$, the pile would consist of one stone and also consist of five stones, and this cannot be correct. To make our model in logic correspond to a real game of Nim, we need constraints on the relationships between the variables.

The game restricts the number of stones in a pile to between zero and ten. We can account for this restriction by requiring the following formula to have the value $True$:[20]

$$x0 \lor x1 \lor x2 \lor \cdots \lor x10$$

This rule would avoid some impossible situations in a real game of Nim. Another restriction is that there cannot be both zero stones in the pile and one stone in the pile nor both zero stones and two stones, and so on. We can account for this restriction by asserting that the following formula has the value $True$:

$$(\neg(x0 \land x1)) \land (\neg(x0 \land x2)) \land \cdots \land (\neg(x0 \land x10))$$

Of course, we also need to eliminate the possibility of having both one stone and two stones, one stone and three stones, and so on.

$$(\neg(x1 \land x2)) \land (\neg(x1 \land x3)) \land \cdots \land (\neg(x1 \land x10))$$

---

[20] To save space, part of the formula has been elided, but we think you can fill in the missing parts. Many of the formulas in our model of Nim will be abbreviated in this way.

2.8 Boolean Models

This new rule could have also included the $(\neg(x0 \wedge x1))$ formula, but we are going to require all our rules to hold, so we don't need to repeat a restriction already made by the first rule.

We could carry on in this vein, but to capture the play of an entire game of Nim, we would need to be able to deal with changes in the pile of stones as the game progresses. Our considerations so far have been limited to an unchanging pile of stones. We need to add a time component to the model. To do that, we will need to have many more variables, and it will help if we name them in a way that helps us remember what they mean in the game.

Our new scheme will use the variable x_{10}^3 to mean "there are 10 stones in the pile after 3 moves." The variable x_5^2 would have the value $True$ for the game summarized in figure 2.28 (page 64) because there are five stones remaining after Bob makes the second move of the game. It is important to recognize that a Boolean variable like x_{10}^3 is no different, except in name, from Boolean variables such as y or z. Both the subscript and the superscript are just part of the name. It is tempting to believe that since we have a variable called x_{10}^3, we have variables called x_{12}^7 and x_3^{26}. However, there are no such variables in our model because there cannot be twelve stones in the pile nor can there be a move number twenty-six.

The subscript part of the name stands for the number of stones remaining, so it must be between zero and ten, and the superscript stands for the move number, so it also must be between zero and ten because, since each move removes at least one stone, no game can have more than ten moves. The numbered portions of the name tell us at a glance how many stones are in the pile and how many moves have been made. There are a lot of things to keep straight. We need 121 variables in all: 11 pile sizes (zero stones through 10 stones) times 11 move numbers (move 0 through move 10, but no move 11 because at least one stone is removed in each move). Our naming scheme makes the model easier to comprehend.

Initially, there are ten stones in the pile, so the variable x_{10}^0 has the value $True$ and the variables $x_0^0, x_1^0, x_2^0, \ldots x_9^0$ all have the value $False$. This is the "initial condition" for the game of Nim, expressed in terms of the Boolean variables we are using in our model.

Another constraint is that after move number six, the number of stones remaining will still be between zero and ten. That is, the following formula has the value $True$:[21]

$$x_0^6 \vee x_1^6 \vee x_2^6 \vee \cdots \vee x_{10}^6.$$

Our model has ten formulas like this. When we combine all of them with logical-and (referred to as AND, here, to make the description more compact), we get a formula that constrains the number of stones in the pile at each stage in the game. The formula will have the value $True$ for any properly played game of Nim.

---

[21] Actually, the formula could be more restrictive because after move six there cannot be more than four stones remaining. We will account for this constraint later and in a more general way.

Another kind of constraint in a properly played game is one we discussed before we added variables to the model to account for the move number. At every stage, there is a particular number of stones remaining in the pile. After move 3, for example, there cannot be both three stones remaining and five stones remaining. So we need some rules like the following after six moves:

$$(\neg(x_0^6) \wedge (x_1^6)) \wedge (\neg(x_0^6 \wedge x_2^6)) \wedge \cdots \wedge (\neg(x_0^6 \wedge x_{10}^6))$$

This formula disallows piles having both zero and five stones or zero and seven stones, but as before, we need nine additional formulas to disallow piles with both one and two stones, one and three stones, and so on.

$$(\neg(x_1^6 \wedge x_2^6)) \wedge (\neg(x_1^6 \wedge x_3^6)) \wedge \cdots \wedge (\neg(x_1^6 \wedge x_{10}^6))$$

That is a total of ten formulas that must be combined using AND to rule out impossible combinations after move six. And that's just move six. We need a group of rules like this to describe the pile after zero moves, one move, two moves, and so on, up to ten moves. These eleven groups of rules should also be combined using AND, making 11 times 10, or 110, separate rules combined with AND.

What about the rules for removing stones from a pile? For example, if the pile has five stones after move three, then it must have four, three, or two stones after four because the player removes either one, two, or three stones in step four. This can be captured with the following rule:

$$x_5^3 \rightarrow (x_4^4 \vee x_3^4 \vee x_2^4)$$

We need to be careful when we get down to piles with fewer than three stones. For example, for the case when there are only two stones left in the pile, we need a rule of the following form:

$$x_2^3 \rightarrow (x_1^4 \vee x_0^4)$$

And what do we do when we run out out of stones? One way to proceed is to continue the pattern and just ensure that there are no stones at the next turn.

$$x_0^3 \rightarrow x_0^4$$

Of course, many such rules will be needed, one rule for each possible combination of stones left and number of moves made.

Now consider a single formula combining all the constraints.

Nim-constraints (a formula that is True for all Nim games)
---
the initial pile containing ten stones AND

the possible legal descriptions of piles at all times AND

the legal ways in which a pile transforms from one move to the next

2.8 Boolean Models

The Nim-constraints formula has the value *True* for all properly played games of Nim. Every Nim game corresponds to a particular combination of values of the Boolean variables in the model. The combination must ensure that the Nim-constraints formula is *True*, and the Nim game corresponding to that combination can be reconstructed from an inspection of the values of the Boolean variables. One combination will correspond to the game in which Alice removes two stones, then Bob removes three, then Alice one more, and so on.

But how do we know who wins? Since Alice moves first, her move numbers are odd (1, 3, 5, ...), and Bob's are even (2, 4, 6, ...). According to the rules, the person who picks up the last stone loses. So, we can characterize a win by one player or the other by the following formulas:

Alice-wins: $x_0^2 \vee x_0^4 \vee x_0^6 \vee x_0^8 \vee x_0^{10}$
Bob-wins: $x_0^1 \vee x_0^3 \vee x_0^5 \vee x_0^7 \vee x_0^9$

If the Alice-wins formula is *True* for a combination of values of the variables that represents a particular game, then Alice won the game. If it is *False*, Bob won. If we were to AND the Nim-constraints formula with the Alice-wins formula and find it to have the value *True*, then we could conclude that Alice always wins the game. If it were *False*, we could conclude that Alice never wins. But there is a third possibility. It may be that for some values of the Boolean variables, the Alice-wins formula is *True* and for other values it is *False*.

Each possible combination of the Boolean variables corresponds to a single game of Nim. Furthermore, for any combination of values of the variables that makes the Nim-constraints formula true, the values of the individual Boolean variables show how to reconstruct the game. So, this collection of Boolean formulas provides a model of the game of Nim that can be used to analyze the play and figure out which player won.

Now here is something that may surprise you. It is one of the central themes of computer science that models of computer programs can take a form similar to the model we discussed for the game of Nim but only if we have an *upper bound on the number of steps in the computation the program represents*. To do this on the scale of a computer program, the formula would need to combine the following specifications:

1. The initial configuration of the computer, AND
2. the possible legal configurations of the computer at all times, AND
3. the legal transitions of the computer from one step to the next, AND
4. a formula that indicates the completion of the computation.

Since digital computers store everything as sequences of ones and zeros, the initial configuration is simply the initial values of those bits. The legal configurations rule out the possibility that a given bit is both a one and a zero at the same time. The legal transitions correspond to the actual program and how it manipulates information over time. And a traditional way to specify that the program is finished is to choose a specific bit in the model, to include in the initial conditions that the chosen bit is zero, and to insist that the bit remains zero until, as its last act, the program sets the bit to the value of one.

So there you have it. If you know that a given program will execute in, say, 10 million or fewer steps, then (in principle) you can write down a Boolean formula that completely captures the computation that the program represents. In this sense, Boolean formulas are sufficient to represent any computer program as long as you can state ahead of time how long it will run. Generally speaking, this kind of analysis isn't feasible because the formulas have way too many variables. However, for small components, specifications of this kind are feasible, and they can be combined to analyze larger systems. The key is modularity: keeping the components small and manageable and combining them in ways that limit a sudden explosion of complexity.

Exercises

1. Did you notice the mistake in the way we described the game of Nim? Suppose that Alice wins in step 5, so that x_0^6 is true. According to our rules, that means that x_0^7 must also be true, but then from the description of "Bob wins," you would conclude that both Alice and Bob won. That is not at all what we meant! Modify the formula that says "Alice wins" to take care of this complication.

2. There are many variants of the game of Nim. In another variant, there are three piles of stones, and each player can remove one or more stones from any single pile. The player who removes the last stone loses the game. Discuss some ways in which a Boolean model of this game using Boolean variables and formulas would differ from the Boolean model of Nim that was discussed in this section.

2.9 More General Models with Predicates and Quantifiers

Boolean formulas and equations can be used to describe and analyze complicated, real-world artifacts. That gives the Boolean domain greater scope than is apparent at first glance, but using it, even for artifacts of modest size, can be cumbersome to the point of infeasibility. Hundreds, thousands, or millions of Boolean variables are needed, and they are difficult to deal with in an understandable fashion.

The Boolean model of the game of Nim in section 2.8 defined 121 variables to deal with various aspects of the game. The variable x_s^t was *True* if after move t there were s stones left in the pile. Let's see how this model would work out if we used a predicate X instead of 121 individual variables. The predicate will have two indexes: (1) an integer s between 0 and 10 representing the number of stones in the pile and (2) an integer t between 0 and 10 indicating the move number. The universe of discourse for the first index is the same as that for the second index, namely, the set of integers between 0 and 10.

The proposition $X(s, t)$ in predicate X has the same meaning as the variable x_s^t in the propositional model. Since the variables are organized in a predicate, we can use quantifiers to make assertions with compact and understandable formulas. In any game of Nim, there is a time at the end when no stones remain in the pile, so the formula $(\exists t.X(0,t))$ is *True*. The formula means that the game of Nim always ends in ten or fewer moves. The Alice-wins formula from the Nim model (page 67) would be $\exists t.(Even(t) \wedge X(0,t))$, where $Even(t)$ is the predicate that detects even numbers (page 56). The formula $\forall s.((s < 10) \rightarrow (\neg X(s,0)))$ is *True* because there are 10 stones in the pile at move 0.

In the propositional model, we specified that the pile after six moves cannot have both no stones and one stone, nor zero stones and two stonres, and so on.

$$(\neg(x_0^6 \wedge x_1^6)) \wedge (\neg(x_0^6 \wedge x_2^6)) \wedge \cdots \wedge (\neg(x_0^6 \wedge x_{10}^6))$$

Using quantifiers, we can write this more succinctly.

$$(\forall s.((s \neq 0) \rightarrow (\neg(X(0,6) \wedge X(s,6)))))$$

Using propositions, we needed to consider ten different cases, one for each possible number of stones other than zero. With the universal quantifier, we need only one formula to describe all of those possibilities. In fact, we can cover more ground. The propositional model also described the fact that no pile can have both one and two stones at the same time, or one stone and three stones, and so on. All of those expressions had to be combined using AND. Using universal quantification ($\forall$), we can cover all possible combinations for move number 6 with a single formula.

$$(\forall s_1.(\forall s_2.((s_1 \neq s_2) \rightarrow ((\neg(X(s_1,6) \wedge X(s_2,6))))))) $$

To specify this constraint for the entire game, not just move number 6, we need ten more similar formulas combined with AND. This can be expressed with a universal quantification spanning the move numbers ($\forall t \ldots$). This single formula is equivalent to over 500 formulas that would have been necessary were we to build the model without using predicates and quantifiers.

$$(\forall s_1.(\forall s_2.(\forall t.((s_1 \neq s_2) \rightarrow (\neg(X(s_1,t) \wedge X(s_2,t)))))))$$

In the game of Nim, the propositional formula specifying that the pile must contain between zero and ten stones after six moves was written as follows:

$$(x_0^6 \vee x_1^6 \vee x_2^6 \vee \cdots \vee x_{10}^6).$$

An existential quantification makes it more succinct: $(\exists s.X(s,6))$. As before, we can use universal quantification to extend this requirement across all the moves: $(\forall t.(\exists s.(X(s,t))))$. The order of the quantifiers matters. This expression says that at any given time, there is a specific number of stones in the pile. If the quantifiers were reversed, the statement would indicate that the pile has some fixed number of stones at all possible points in time, which is false because the size of the pile changes with every move.

> **Box 2.11**
> Other Quantifiers
>
> Universal (∀) and existential (∃) quantifiers are not the only possibilities. A "for most" quantifier (≈∀) might be perfectly sensible. In some branches of mathematics, it is common to use the quantifiers "almost everywhere" and "almost nowhere." They sound fuzzy but they're not. They have precise mathematical definitions.

Quantifiers make formulas using predicates more compact than those that stay in the domain of propositions. We saw in the previous section a way to characterize the game of Nim using just Boolean variables, with a formula that specifies the following conditions:

1. The initial pile containing ten stones, AND
2. the possible legal descriptions of piles at all times, AND
3. the legal ways in which a pile transforms from one step to the next, AND
4. the formula that "Alice wins."

The formulas in this section specify the same model but in a more understandable way and in less space. However, predicate logic can do more than simplify the formulas. It can express bigger ideas. Consider the formula with the meaning "Alice wins." In the propositional model this was built by taking together the more specialized formulas "Alice wins at time 2," "Alice wins at time 4," ..., "Alice wins at time 10" and joining them all with logical-or. Using predicates and quantifiers, the formula $(\exists t.(AliceWinsAtTime(t)))$ expresses the same idea.

This is not just a simple syntactic change. We have gained much more than the ability to collect all possible "Alice wins at time ..." phrases into a single statement. The variable t can potentially stand not just for the numbers zero through ten but for any possible number of moves, so this same approach applies to open-ended games like chess. Using propositions, we could (in principle) completely describe the game of chess, but only if we limit chess games to, say, 200 moves. With predicates, there is no such limitation. We can refer to an arbitrary time at which black wins because the universe of discourse can be infinite.

The same is true for models of computer programs. Before, to build a model we had to limit the number of computational steps to a fixed upper bound, such as a million steps. Then (in principle) a Boolean formula could be constructed that was *True* if and only if the program produced a given answer. With predicates, we can do away with the upper bound on computational steps. The computation can be open-ended because the universe of discourse for computational steps can have an infinite number of elements. It could be the set of natural numbers, for example. It is possible, in principle, to describe *all* possible computer programs using formulas in predicate logic. Our primary use of predicate logic will be in the analysis of computer programs and digital circuits, which will involve describing models of programs and circuits.

3 Software Testing and Prefix Notation

We will be using a software development environment that automates parts of the software testing process. To use it, we can define tests that check results corresponding to inputs provided directly in the test. Or, we can ask the system to generate random data and check results against logic formulas specifying properties that we expect the results to have.

Box 3.1
Proof Pad and ACL2

Proof Pad is a tool that provides a convenient way to write, test, and reason about programs written in the ACL2 dialect of a programming language called Lisp. ACL2 includes a mechanized logic that supports rigorous checking of the reasoning process, and we will use it for that purpose starting in this chapter.

We will begin to use Proof Pad in this chapter, so you should install it on your computer. Get it at the following URL. It's free.
```
                    http://proofpad.org
```

To get the syntax straight, let's specify some direct tests that Proof Pad can carry out for the numeric addition operator. The following tests check some equations that we expect to be true: 2 + 2 = 4, 5 + 7 = 10 + 2, etc.

```
(include-book "testing" :dir :teachpacks)
(check-expect (+ 2 2) 4)
(check-expect (+ 5 7) (+ 10 2))
(check-expect (+ 27 6) (+ (+ 23 5) (+ 2 2)))
```

This notation looks strange, with formulas like (+ 2 2) instead of 2 + 2. ACL2 expresses formulas with a prefix notation rather than the customary infix notation. In ACL2's prefix notation, a computational formula starts with a parenthesis, followed by the operator it invokes. The operator comes first (as a "prefix"), then the operands follow. Infix notation places the operator between the operands but prefix notation puts it before the operands.

That seems simple enough, but it gets gradually more unfamiliar as the formulas get more complicated. For example, a formula for the sum of x and three times y, $(x + 3 * y)$, comes out as $(+\ x\ (*\ 3\ y))$ in prefix form. This will take a little getting used to, but it is a lot easier to get good at than most of the stuff you have been studying. We hope you will be willing to accept it as just another way of writing formulas. One advantage is that the formulas are, by default, fully parenthesized, so there is never any question about what operands go with which operators. Some people end up liking the prefix format better than infix. We will use both, in different contexts, depending on whether we're doing paper-and-pencil reasoning or formal, mechanized logic.

Another important element is the include-book directive. The facility for tests of this kind resides in an ACL2 book known as testing, which in turn resides in a directory known as :teachpacks (with a leading colon). The include-book directive tells the system the name of the testing package and where to find it. This boilerplate is needed (including the keyword :dir, with the leading colon) whenever you want to use check-expect testing, even though we won't repeat it in all of our examples.

If you install Proof Pad and put these tests in a file with a .lisp extension (plusoptests.lisp, for example), you can use Proof Pad to run the tests. The first two tests pass without fanfare, but the third test has a mistake. We should not expect that $27 + 6 = (23 + 5) + (2 + 2)$. The system reports that the assertion failed. If you change the 6 to 5, the test will pass, as it should. Normally, when we run tests, we want to know whether the operator is doing the right thing, but in this case the operator (+) was fine. It was the test that was wrong.

Direct, one-off, check-expect tests are a good way to get started, but we will want to test more general properties that we specify with logic formulas. Proof Pad facilitates that kind of testing with a package called DoubleCheck that generates random data to run against a predicate (that is, a formula representing a true/false value). To demonstrate the idea, consider testing the addition operator to see if it conforms to the associative law for addition: $(x + (y + z)) = ((x + y) + z)$ ({+ associative}, figure 2.1, page 16). To do that, we define a property as an equation and direct Proof Pad to generate some random numbers and report whether or not the equation holds for all of the random data. The Proof Pad "doublecheck" package, like the "testing" package, resides in the teachpacks directory. An include-book directive is required to make it accessible for use.

```
(include-book "doublecheck" :dir :teachpacks)
(defproperty +associative-test
  (x :value (random-integer)
   y :value (random-integer)
   z :value (random-integer))
  (= (+ x (+ y z))
     (+ (+ x y) z)))
```

A property definition with DoubleCheck begins with a parenthesis, then the keyword defproperty and a name for the property to be tested, which is +associative-test in this example.[22] A sequence of data generator specifications follows the name of the property. The entire sequence is enclosed in parentheses, and each specification has three parts: a name, the keyword :value (with the leading colon), and a generator, also enclosed in parentheses. In this example, the definition says to use random integers. DoubleCheck is able to generate many kinds of random data, and we will gradually introduce different data generators as the story progresses. Finally, a predicate formula specifies the test. In this case, the predicate is an equation, written of course in prefix notation.

Let's look at another example. ACL2 can deal with a collection of data elements by keeping them in an ordered sequence called, and this is a technical term, a *list*. There is an operator, list, that creates a list of its operands (any number of operands). Another operator, append, concatenates two lists. A few check-expect tests may clarify what the append and list operators do. Later, we will describe them formally.

```
(check-expect (append (list 1 2 3 ) (list 4 5)) (list 1 2 3 4 5))
(check-expect (append (list 9 8 7) (list 6)) (list 9 8 7 6))
(check-expect (append (list 11 7) (list 2 5 3)) (list 11 7 2 5 3))
(check-expect (append (list 2 0) (list 1 8)) (list 2 0 1 8))
```

We would anticipate that the length of a concatenation of two lists would be the length of the first list plus the length of the second. Using the len operator, which computes the length of a list, we can state this property in a DoubleCheck test. Later, we will prove this relationship between append and len, but for the moment we confine ourselves to stating properties and asking Proof Pad to check them against random data.

```
(defproperty additive-law-of-concatenation-test
  (xs :value (random-list-of (random-integer))
   ys :value (random-list-of (random-integer)))
  (= (len (append xs ys))
     (+ (len xs) (len ys))))
```

Box 3.2
Natural Numbers

A *natural number* is an integer that is zero or more. Some treatments start the naturals with one rather than zero, but in the computing domain, starting with zero simplifies some formulas.

When we discuss computer arithmetic, we will need to know something about arithmetic on finite sets of numbers, and the mod operator will play a crucial role. The mod operator takes two operands and delivers the remainder in the division of the first operand by the

---

[22] The rules for naming things in ACL2 are lax compared to many programming languages. Names can be made up of letters and digits, of course, but they can also include some special characters, such as plus signs, hyphens, and colons (not semicolons or commas, though).

second. ACL2 does not restrict the operands to the set of natural numbers, but in this discussion, we are going to stick with that domain. Think divisor, dividend, quotient, remainder, as in long division: third-grade stuff.

Let's test drive the mod operator, starting with some simple checks, such as using mod to compute the remainder when dividing by two, which will produce zero for even numbers and one for odd numbers.

```
(check-expect (mod 12 2) 0)
(check-expect (mod 27 2) 1)
```

Here are a few more sanity checks, this time with three as the divisor:

```
(check-expect (mod 14 3) 2)
(check-expect (mod  7 3) 1)
(check-expect (mod 18 3) 0)
```

Box 3.3
Modular Arithmetic

Modular arithmetic (aka clock arithmetic) produces integers in a fixed range: $0, 1, \ldots (m-1)$, where $m > 0$ is a natural number known as the *modulus*. If x is an integer, the formula (x mod m) denotes the remainder in the division of x by m ($0 \le (x \bmod m) < m$). Modular addition, subtraction, and multiplication are consistent with ordinary arithmetic.

$$((x + y) \bmod m) = (((x \bmod m) + (y \bmod m)) \bmod m)$$
$$((x - y) \bmod m) = (((x \bmod m) - (y \bmod m)) \bmod m)$$
$$((x \times y) \bmod m) = (((x \bmod m) \times (y \bmod m)) \bmod m)$$

The ACL2 operation (mod x m) delivers the remainder in $x \div m$. Our discussion restricts the operands to natural numbers but ACL2 doesn't.

We can use Proof Pad to put mod through its paces on a large number of tests using DoubleCheck, but for that we will need to come up with relationships that express more general properties of division and remainders. One such property is that the remainder doesn't change when the divisor is added to the dividend.

```
(defproperty mod-invariant
  (divisor-minus-1 :value (random-natural)
   dividend        :value (random-natural))
  (let* ((divisor (+ divisor-minus-1 1))) ; avoid zero divisor
    (= (mod dividend divisor)
       (mod (+ dividend divisor) divisor))))
```

Generating random data is an art. In this example, we have made sure the divisor isn't zero by adding one to a natural number. Since negative numbers aren't natural numbers, adding one to a natural number ensures that the sum is nonzero. The definition of the mod-invariant property uses a let* formula, which provides a way to attach names to values temporarily (box 3.4). In this case, the let∗ says that the name divisor will stand for the value represented by the formula (+ divisor-minus-1 1).

Box 3.4
Local Names for Values: let∗

A let∗ formula attaches values to names. The scope of the let∗ opens with a parenthesis, then the keyword let∗. The scope continues with a sequence of name/value bindings and ends with a formula for the value to be delivered by the let∗. A final parenthesis closes the scope.

Each binding has the form (h v), where h is a name and v is a formula specifying a value to be associated with that name (h does not denote an operator). The parentheses around each binding are required. Any of the names attached to values in the bindings denote those values at any subsequent point within the scope of the let∗, but the name/value associations do not extend outside that scope. There can be any number of bindings. The most common use of let∗ is to give names to values needed more than once in a computation, but sometimes let∗ is used to provide mnemonic names to make a formula more easily understood.

The parentheses in a let∗ formula are tricky. Each individual binding is enclosed in parentheses, and the whole sequence of bindings is enclosed in another set of parentheses. In addition, the entire let∗ formula is enclosed in parentheses. A let∗ formula delivers a value v_{let*} specified by a formula positioned after name/value bindings.

$$(\text{let}* \ ((h_1 \ v_1) \ (h_2 \ v_1) \ldots (h_n \ v_n)) \ v_{let*})$$

Since (mod x m) < m when $m > 0$, the mod operator will pass the following test:

```
(defproperty mod-upper-limit-test
  (divisor-minus-1 :value (random-natural)
   dividend        :value (random-natural))
  (let* ((divisor (+ divisor-minus-1 1)))  ; avoid zero divisor
    (< (mod dividend divisor) divisor)))
```

In this test, the property is not expressed as an equation but as an inequality specified with the less-than operator (<). As always, the formula puts the operator in the prefix position, in front of its operands. For practice, add this property to the .lisp file with the other tests and use Proof Pad to run it.

When one natural number is divided by another, the remainder is also a natural number. We can use the logical-and operator to combine the upper-limit test with the natural-number test in one property definition ("and" is the ACL2 name for ∧). The value of the formula (natp x) is true if x is a natural number and false if it isn't.

Axiom {*natp*}

(natp x) = x *is a natural number* = (and (integerp x) (>= x 0))
Note: (integerp x) is true if x is a whole number and false otherwise.

```
(defproperty mod-range-test
  (divisor-minus-1 :value (random-natural)
   dividend        :value (random-natural))
  (let* ((divisor (+ divisor-minus-1 1)))  ; avoid zero divisor
    (and (natp (mod dividend divisor))
         (< (mod dividend divisor) divisor))))
```

There are two parts to the result of dividing one number by another: the quotient and the remainder. The **mod** operator delivers the remainder and an operator called **floor** (box 3.5, page 77) delivers the quotient. The quotient is always strictly smaller than the dividend when the divisor is bigger than one and the dividend is a nonzero, natural number. The following test checks for that property. The random-value generator for the divisor makes sure the divisor exceeds one by adding two to a natural number. Similarly, we make sure the dividend isn't zero by adding one.

```
(defproperty quotient-less-than-dividend-test
  (divisor-minus-2   :value (random-natural)
   dividend-minus-1  :value (random-natural))
  (let* ((divisor  (+ divisor-minus-2 2))   ; divisor > 1
         (dividend (+ dividend-minus-1 1))) ; dividend > 0
    (< (floor dividend divisor)
       dividend)))
```

Checking the result of a division is a matter of multiplying the quotient by the divisor and adding the remainder. If this fails to reproduce the dividend, something has gone wrong in the division process. The following property tests this relationship between the **mod** and **floor** operators. It needs to use the multiplication operator (*).

```
(defproperty division-check-test
  (divisor-minus-1 :value (random-natural)
   dividend        :value (random-natural))
  (let* ((divisor (+ divisor-minus-1 1)))  ; avoid zero divisor
    (= (+ (* divisor (floor dividend divisor))
          (mod dividend divisor))
       dividend)))
```

We hope by now you are starting to get comfortable with prefix notation and using Proof Pad to run tests. The exercises will give you a chance to practice.

Exercises

1. Define a test of the floor operator that checks to make sure its value is a natural number when its operands are natural numbers and the divisor (second operand) is not zero. Use Proof Pad to run the test.

2. The max operator chooses the larger of two numbers: (max 2 7) is 7, (max 9 3) is 9. Define a DoubleCheck property that tests to make sure (max x y) is greater than or equal to (>=) both x and y. Use Proof Pad to run tests of the property.

Exercises

3. Define a DoubleCheck property to test the distributive law of arithmetic ({distributive law}, figure 2.1, page 16). Use Proof Pad to run your test.

4. Define DoubleCheck properties to test consistency between clock arithmetic and ordinary arithmetic as described in box 3.3 (page 74). Use Proof Pad to run your tests.

5. The ACL2 operator reverse delivers a list whose elements are in the reverse order of those in its operand. For example, (reverse (list 1 2 3)) is (list 3 2 1). Find an equation that expresses the value of (reverse (append *xs ys*)) in terms of (reverse *xs*) and (reverse *ys*). Define a property based on your equation and use Proof Pad to test it.

Box 3.5
Floor and Ceiling Operators, Floor and Ceiling Brackets

ACL2 provides two operators, ceiling and floor, that divide one number by another, then round the result up or down to an integer value. In algebraic formulas, special brackets known as floor and ceiling brackets indicate the direction of rounding. Ceiling brackets (like square brackets with the bottom chopped off) indicate the next integer that is the same as or larger than the enclosed value. Floor brackets (square brackets without the top part) go in the other direction, down to the next integer that is the same as or smaller than the enclosed value.

$$(\text{floor } x \ y) = \lfloor x \div y \rfloor \text{ quotient rounded down}$$
$$(\text{ceiling } x \ y) = \lceil x \div y \rceil \text{ quotient rounded up}$$

---

4 Mathematical Induction

4.1 Lists as Mathematical Objects

A sequence is an ordered list of elements. In fact, for our purposes, the terms "list" and "sequence" are synonyms. Many things that computers do come down to keeping track of lists, so lists are an important class of mathematical objects. We will need a formal notation, including an algebra of formulas, to discuss lists with the level of mathematical precision required in specifications of computer hardware and software.

Informally, we will write lists as sequences of elements separated by spaces, with square brackets marking the beginning and end of the list. For example, [8 3 7] denotes a list with first element 8, second element 3, and third element 7, and [9 8 3 7] denotes a list with the same elements plus an additional element 9 at the beginning. We use the symbol nil for the empty list (that is, the list with no elements). We use square brackets rather than round ones in formulas specifying lists, to avoid confusion with formulas that invoke operators.[23] For example, [4 7 9] denotes a three-element list, whereas (+ 7 9) is a numeric formula representing the value 16. However, ACL2 does not employ this square-bracket notation. When it displays the list [4 7 9], it uses round brackets: (4 7 9). The square-bracket notation helps keep data in computational formulas straight in written discussion and also makes some formulas more compact in writing, but square-bracket notation for lists is not ACL2 notation.

One of the basic operators in the algebra of lists is the construction operator cons, which inserts a new element at the beginning of a list. Formulas using cons, like all formulas in the mathematical notation we have been using to discuss software concepts, are written in prefix form. So, the formula (cons x xs) denotes a list with the same elements as the list xs plus an additional element x inserted at the beginning. If x stands for the number 9 and xs stands for the list [8 3 7], then (cons x xs) constructs the list [9 8 3 7].

---

[23] To *invoke* an operator is to apply it to its operands to make a computation. An *invocation* is a formula that invokes an operator.

$[x_1 \; x_2 \ldots x_n]$ = (list $x_1 \; x_2 \ldots x_n$) = (cons x_1 (cons x_2 ... (cons x_n nil) ...))

[1 2]   = (list 1 2)   = (cons 1 (cons 2 nil))
[16 256 4096] = (list 16 256 4096) = (cons 16 (cons 256 (cons 4096 nil)))
[1 9 4 7]   = (list 1 9 4 7)   = (cons 1 (cons 9 (cons 4 (cons 7 nil))))

Figure 4.1
Shorthand for nested cons: list.

Box 4.1
Square Bracket Notation for Lists: Paper-and-Pencil Only

Most of the time, we will use square bracket notation for lists to distinguish them from computational formulas. However, ACL2 does not display lists with square brackets. It uses round parentheses both for lists and for computational formulas.

paper-and-pencil only   *formal ACL2 notation*
[1 2 1 5]   (list 1 2 1 5)

Any list can be constructed by starting from an empty list and using the construction operator to insert the elements of the list, one by one. The empty list, nil, which is intrinsic in ACL2, needs no construction. Nonempty lists are constructed using the cons operator. The formula [8 3 7] is paper-and-pencil shorthand for (cons 8 (cons 3 (cons 7 nil))). ACL2 also has shorthand for nested cons operations (figure 4.1): (list 8 3 7) is another way to write the formula (cons 8 (cons 3 (cons 7 nil))).

Box 4.2
Equal by Definition: $\equiv$

The three-line variation of the equals sign indicates that the term on the left stands for the formula on the right *by definition*.

$term \equiv \ldots \textit{some formula} \ldots$ *definition of term*
$P(xs, y, ys) \equiv (xs = (\text{cons } y \; ys))$ $P(xs, y, ys)$ means $(xs = (\text{cons } y \; ys))$

Suppose we take $P(xs, y, ys)$ as shorthand for the equation $xs = (\text{cons } y \; ys)$.

$$P(xs, y, ys) \equiv (xs = (\text{cons } y \; ys))$$

Given a particular list xs, together with a value y, we can view the equation $P(xs, y, ys)$ as a set of propositions indexed by the variable ys whose universe of discourse is the set of

4.1 Lists as Mathematical Objects

Axiom {*consp*}

$(\text{consp } xs) = (\exists y.(\exists ys.(xs = (\text{cons } y\ ys))))$

Figure 4.2
Nonempty list predicate: consp.

Axiom {*nlst*}

$[x_m\ x_{m+1}\ \ldots\ x_n]$ *denotes a list with* $n - m + 1$ *elements* {*nlst*}

Note: Denotes nil, *the empty list, if* $m > n$

Figure 4.3
Numbered list notation.

lists that can be constructed by ACL2. In this set of propositions, the one corresponding to the list ys is the equation that $P(xs, y, ys)$ stands for: $(xs = (\text{cons } y\ ys))$. If that equation holds, the value of the proposition $P(xs, y, ys)$ is true. Otherwise, it's false. For example, if xs denotes the list [1 2 3] and y denotes the natural number 1, then $P(xs, y, ys)$ is $P([1\ 2\ 3], 1, ys)$, which stands for an equation involving the variable ys. There is one such equation for each different list ys. Taken all together, those equations comprise a predicate whose universe of discourse is ACL2 lists.

The operator consp checks for nonempty lists. That is, the formula (consp xs) delivers true if xs is a nonempty list and false otherwise. The {*consp*} axiom (figure 4.2) formally asserts that all nonempty lists are constructed with the cons operator.

The formula $(\exists ys.P([1\ 2\ 3], 1, ys))$ is true because [1 2 3] = (cons 1 [2 3]), so there is a value of ys, namely, $ys = [2\ 3]$, for which $P([1\ 2\ 3], 1, ys)$ is true. If there were no list that made the equation valid, the formula $(\exists ys.P([1\ 2\ 3], 1, ys))$ would be false.

If, on the other hand, xs were the list [1 2 3] and y were the number 2, there would be no list ys that would make the equation [1 2 3] = (cons 2 ys) valid because the list on the left-hand side of the equation starts with 1 and the list on the right-hand side starts with 2. So, the formula $(\exists ys.P([1\ 2\ 3], 2, ys))$ is false.

Now, let's take a step back. We can view the formula $(\exists ys.\ (xs = (\text{cons } y\ ys)))$ as a set of propositions, one for each object y that ACL2 can represent. The formula $(\exists ys.P(xs, y, ys))$ is one way to represent that set of propositions. Since any set of propositions is a predicate, we can view $(\exists ys.P(xs, y, ys))$ as a predicate indexed by the set of ACL2 objects y.

We can convert the predicate $(\exists ys.P(xs, y, ys))$ into a true/false value (that is, convert it to a proposition) by applying the $\exists$ quantifier again, but this time with y as the bound variable: $(\exists y.(\exists ys.P(xs, y, ys)))$. When xs is a list for which this formula has the value true, then (consp xs) is true. That is, consp is the ACL2 predicate that means $(\exists y.(\exists ys.P(xs, y, ys)))$. The universe of discourse of the predicate consp is the set of all objects that ACL2 can

Axioms $\{cons\}$, $\{first\}$, and $\{rest\}$	
$[x_1\ x_2\ \ldots x_{n+1}] = (\text{cons } x_1\ [x_2 \ldots x_{n+1}])$	$\{cons\}$
(first (cons x xs)) = x	$\{fst\}$
(rest (cons x xs)) = xs	$\{rst\}$
(first nil) = nil	$\{fst0\}$
(rest nil) = nil	$\{rst0\}$

Figure 4.4
List constructor and deconstructors: cons, first, rest.

represent. That specification of consp is expressed in the $\{consp\}$ axiom (figure 4.2, page 81). So, (consp xs) is a way to write the formula $(\exists y.(\exists ys.(xs = (\text{cons } y\ ys))))$ in ACL2.

When we know that a list ys is nonempty, we can cite the $\{consp\}$ axiom to rewrite ys in the form (cons x xs). When we do this, we choose the symbols x and xs carefully to avoid conflicts with other symbols that appear in the context of the discussion.

The $\{consp\}$ axiom refers to cons, so we will need a $\{cons\}$ axiom. The $\{cons\}$ axiom uses numbered list notation (figure 4.3, page 81) to specify that cons delivers a nonempty list, $[x_1\ x_2\ \ldots x_{n+1}]$, where n stands for a natural number. Because that list has $n + 1$ elements and $n + 1$ is at least one when n is a natural number, the list cannot be empty. Therefore, the list can be constructed by cons.

The construction operator, cons, cannot be the whole story, of course. To compute with lists, we need to be able to construct them, but we also need to be able to take them apart. There are two basic operators for taking lists apart: first and rest. We express the relationship between these operators and the construction operator in the form of equations, $\{fst\}$ and $\{rst\}$, that we take as axioms (figure 4.4).

The $\{fst\}$ axiom states formally that the operator first delivers the first element from a nonempty list. The $\{rst\}$ axiom states that the operator rest delivers a list like its operand but without the first element. Note that the lists to which the operators first and rest are applied in the axioms have at least one element because those lists are constructed by the cons operator. The axioms $\{fst0\}$ and $\{rst0\}$ provide an interpretation of the formulas (first nil) and (rest nil) when the operand is a list with no elements.

We will use equations like the ones in these axioms in the same way we used the logic equations in figure 2.2 (page 18) and the arithmetic equations of figure 2.1 (page 16). That is, whenever we see a formula like (first (cons x xs)), no matter what formulas x and xs stand for, we will be able to cite equation $\{fst\}$ to replace (first (cons x xs)) by the simpler formula x. Equations go both ways, so we can also cite equation $\{fst\}$ to replace any formula x by the more complicated formula (first (cons x xs)), where xs stands for any formula we care to make up, as long as it is grammatically correct.

Similarly, we can cite the equation $\{rst\}$ to justify replacing the formula (rest (cons x xs)) by xs and vice versa, regardless of what formulas the symbols x and xs stand for. In

4.1 Lists as Mathematical Objects

other words, these are ordinary algebraic equations. The only new factors are (1) the kind of mathematical object they denote and (2) the syntactic quirk of prefix notation, instead of the more familiar infix notation. All properties of lists, as mathematical objects, derive from the {cons}, {fst}, and {rst} axioms.

The operator len (page 73) delivers the number of elements in a list. We can use check-expect to test len in some specific cases.

```
(check-expect (len (cons 8 (cons 3 (cons 7 nil)))) 3)
(check-expect (len nil) 0)
```

We can use the DoubleCheck facility for more general tests. For example, we expect that the number of elements in a list constructed by the cons operation is one more than the number of elements in its second operand. The following property tests this expectation:

```
(defproperty len-cons-test
   (x  :value (random-natural)
    xs :value (random-list-of (random-natural)))
   (= (len (cons x xs))
      (+ 1 (len xs))))
```

By the same token, we expect that a list always has one more element than it would have if its first element were removed: (len xs) = 1+ (len (rest xs)). However, that is true only if the list xs has some elements. It would not be true if xs were nil. What we want to test is an implication: (consp xs) → ((len xs) = 1 + (len (rest xs))). The ACL2 name for the implication operator is implies, and we can use that operator to specify a test that constrains xs to nonempty lists and thereby makes the length equation in the test true.

```
(defproperty len-rest-test
   (xs :value (random-list-of (random-natural)))
   (implies (consp xs)
            (= (len xs)
               (+ 1 (len (rest xs))))))
```

The equation in the len-rest test can serve as an axiom for the len operator in the case when its operand is a nonempty list. The axiom for the empty case is simpler. Figure 4.5 (page 84) states these two axioms for the len operator. Axiom {len1} applies when the operand is nonempty and the other axiom applies in all other circumstances.[24]

---

[24] Normally the operand of len will be either a nonempty list or nil, the empty list. That is a good way to think of it for now, but the operand may not be a list at all, and according to axiom {len0}, its value is zero in that case. So, (len nil) = 0 but (len 3) = 0 too since 3 is not a list. Later, we will say more about this kind of axiom.

Axioms {len}

(len (cons x xs)) = (+ 1 (len xs))   {len1}
(len e) = 0   {len0}

Note: Cite {len0} only if {len1} doesn't match.

Figure 4.5
Length of list: len.

We expect the len operator to deliver a natural number, regardless of the value of its operand. For the record, we state this property as a theorem. Later, you will have a chance to derive this theorem from the {len} axioms. The theorem refers to the natp operator (page 75), which delivers true if its operand is a natural number and false otherwise.

$$\text{Theorem } \{len\text{-}nat\}: \forall xs.(\text{natp (len } xs))$$

A related fact is that the length of a nonempty list is strictly positive. One way to state that fact is to observe that the formula (consp xs) is true if (> (len xs) 0) and vice-versa. This theorem too can be derived from the axioms for len, consp, and cons. For the moment, we state the theorem without proof.

$$\text{Theorem } \{consp = (len > 0)\}: \forall xs.((\text{consp } xs) = (> (\text{len } xs) \, 0))$$

Box 4.3
Suppressing Computation with Single-Quote

To specify the list [1 2 3 4] in an ACL2 formula rather than in a paper-and-pencil formula, we can, of course, use the cons operator to construct it: (cons 1 (cons 2 (cons 3 (cons 4 nil)))). Or, we can use the list operator (page 73) to write it more compactly: (list 1 2 3 4).

However, the single-quote trick provides a less bulky ACL2 formula for lists whose elements are numbers (or literals denoting other ACL2 constants). The formula '(1 2 3 4) has the same meaning as (list 1 2 3 4). Normally, ACL2 interprets the first symbol after a left-parenthesis as the name of an operator. However, the single-quote mark suppresses that interpretation and delivers a list made up of the elements in the parentheses. Without the single-quote mark, the formula would make no sense because 1 is not the name of an operator.

Exercises

1. Prove theorem {rst1}: (rest (list x)) = nil.
   Hint: Cite some equations from figure 4.4 (page 82) and figure 4.1 (page 80).

Axioms {*if*}

(if *p x y*) = *x*, *if p* ≠ nil	{*if-true*}
(if *p x y*) = *y*, *if p* = nil	{*if-false*}

Figure 4.6
Formula selector: if.

4.2 Mathematical Induction

The cons, first, and rest operators form the basis for computing with lists, but there are lots of other operators for lists. The operator append, previously discussed in terms of some check-expect tests (page 73), concatenates two lists, as illustrated in the following check-expect tests, which use the single-quote notation (box 4.3, page 84) to make them more compact:

```
(check-expect (append '(1 2 3 4) '(5 6 7)) '(1 2 3 4 5 6 7))
(check-expect (append '(1 2 3 4 5) nil) '(1 2 3 4 5))
```

The numbered-list notation (figure 4.3, page 81) provides a way to define the append operator informally. In this form, the definition implicitly reveals the number of elements in the lists involved in the concatenation. In the following list schematics, the *x* list has *m* elements, the *y* list has *n* elements, and the concatenated list has *m* + *n* elements:

$$(\text{append } [x_1 \ x_2 \ \ldots \ x_m] \ [y_1 \ y_2 \ \ldots \ y_n]) = [x_1 \ x_2 \ \ldots \ x_m \ y_1 \ y_2 \ \ldots \ y_n]$$

Let's analyze the concatenation (append *xs ys*). If *xs* is the empty list, then we expect the concatenation to deliver the list *ys*. This is the {*app0*} case: (append nil *ys*) = *ys*. If *xs* is not empty, then we expect the concatenation to start with the first element of *xs*, that is, (first *xs*), and to continue with the remaining elements of *xs*, that is, (rest *xs*), and then the elements of *ys* come after that. Put another way, when *xs* is not empty, the result is the concatenation of (rest *xs*) and *ys*, with (first *xs*) inserted at the front. This is the {*app1*} case: (append *xs ys*) = (cons (first *xs*) (append (rest *xs*) *ys*)).

We would like to express our expectations formally. To do so, we use a special ACL2 operator called if (figure 4.6), which has three operands. It delivers its second operand if its first operand is true (that is, not nil) and selects its third operand if its first operand is false (that is, nil). So, the if operator provides a way to select between the two formulas supplied as its second and third operands.

With the if operator, we can use DoubleCheck to test our expectations of (append *xs ys*) by comparing (append *xs ys*) to the {*app1*} formula if *xs* is nonempty and to the {*app0*} formula if *xs* is empty. Figure 4.7 (page 86) displays a property definition that formalizes this idea. It uses an operator called equal (box 4.4, page 86) to compare (append *xs ys*) to a formula selected by the if operator.

```
(defproperty append-test
  (xs :value (random-list-of (random-natural))
   ys :value (random-list-of (random-natural)))
  (equal (append xs ys)
         (if (consp xs)
             (cons (first xs)
                   (append (rest xs) ys))
             ys)))
```

Figure 4.7
DoubleCheck test of append.

Axioms {*append*}

(append (cons *x xs*) *ys*) = (cons *x* (append *xs ys*))   {*app1*}
(append nil *ys*) = *ys*   {*app0*}

Figure 4.8
List concatenation: append.

Box 4.4
"equal" vs "="

Why is it (equal (append *xs ys*) ...)? Why not (= (append *xs ys*) ...)? The "=" operator is restricted to numbers, whereas the operator equal can check for equality between any two values: numbers, lists, whatever.

The = symbol reminds people who are reading the formula that it compares numbers. It reminds the computer, too, and that could simplify mechanized reasoning or promote efficient comparison. Mostly, though, it's a matter of taste.

The append-test property might not be the first test you would think of, but if the test failed to pass, you would know for sure that something was wrong with the append operator. The append-test property is so plainly correct that we are going to state it in the form of equations that we accept as axioms (figure 4.8). As in the {*len*} theorem, there are two {*append*} equations, which specify the meaning of the append operation in different situations. One of them specifies the meaning when the first operand is a nonempty list, the other specifies the meaning under all other circumstances.

These equations state simple properties of the append operation, and it turns out that lots of other properties of the append operation can be derived from them. For example, we can prove that the length of the concatenation of two lists is the sum of the lengths of the lists, as tested by a DoubleCheck property in chapter 3 (page 73). We call this theorem the *additive law of concatenation*. Let's see how a proof of this law could be carried out.

4.2 Mathematical Induction

First, let's break it down into some special cases. We will use L(*n*) as shorthand for the proposition that (len (append [x_1 x_2 ... x_n] *ys*)) is the sum of (len [x_1 x_2 ... x_n]) and (len *ys*). That makes L a predicate whose universe of discourse is the natural numbers.

L(*n*) ≡ (len (append [x_1 x_2 ... x_n] *ys*)) = (+ (len [x_1 x_2 ... x_n]) (len *ys*))

For the first few values of *n*, L(*n*) would stand for the following equations:

L(0) ≡ (len (append nil *ys*)) = (+ (len nil) (len *ys*))
L(1) ≡ (len (append [x_1] *ys*)) = (+ (len [x_1]) (len *ys*))
L(2) ≡ (len (append [x_1 x_2] *ys*)) = (+ (len [x_1 x_2]) (len *ys*))
L(3) ≡ (len (append [x_1 x_2 x_3] *ys*)) = (+ (len [x_1 x_2 x_3]) (len *ys*))

We can derive L(0) from the {*append*} and {*len*} axioms as follows. We start from the left-hand side of the equation that L(0) stands for and cite some axioms about append, len, and numeric algebra, one by one. We end up with the right-hand side of the equation L(0).

Proof of L(0), citing axioms

	(len (append nil *ys*))	
=	(len *ys*)	{*app0*} (page 86)
=	(+ (len *ys*) 0)	{+ identity} (page 16)
=	(+ 0 (len *ys*))	{+ commutative} (page 16)
=	(+ (len nil) (len *ys*))	{*len0*} (page 84)

That takes care of L(0). How about L(1)?

Proof of L(1), citing axioms and proven equations

	(len (append [x_1] *ys*))	
=	(len (append (cons x_1 nil) *ys*))	{*cons*} (page 82)
=	(len (cons x_1 (append nil *ys*)))	{*app1*}
=	(+ 1 (len (append nil *ys*)))	{*len1*}
=	(+ 1 (+ (len nil) (len *ys*)))	{L(0)} Note: L(0) already proved
=	(+ (+ 1 (len nil)) (len *ys*))	{+ associative} (page 16)
=	(+ (len (cons x_1 nil)) (len *ys*))	{*len1*}
=	(+ (len [x_1]) (len *ys*))	{*cons*}

That was a little harder. Will proving L(2) be still harder? Let's try it.

Proof of L(2), citing axioms and proven equations

	(len (append [x_1 x_2] ys))		
=	(len (append (cons x_1 [x_2]) ys))	{*cons*}	
=	(len (cons x_1 (append [x_2] ys)))	{*app1*}	
=	(+ 1 (len (append [x_2] ys)))	{*len1*}	
=	(+ 1 (+ (len [x_2]) (len ys)))	{L(1)}	Note: *L(1) already proved*
=	(+ (+ 1 (len [x_2])) (len ys))	{+ *associative*}	
=	(+ (len (cons x_1 [x_2])) (len ys))	{*len1*}	
=	(+ (len [x_1 x_2]) (len ys))	{*cons*}	

Proving L(2) was no harder than proving L(1). In fact, the two proofs cite exactly the same equations all the way through except in one place. Where the proof of L(1) cited the equation L(0), the proof of L(2) cited the equation L(1). Maybe the proof of L(3) will work the same way.

Proof of L(3), citing axioms and proven equations

	(len (append [x_1 x_2 x_3] ys))		
=	(len (append (cons x_1 [x_2 x_3]) ys))	{*cons*}	
=	(len (cons x_1 (append [x_2 x_3] ys)))	{*app1*}	
=	(+ 1 (len (append [x_2 x_3] ys)))	{*len1*}	
=	(+ 1 (+ (len [x_2 x_3]) (len ys)))	{L(2)}	Note: *L(2) already proved*
=	(+ (+ 1 (len [x_2 x_3])) (len ys))	{+ *associative*}	
=	(+ (len (cons x_1 [x_2 x_3])) (len ys))	{*len1*}	
=	(+ (len [x_1 x_2 x_3]) (len ys))	{*cons*}	

By now it's easy to see how to derive L(4) from L(3), then L(5) from L(4), and so on. If you had the patience, you could prove L(100), L(1000), or even L(1000000) by following the established pattern. It would not be hard to write a program to print out the proof of L(n) given any natural number n. Since we know how to prove L(n) for any natural number n, it seems fair to say that we know all those equations are true. That is, we think we know that the formula ($\forall n$.L(n)) is true. However, to prove that in a formal sense, we need a rule of inference that allows us to make conclusions from patterns like those we observed in proving L(1), L(2), and so on. That rule of inference is known as *mathematical induction*.

Mathematical induction provides a way to prove that formulas like ($\forall n$.P(n)) are true when P is a predicate whose universe of discourse is the natural numbers. If for each natural number n, P(n) stands for a proposition, then mathematical induction is an inference rule that may be useful in a proof that ($\forall n$.P(n)) is true. That is not to say that such a proof can always be constructed. It's just that mathematical induction might provide some help in

4.2 Mathematical Induction

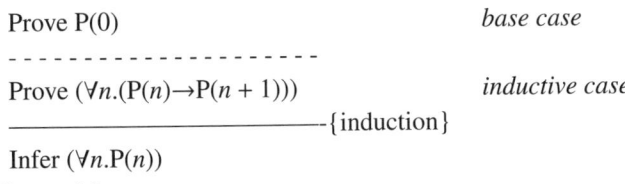

Figure 4.9
Mathematical induction: a rule of inference.

the process. The inverse is also true: mathematical induction cannot help if the universe of discourse is not the natural numbers.[25]

The rule goes as follows. Infer the truth of $(\forall n.P(n))$, citing {induction} as a rule of inference, from proofs of two propositions: $P(0)$ and $(\forall n.(P(n) \to P(n + 1)))$.

$$((P(0) \land (\forall n.(P(n) \to P(n+1)))) \to (\forall n.P(n))) = True \quad \{induction\}$$

It's a very good deal if you think about it. A direct proof of $(\forall n.P(n))$ would require a proof of proposition $P(n)$ for each value of n (0, 1, 2, ...). But, in a proof by induction, the only proposition that needs to be proved on its own is $P(0)$. To justify any step in the proof of proposition $P(n + 1)$ (that is, any proposition in the predicate P with a nonzero index: $P(1), P(2), P(3), \ldots$), you are allowed to cite $P(n)$ as if it were a known theorem. The proof of $P(0)$ is known as the *base case* in the proof by induction, and the proof of $(P(n) \to P(n + 1))$ is known as the *inductive case*.

The reason you can assume that $P(n)$ is true in the proof of $P(n+1)$ (that is, in the inductive case) is because the goal, according to part two of the inference rule for induction, is to prove that the implication $P(n) \to P(n+1)$ is true. When $P(n)$ is false, we know already from the truth table of the implication operator (page 24) that the implication $P(n) \to P(n + 1)$ is true. There is no need to prove that again. Therefore, in the proof of $P(n + 1)$, we can assume $P(n)$ is true. The assumption of $P(n)$ is known as the *induction hypothesis*. It gives you a leg up in the proof of $P(n + 1)$.

Figure 4.9 states the {induction} inference rule in the form of natural deduction, a formal method of logic discussed in section 2.5. You don't need to have studied that section to understand inductive proofs, but in case you did study natural deduction, figure 4.9 puts proof by induction in that context.

Now, let's apply mathematical induction to prove the additive law of concatenation. Here, the predicate that we will apply the method to is L (page 87). We have already proved L(0), so we have already completed one of the two proofs required to cite the math-

---

[25] Mathematical induction is not the only form of proof by induction, but all the other forms (other than transfinite induction, which is a different animal) can be contorted into proofs by mathematical induction. We will stick with classical mathematical induction and leave the variations for another time. They are easy to learn for people who understand ordinary mathematical induction.

ematical induction inference rule. All that is left is to prove ($\forall n.(L(n) \rightarrow L(n + 1))$). That is, we have to derive $L(n + 1)$ from $L(n)$ for an arbitrary natural number n. Fortunately, we know how to do this. Just copy the derivation of, say, $L(3)$ from $L(2)$, but start with an append formula in which the first operand is a list with $n + 1$ elements and cite $L(n)$ where we would have cited $L(2)$.

Proof of L(n+1), citing axioms, proven equations, and L(n): $L(n) \rightarrow L(n+1)$

	(len (append [x_1 x_2 ... x_{n+1}] ys))	
=	(len (append (cons x_1 [x_2 ... x_{n+1}]) ys))	{*cons*}
=	(len (cons x_1 (append [x_2 ... x_{n+1}] ys)))	{*app1*}
=	(+ 1 (len (append [x_2 ... x_{n+1}] ys)))	{*len1*}
=	(+ 1 (+ (len [x_2 ... x_{n+1}]) (len ys)))	{L(n)} Note: induction hypothesis
=	(+ (+ 1 (len [x_2 ... x_{n+1}])) (len ys))	{+ associative}
=	(+ (len (cons x_1 [x_2 ... x_{n+1}])) (len ys))	{*len1*}
=	(+ (len [x_1 x_2 ... x_{n+1}]) (len ys))	{*cons*}

This completes the mathematical induction proving the additive law of concatenation.

Theorem {additive law of concatenation}

$\forall n.((\text{len (append } [x_1\ x_2\ \ldots\ x_n]\ ys)) = (+ (\text{len } [x_1\ x_2\ \ldots\ x_n])\ (\text{len } ys)))$

An important point to notice in this proof is that we could not cite the {*cons*} equation to replace [x_2 ... x_{n+1}] with (cons x_2 [x_3 ... x_{n+1}]). The reason we could not do this is that we are trying to derive $L(n + 1)$ from $L(n)$ without making any assumptions about n other than the fact that it is a natural number. Since zero is a natural number, the list [x_2 ... x_{n+1}] could be empty,[26] and the cons operator cannot deliver an empty list as its value.

In the next section, we will prove several properties of **append** and its relationships with other operators. These properties, and in fact all properties of the **append** operator, can be derived from the {*append*} axioms (figure 4.8, page 86). Those axioms state properties of the **append** operation in two separate cases: (1) {*app0*}: the first operand is the empty list and (2) {*app1*}: the first operand is a nonempty list. When the first operand is the empty list, the result must be the second operand, no matter what. When the first operand is not empty, it must have a first element, and that element must be the first element of the concatenation. The other elements of the concatenation are the ones you would get if you appended the second operand to the rest of the first operand.

Both of these properties are so straightforward and easy to believe that we would probably be willing to accept them as axioms with no proof at all, so it might seem surprising that all of the other properties of the **append** operation can be derived from the two sim-

---

[26] The list [x_2 ... x_{0+1}] is empty ({*nlst*} numbered list notation, figure 4.3, page 81).

Complete	All possible combinations of operands are covered by at least one equation in the definition.
Consistent	Combinations of operands covered by two or more equations define the same value for the operation.
Computational	1. *Noninductive Equation*: In at least one equation, the operator being defined appears only on the left-hand side. 2. *Reduced Computation*: On the right-hand side of an inductive equation, each invocation of the operator being defined has operands that are closer to the operands on the left-hand side of a noninductive equation than to those on the left-hand side of the inductive equation.

Figure 4.10
The three C's: a guide to inductive definitions.

ple properties {*app0*} and {*app1*}. That is the power of mathematical induction. The two equations of the {*append*} axioms amount to an inductive definition of the append operator.

An inductive definition is circular in the sense that some of the equations in the definition refer to the operator on both sides of the equation. Most of the time, people think of circular definitions as unhelpful at best, but they can be useful in mathematics. You will see a lot of them and gradually learn how to recognize and create useful, circular (that is, inductive) definitions.

It turns out that all operators that can be defined in software have inductive definitions in the manner of the equations of the {*append*} axioms. The keys to an inductive definition of an operator are listed in figure 4.10. All of the software we will discuss will take the form of inductive definitions of operators. That makes it possible to use mathematical induction as a fundamental tool in verifying, to a logical certainty, properties of that software.

Exercises

1. Prove $\forall xs.(\text{natp } (\text{len } xs))$. You may cite {natp0} and {natp1}, defined as follows. (*Note*: This definition will be adequate for this exercise but it is not the full definition of natp.)

$$\frac{(\text{natp } 0) \qquad \{\text{natp0}\}}{\forall x.((\text{natp } x) \to (\text{natp } (+\ x\ 1))) \quad \{\text{natp1}\}}$$
$\forall x$ *universe of discourse: numbers*

2. Prove: $\forall n.(\forall x.((\text{expt } x\ n) = x^n))$ {*expt*}
   $\forall n$ universe of discourse: natural numbers; $\forall x$ universe of discourse: non-zero numbers
   Assume the equations {*expt0*} and {*expt1*} are true.

$$\frac{(\text{expt } x\ 0) = 1 \qquad \{expt0\}}{(\text{expt } x\ (+\ n\ 1)) = (*\ x\ (\text{expt } x\ n)) \qquad \{expt1\}}$$
$$(*\ x\ y) = x \times y$$

4.3 Defun: Defining Operators in ACL2

Now, we are going to let you in on a little secret. The axioms we wrote for the **append** operator are very close to an ACL2 definition of **append**. Operators are defined in ACL2 with a **defun** command, which has four parts.

(defun f $(x_1\ x_2\ \ldots\ x_n)$ $formula$)	
defun	keyword
f	name for the operator being defined
$(x_1\ x_2\ \ldots\ x_n)$	names designating operands, surrounded by parentheses
$formula$	ACL2 formula specifying the value the operator will deliver

Most of the time, the formula for the value the operator delivers will have subformulas specifying alternative values for different cases. Formulas interpreted as predicates select one of the subformulas to compute the value.

In section 4.2 (figure 4.8, page 86), we defined the **append** operator with two equations, one for the case when the first operand was a nonempty list and the other for all other possibilities. The ACL2 definition, following that pattern, has two subformulas, one for each case. The **if** operator (figure 4.6, page 85) selects the appropriate formula.

Box 4.5
Why Inductive Definitions?

Inductive definitions are one of many ways to specify software, and other methods are more common in practice. However, proving properties of software written using conventional methods is clumsy at best, especially in the framework of classical logic. So, in terms of understanding what computers do and how they do it, inductive definitions provide solid footing. Since reasoning is a central theme in this presentation, it focuses on software written in the form of inductive equations.

Figure 4.11 (page 93) repeats the axioms for the **append** operator from figure 4.8 (page 86) and also displays a definition of **append** using **defun**, which expresses the axioms in ACL2 notation. As it happens, the **append** operator is an ACL2 intrinsic. It is defined by the ACL2 system, so the definition in figure 4.11 is redundant, and the ACL2 system will

Axioms {*append*}	
(append (cons *x xs*) *ys*) = (cons *x* (append *xs ys*))	{*app1*}
(append nil *ys*) = *ys*	{*app0*}

```
(defun append (xs ys)     ; intrinsic operator, so defun is redundant
  (if (consp xs)                          ; select formula
      (cons (first xs) (append (rest xs) ys))  ; {app1}, xs not empty
      ys))                                ; {app0}, xs is empty
```

Figure 4.11
Defining concatenation: **append**.

tell you that if you try to define it. Shortly, we will begin to define operators that are not intrinsic, so those definitions won't be redundant. This familiar example is just a starting point to get on the right track.

So, now you know. We've been writing programs on the sly, passing them off as axioms. Why? Because that's how we want you to think of them. A program is a collection of axioms expressed as equations that specify properties that you want operators to have. You can reason from those equations using the same methods you have used in reasoning about Boolean equations or numeric equations. The program is written in the syntax of a programming language, which makes it look a bit stilted. That is always the case in programming languages because they have their own syntax, and you have to conform to it. It pays off, though. If you stick with the program, you can get the computer to carry out computations, and you will have some confidence in the results of those computations.

Since definitions must use the ACL2 syntax, they don't look much like equations, but if they are complete, consistent, and computational (figure 4.10, page 91), the values they deliver will have the properties you derive from the equations. They may not have all the properties you expected. They may have bugs. But you'll have a good chance of fixing them with automated testing (**defproperty**) and the reasoning assistance of the ACL2 mechanized logic.

Going forward, we will define operators that aren't intrinsic in ACL2 (and, therefore, need definitions) both in the form of an ACL2 **defun** and in the form of equations for paper-and-pencil reasoning. The corresponding ACL2 definitions make it possible to apply automated testing and mechanized logic to confirm some of the properties of those operators. When the operators we are discussing are intrinsic, we will sometimes not provide **defun** versions of them, just axioms for paper-and-pencil reasoning. Automated testing and mechanized logic will make use of their intrinsic definitions in the ACL2 system.

4.4 Concatenation, Prefixes, and Suffixes

If you concatenate two lists, *xs* and *ys*, you would expect to be able to retrieve the elements of *ys* by dropping some of the elements of the concatenated lists. How many elements would you need to drop? That depends on the number of elements in *xs*. If there are *n* elements in *xs* and you drop *n* elements from (append *xs ys*), you expect the result to be identical to the list *ys*. To express that expectation, we can use an intrinsic operator in ACL2 with the arcane name nthcdr. The nthcdr operator has two operands: a natural number and a list. The formula (nthcdr *n xs*) delivers a list like *xs* but without its first *n* elements. If *xs* has fewer than *n* elements, then the formula delivers the empty list. In any case, nthcdr delivers a suffix of the list supplied as its second operand.

If the first operand (the number of elements to be dropped) is zero, you would expect nthcdr to deliver the entire list, having dropped no elements. If the second operand has no elements, you would expect nthcdr to deliver a list just like that (that is, a list with no elements). Combining these two observations, we find that *xs* would be the appropriate value for (nthcdr *n xs*) if either *n* is zero or *xs* has no elements.

Box 4.6
Natural Number Predicates: Zero (zp) and Nonzero (posp)

The predicate posp is used to test for nonzero values in the domain of natural numbers. In the ACL2 logic, the formula (posp *n*) has the value true if *n* is a nonzero, natural number (that is, a strictly positive integer). The value of (posp *n*) is false if *n* is not a natural number or if *n* is zero. That makes posp especially useful in definitions that are inductive on the natural numbers. You might think that (> *n* 0) would work the same way but it doesn't because that formula does not constrain *n* to the natural numbers, and many inductive definitions rely on that constraint.

The predicate zp imposes the same constraint, but it is true when its operand is zero and false otherwise. Both zp and posp are useful for inductive definitions that rely on the domain of natural numbers to ensure that the defined operator terminates.

The other possibility is that *n* is not zero and *xs* has some elements. Since the first operand is a natural number, being nonzero is the same as being one or more. In that case you would expect (nthcdr *n xs*) to deliver the same list that it would deliver if you dropped the first element of *xs* and then, in addition, dropped ($n - 1$) more elements. Together, these two actions would drop *n* elements. The axioms in figure 4.12 (page 95) express these observations as equations.

The figure also contains an ACL2 definition of nthcdr, which is of course redundant because nthcdr is intrinsic in ACL2.[27] The definition uses the predicate consp (figure 4.2, page 81) to find out whether the list contains some elements and uses the predicate posp to determine whether the number of elements to be dropped is one or more. It combines

---

[27] If you submit a definition of nthcdr, the system will inform you of the redundancy.

4.4 Concatenation, Prefixes, and Suffixes

Axioms {*nthcdr*}	
(nthcdr (*n* + 1) (cons *x xs*)) = (nthcdr *n xs*)	{*sfx1*}
(nthcdr *n xs*) = *xs*	{*sfx0*}
Note 1: Cite {*sfx0*} only if {*sfx1*} doesn't match.	
Note 2: *n* is a natural number.	

```
(defun nthcdr (n xs)      ; intrinsic operator, so defun is redundant
   (if (and (posp n) (consp xs))   ; select formula
       (nthcdr (- n 1) (rest xs))  ; {sfx1}
       xs))                        ; {sf0}
```

Figure 4.12
Defining list suffix extractor: nthcdr.

these predicates with the and operator, which is the ACL2 notation for the logical-and (∧). The value (and *a b*) is false (nil) if either *a* or *b* is false and true otherwise.

The equations in figure 4.12 cover all combinations of values that the operands of nthcdr can have. The first operand is a natural number, so it's either zero or bigger than zero. The second operand, a list, either has some elements or it doesn't. So the definition is complete, having covered all the cases. The cases do not overlap, so we don't need to worry about consistency between the axioms.

Box 4.7
Fall-Through Axioms

There is a subtlety in the axioms for nthcdr (figure 4.12) that needs to be discussed. The operand prototypes in the {*sfx1*} axiom match whenever the first operand is a nonzero natural number (*n* + 1 cannot be zero when *n* is a natural number) and the second operand is a nonempty list. However, the operand prototypes in the {*sfx0*} axiom match anything. There is a note restricting citations of the second axiom to cases where the first axiom does not apply. The definition of the len operator had a fall-through axiom like this too (figure 4.5, page 84).

Usually we state axioms with operand prototypes that constrain the operands to specific forms, but in the case of nthcdr, Note 2 conforms to the meaning of the (if *p a b*) formula, which chooses formula *b* only if *p* has the value nil (representing false). Since {*sfx0*} only applies if the operand prototypes in {*sfx0*} don't match the operands in a formula that refers to nthcdr, the two axioms do not share any combination of operands, so they cannot cause an inconsistency in the specified results.

That covers two of the three C's guidelines for inductive definitions of operators (figure 4.10, page 91). The equations are complete and consistent. The third guideline (computational) has two parts, one of which is a requirement that at least one axiom must be noninductive. The {*sfx0*} equation is not inductive because the nthcdr operator is not invoked on the right-hand side of the equation. In that case, nthcdr just delivers its second operand as is. So, the axioms pass muster on that part of the computational guideline.

With regard to the inductive axiom {*sfx1*}, the operands on the right-hand side of the equation are smaller and shorter than the operands on the left-hand side, which makes them closer to the noninductive case since that axiom, {*sfx0*}, will apply if either the first operand is zero or the second one doesn't have any elements. Therefore, the equations conform to the three C's guidelines, and we can conclude that they define an operator.

At this point, we are in a position to verify the relationship between the **append** and **nthcdr** operators that started this discussion. Namely, we want to prove that if the lists *xs* and *ys* are concatenated and then (len *xs*) elements are dropped from the beginning of the concatenation, the result will be the list *ys*. We will use $S(n)$ as a shorthand for this property when *xs* has n elements.

$$S(n) \equiv ((\text{nthcdr (len } [x_1\ x_2\ \ldots\ x_n])\ (\text{append } [x_1\ x_2\ \ldots\ x_n]\ ys)) = ys$$

S is a predicate indexed by the natural numbers, so the formula $\forall n.S(n)$ is a candidate for proof by induction. A proof of this formula by mathematical induction (figure 4.9, page 89) requires two proofs: (1) the formula $S(0)$ is true and (2) the formula $S(n + 1)$ is true under the assumption that $S(n)$ is true, regardless of what natural number n stands for.

Let's do those two proofs. First, we prove $S(0)$. When n is zero, the list $[x_1\ x_2\ \ldots\ x_n]$ is empty: $[x_1\ x_2\ \ldots\ x_0]$ = nil, according to the {*nlst*} axiom (figure 4.3, page 81). So, $S(0)$ stands for the following equation:

$$S(0) \equiv ((\text{nthcdr (len nil)}\ (\text{append nil } ys)) = ys$$

As usual in a proof of an equation, we start with the formula on one side of the equation and use known equations to gradually transform that formula into the one on the other side.

	(nthcdr (len nil) (append nil *ys*))	
=	(nthcdr (len nil) *ys*)	{*app0*} (page 93)
=	(nthcdr 0 *ys*)	{*len0*} (page 84)
=	*ys*	{*sfx0*} (page 95)

That takes care of $S(0)$. Figure 4.13 (page 97) displays a proof of $S(n + 1)$, assuming that the induction hypothesis $S(n)$ is true. The last step in the proof is justified by citing $S(n)$. This is a little tricky because the formula that $S(n)$ stands for is not exactly the same as the formula in the next-to-last step of the proof. We interpret the formula $[x_1\ x_2\ \ldots\ x_n]$ in the definition of $S(n)$ to stand for any list with n elements. The elements in the list $[x_2\ \ldots\ x_{n+1}]$ are numbered 2 through $n + 1$, which means there must be exactly n of them ({*nlst*}, figure 4.3, page 81). With this interpretation, the formula in the next-to-last step matches the formula in the definition of $S(n)$, which makes it legitimate to cite $S(n)$ to justify the transformation to *ys* in the last step of the proof. We will cite axiom {*nlst*}, the numbered-list interpretation, frequently in proofs about lists.

4.4 Concatenation, Prefixes, and Suffixes

$$S(n + 1) \equiv ((\text{nthcdr } (\text{len } [x_1\ x_2\ \ldots\ x_{n+1}])\ (\text{append } [x_1\ x_2\ \ldots\ x_{n+1}]\ ys)) = ys)$$

	(nthcdr	(len $[x_1\ x_2\ \ldots\ x_{n+1}]$)	
		(append $[x_1\ x_2\ \ldots\ x_{n+1}]\ ys$))	
=	(nthcdr	(len (cons $x_1\ [x_2\ \ldots\ x_{n+1}]$))	{*cons*} (page 82)
		(append (cons $x_1\ [x_2\ \ldots\ x_{n+1}]$) ys))	{*cons*}
=	(nthcdr	(+ 1 (len $[x_2\ \ldots\ x_{n+1}]$))	{*len1*} (page 84)
		(cons x_1 (append $[x_2\ \ldots\ x_{n+1}]$) ys))	{*app1*} (figure 4.11, page 93)
=	(nthcdr	(+ (len $[x_2\ \ldots\ x_{n+1}]$) 1)	{+ commutative} (page 16)
		(cons x_1 (append $[x_2\ \ldots\ x_{n+1}]$) ys))	
=	(nthcdr	(len $[x_2\ \ldots\ x_{n+1}]$)	{*sfx1*}
		(append $[x_2\ \ldots\ x_{n+1}]\ ys$))	
=	ys		{$S(n)$} (induction hypothesis)

Figure 4.13
Inductive case: $S(n) \rightarrow S(n + 1)$.

At this point, we know that (append $xs\ ys$) delivers a list that has the right elements at the end. How about the beginning? We expect the concatenation to start with the elements of the list xs, so if we extract the first n elements of (append $xs\ ys$), where n is (len xs), we would expect to get a list identical to xs. To express this expectation formally, we need an operator that, given a number n and a list xs, delivers the first n elements of xs. Let's call that operator **prefix** and think about properties it would have to satisfy.

Of course, if n is zero or if xs is empty, (prefix $n\ xs$) must be the empty list. If n is a nonzero natural number and xs is not empty, then the first element of (prefix $n\ xs$) must be the first element of xs and the other elements must be the first $n - 1$ elements of (rest xs). Figure 4.14 (page 98) displays equations that define the **prefix** operator. We can derive the prefix property of the **append** operator from those equations and the axioms of the **append** operator (figure 4.11, page 93). We will prove $\forall n.P(n)$ by induction, where the predicate P is defined as follows:

$$P(n) \equiv ((\text{prefix } (\text{len } [x_1\ x_2\ \ldots\ x_n])\ (\text{append } [x_1\ x_2\ \ldots\ x_n]\ ys)) = [x_1\ x_2\ \ldots\ x_n])$$

As required in proofs by induction, we will prove that P(0) is true and also that P($n + 1$) is true whenever P(n) is true. Those two proofs will allow us to conclude that $\forall n.P(n)$ is true, citing {induction}.

As in the proof of the append suffix theorem, we start with the formula on one side of the P(0) equation and use known equations to gradually transform that formula into the one on the other side of the equation.

Axioms {*prefix*}	
(prefix (*n* + 1) (cons *x* *xs*)) = (cons *x* (prefix *n* *xs*))	{*pfx1*}
(prefix *n* *xs*) = nil	{*pfx0*}

Note 1: Cite {*pfx0*} only if {*pfx1*} doesn't match.

Note 2: *n* is a natural number.

```
(defun prefix (n xs)
  (if (and (posp n) (consp xs))
      (cons (first xs) (prefix (- n 1) (rest xs)))   ; {pfx1}
      xs))                                            ; {pfx0}
```

Figure 4.14
Defining list prefix extractor: prefix.

$$P(0) \equiv ((\text{prefix (len nil) (append nil } ys)) = \text{nil})$$

$$\begin{aligned}
&(\text{prefix (len nil) (append nil } ys)) \\
={} &(\text{prefix 0 (append nil } ys)) \quad \{len0\} \text{ (page 84)} \\
={} &\text{nil} \quad \{pfx0\}
\end{aligned}$$

That takes care of P(0). Figure 4.15 (page 99) displays a proof of P(*n*) → P(*n* + 1). At this point we know three important facts about the **append** operator:

Additive-length theorem: (len (append *xs* *ys*)) = (+ (len *xs*) (len *ys*))
Append-prefix theorem: (prefix (len *xs*) (append *xs* *ys*)) = *xs*
Append-suffix theorem: (nthcdr (len *xs*) (append *xs* *ys*)) = *ys*

Together, these theorems provide some assurance that **append** does what we would expect for a concatenation operator. We could think of them as correctness properties for **append**. Of course, the **append** operator has an infinite variety of other properties too. Their relative importance depends on how we are using the operation. A property that is sometimes important to know is that concatenation is associative, like addition and multiplication in numeric algebra (figure 2.1, page 16). That is, if there are three lists to be concatenated, you could concatenate the first list with the concatenation of the last two. Or, you could concatenate the first two, then append the third list at the end.

Theorem {*app-assoc*}: (append *xs* (append *ys* *zs*))=(append (append *xs* *ys*) *zs*)

The associative property of **append** can be proved by mathematical induction, starting from the following predicate with the natural numbers as its universe of discourse.

The goal would be to prove that the formula ($\forall n.A(n)$) is true. We leave the proof as an exercise.

$A(n) \equiv$ (append [$x_1\ x_2\ \ldots\ x_n$] (append *ys* *zs*))=(append(append [$x_1\ x_2\ \ldots\ x_n$] *ys*) *zs*)

```
   (prefix (len      [x_1 x_2 ... x_{n+1}])
           (append [x_1 x_2 ... x_{n+1}] ys))
= (prefix (len     (cons x_1 [x_2 x_3 ... x_{n+1}]))         {cons} (page 82)
          (append (cons x_1 [x_2 x_3 ... x_{n+1}]) ys))
= (prefix (+  1    (len [x_2 x_3 ... x_{n+1}]))              {len1} (page 84)
          (cons x_1 (append [x_2 x_3 ... x_{n+1}] ys)))      {app1} (page 86)
= (cons  (first    (cons x_1 [x_2 x_3 ... x_{n+1}]))
         (prefix  (- (+ 1 (len [x_2 x_3 ... x_{n+1}])) 1)    {pfx1}
                  (rest (cons x_1 (append [x_2 x_3 ... x_{n+1}] ys)))))
= (cons  x_1                                                 {fst} (page 82)
         (prefix  (len [x_2 x_3 ... x_{n+1}])                {arithmetic}
                  (append [x_2 x_3 ... x_{n+1}] ys)))        {rst} (page 82)
= (cons  x_1
         [x_2 x_3  ... x_{n+1}])                             {P(n)}
= [x_1 x_2 ... x_{n+1}]                                      {cons} (page 82)
```

$P(n + 1) \equiv$ ((prefix (len[$x_1 x_2 \ldots x_{n+1}$]) (append [$x_1 x_2 \ldots x_{n+1}$] ys)) = [$x_1 x_2 \ldots x_{n+1}$])

Figure 4.15
Inductive case: $\forall n. P(n) \rightarrow P(n + 1)$.

Exercises

1. Prove the {*app-assoc*} theorem (page 98).

2. Prove $\forall n.$((rest [$x_1 x_2 \ldots x_n$]) = (nthcdr 1 [$x_1 x_2 \ldots x_n$])).

3. Prove: $\forall n.$((len (rep n x)) = n)   {*rep-len*}
   The operator rep is defined as follows:

```
(defun rep (n x)
  (if (posp n)
      (cons x (rep (- n 1) x))    ; {rep1}
      nil))                        ; {rep0}
```

4. Prove theorem {app-nil}: $\forall n.$([$x_1 x_2 \ldots x_n$] = (append [$x_1 x_2 \ldots x_n$] nil))

5. Prove: $\forall n.$((nthcdr (len [$x_1 x_2 \ldots x_n$]) (append [$x_1 x_2 \ldots x_n$] nil)) = nil)

6. Prove the following implication:
   $\forall n.$((member-equal y (rep n x)) $\rightarrow$ (member-equal y (cons x nil)))   {*rep-mem*}
   The operator rep is defined in exercise 3, and the equations {*mem0*} and {*mem1*} define the predicate member-equal.

   (member-equal y (cons x xs)) = (equal y x) $\lor$ (member-equal y xs)   {*mem1*}
   (member-equal y nil) = nil   {*mem0*}

5 Mechanized Logic

Proofs of algebraic equations like those in Chapter 4 depend on matching grammatical elements in formulas against templates in axioms and theorems. A proof starts with a formula on one side of the equation of a theorem, cites a known equation to transform the formula to another one with the same meaning, and moves gradually, step by step, to the formula on the other side of the equation of the theorem. It is easy for people to make mistakes in this detailed, syntax-matching process, but computers can carry it out flawlessly, relieving people from an obligation to focus with monk-like devotion on grammatical details.

Box 5.1
ACL2 Learning Objectives

One of the aims of this book is to help readers learn to verify, using rigorous logic, properties of software that are specified in the form of inductive equations. Another goal is to provide an inkling of how mechanized logic can formalize reasoning of that kind and lead to greater confidence in the results. If you follow the broad outlines of the examples and successfully work your way through some of the exercises, we think you will gain a basic understanding of what can be done with mechanized logic. Becoming an accomplished user of such tools would require substantially more study and experience.

ACL2 is one of several systems of mechanized logic that provide this kind of assistance with proofs. Theorems for the ACL2 proof engine take the same form as properties for the DoubleCheck testing facility in Proof Pad. ACL2 has a built-in way to look for inductive proofs, and for some theorems it succeeds without guidance. Most of the time, however, ACL2 needs some help in finding its way through the morass of strategies. In any case, ACL2 checks the details automatically.

To illustrate how this works, we return to the theorems discussed in chapter 4. The syntax of ACL2 requires prefix notation exclusively, as you would expect, and there are additional issues to be discussed, but the form of the theorems will be familiar from your experience with DoubleCheck.

5.1 ACL2 Theorems and Proofs

The first proof by mathematical induction in chapter 4 verified the additive law of concatenation (page 90). The statement of the theorem asserts that $(\forall n.L(n))$ is true, where $L(n)$ stands for the following equation:

$$L(n) \equiv ((\text{len (append } [x_1 \; x_2 \; \ldots \; x_n] \; ys)) = (+ \; (\text{len } [x_1 \; x_2 \ldots x_n]) \; (\text{len } ys)))$$

In chapter 3 (page 73) we defined a DoubleCheck test of this equation.

```
(defproperty additive-law-of-concatenation-test
   (xs :value (random-list-of (random-natural))
    ys :value (random-list-of (random-natural)))
  (= (len (append xs ys))
     (+ (len xs) (len ys))))
```

Box 5.2
An Informality: $[x_1 \; x_2 \; \ldots \; x_n]$ versus xs

$$A(xs, ys) \equiv ((\text{len (append } xs \; ys)) = (+ \; (\text{len } xs) \; (\text{len } ys))) \quad \textit{formal}$$
$$L(n) \equiv ((\text{len (append } [x_1 \; x_2 \; \ldots \; x_n] \; ys)) \quad \textit{informal}$$
$$= (+ \; (\text{len } [x_1 \; x_2 \; \ldots \; x_n]) \; (\text{len } ys)))$$

A proof of $(\forall n.L(n))$ is intended to apply to all lists with n elements, regardless of what those elements are. The formula $(\forall xs.(\forall ys.A(xs, ys)))$ is a more formal statement of the intended theorem, where A is a predicate whose universe of discourse is pairs of lists. A proof of $(\forall n.L(n))$ accomplishes this goal rigorously (but not with full formality) if it avoids depending in any way on the elements in the lists that $[x_1 \; x_2 \; \ldots \; x_n]$ and ys stand for.

Furthermore, the proof assumes that any list of length n has a representation in the form $[x_1 \; x_2 \; \ldots \; x_n]$. The following axiom $\{lst\}$ states this assumption in terms of the bound variables xs and n, with lists and natural numbers, respectively, as their universes of discourse. The formula $[x_1 \; x_2 \; \ldots \; x_n]$ stands for a list of n elements from specifiable domains.

$$\text{Axiom } \{lst\}: (\forall xs.(\exists n.(\exists [x_1 \; x_2 \; \ldots \; x_n].(xs = [x_1 \; x_2 \; \ldots \; x_n]))))$$

Of course, the DoubleCheck specification cannot employ the informal notation of numbered lists (page 81). Instead, the property uses the variable xs to designate the list. The ACL2 statement of the theorem corresponding to this property is the same as the property specification except for the keyword **defproperty**, which becomes **defthm**, and the random data generators (introduced by the :value keyword), which are not present in the

5.2 Using Books of Proven Theorems

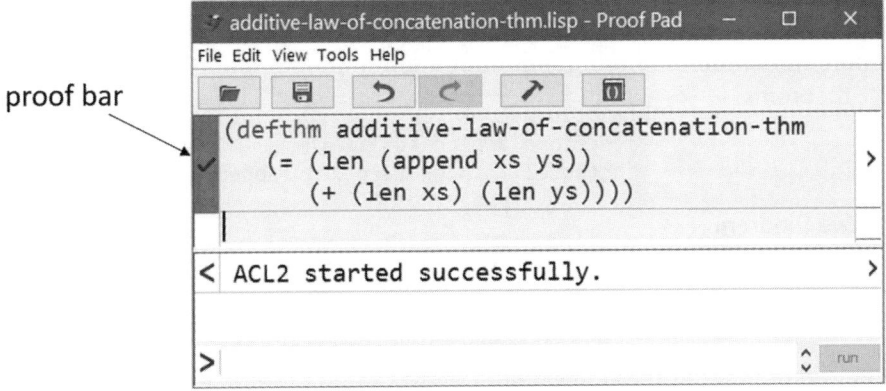

Figure 5.1
Proof Pad session with proof bar.

theorem. Theorem statements in ACL2 start with the **defthm** keyword. After that comes a name for the theorem and then the Boolean formula that states the theorem.

```
(defthm additive-law-of-concatenation-thm
  (= (len (append xs ys))
     (+ (len xs) (len ys))))
```

The mechanized logic of ACL2 fully automates the proof of this theorem. It follows its own heuristic procedures to find an induction scheme and pushes the proof through on its own. To see ACL2 in action, enter the theorem into a Proof Pad session and click in the green proof bar (figure 5.1). This sets ACL2 in motion to prove the theorem. A check mark appears in the proof bar when the mechanized logic succeeds.

5.2 Using Books of Proven Theorems

The append-suffix theorem states that if the first operand of the **append** operator is a list of length n, then dropping n elements from the front of the concatenation reproduces the second operand of **append**. In section 4.4 (page 96), we stated this theorem in the form ($\forall n.S(n)$), where $S(n)$ was shorthand for the following equation:

$$S(n) \equiv ((\text{nthcdr } (\text{len } [x_1\ x_2\ \ldots\ x_n])\ (\text{append } [x_1\ x_2\ \ldots\ x_n]\ ys)) = ys)$$

In ACL2 notation, a definition of this theorem takes the following form:

```
(defthm append-suffix-thm
  (equal (nthcdr (len xs) (append xs ys))
         ys))
```

The mechanized logic can prove this theorem, but the proof depends on some equations from numeric algebra. Fortunately, ACL2 experts have already proved many such theo-

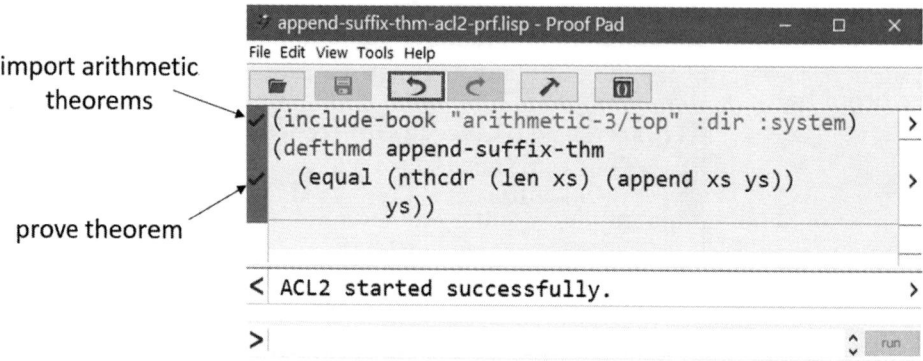

Figure 5.2
Importing theorems to facilitate proof: append-suffix theorem.

rems, and the ACL2 system makes them available in the form of "certified books" (ACL2 terminology for a package of theorems successfully proven by the mechanized logic). A book known as **arithmetic-3/top** contains theorems from algebra that the mechanized logic can cite in a proof of the append-suffix theorem. An **include-book** directive tells ACL2 to import these theorems.

    (include-book "arithmetic-3/top" :dir :system)

To make the theorems in the book available for ACL2 to cite in a proof, the directive precedes the **defthm** command that defines the theorem to be proved, as shown in figure 5.2. Clicking in the proof bar sets the mechanized logic in motion, and a check mark appears in the proof bar when the proof is complete. Try it for yourself in a Proof Pad session. Enter the **include-book** directive and theorem definition of figure 5.2, click in the proof bar, and see what happens.

Exercises

1. Define the associativity property of **append** ({*app-assoc*}, page 98) in ACL2 notation, and use Proof Pad to run it through the ACL2 mechanized logic. If you state the theorem correctly, ACL2 will succeed in proving it.[28]

2. Define the following theorem in ACL2 and get the mechanized logic to prove it:
    $\forall xs.((\text{len (nthcdr (len xs) xs)}) = 0)$

---

[28] According to J Strother Moore, a pioneer in mechanized logic and a principal developer of ACL2, the associativity of **append** was one of the driving examples in early work in mechanized logic and one of the first theorems that such a system proved autonomously.

5.3 Theorems with Constraints

Another of the paper-and-pencil proofs in section 4.4 was the append-prefix theorem (page 98), which referred to the prefix operator.

```
(defun prefix (n xs)
  (if (and (posp n) (consp xs))
      (cons (first xs) (prefix (- n 1) (rest xs)))   ; {pfx1}
      nil))                                           ; {pfx0}
```

The append-prefix theorem, as stated for the paper-and-pencil proof, had the form $\forall n.S(n)$, where $S(n)$ was written in terms of a numbered list. The same theorem in ACL2 terminology uses a variable to designate the list.

Theorem {append-prefix}: $\forall n.S(n)$
where $S(n) \equiv ((\text{nthcdr } (\text{len } [x_1 \ x_2 \ \ldots \ x_n]) \ (\text{append } [x_1 \ x_2 \ \ldots \ x_n] \ ys)) = ys)$

```
(defthm append-prefix-thm-NOT-QUITE-RIGHT
  (equal (prefix (len xs) (append xs ys))
         xs))
```

However, the theorem as stated is not quite correct, and to explain why, we need to mention some things we haven't told you about lists. A *true list* in ACL2 is either nil (the empty list) or a value of the form (cons *x xs*), where *xs* is a true list. The predicate true-listp is true when its operand is a true list and false otherwise.

It can be verified by induction that the prefix operator always delivers a true list. The value (prefix 0 *xs*) is nil. That takes care of the base case. The value of (prefix (*n*+1) *xs*) is either nil (again, a true list) or (cons (first *xs*) (prefix *n* (rest *xs*))). We can assume by the induction hypothesis that (prefix *n* (rest *xs*)) is a true list, which means (by the definition of the term "true list") that (cons (first *xs*) (prefix *n* (rest *xs*))) is a true list. That takes care of the inductive case and completes the proof that (prefix *n xs*) is always a true list.

When the second operand of cons is not a true list, the value it constructs isn't a true list either. For example, (cons 2 1) is not a true list because 1 is not a true list. The formula (cons 3 (cons 2 1)) also is not a true list, again, because the second operand, in this case (cons 2 1), is not a true list. However, (len (cons 3 (cons 2 1))) is 2, which you can work out from the axioms or, easier, just submit the formula to ACL2 and let it do the computation. A similar computation reveals that (prefix (len (cons 3 (cons 2 1))) (cons 3 (cons 2 1))) is (cons 3 (cons 2 nil)), not (cons 3 (cons 2 1)). That is, when *xs* is (cons 3 (cons 2 1)), then (prefix (len *xs*) (append *xs ys*)) is not equal to *xs*. The equation that theorem append-prefix-thm-NOT-QUITE-RIGHT guarantees does not hold when *xs* is (cons 3 (cons 2 1)). Since a theorem cannot have exceptions, the theorem append-prefix-thm-NOT-QUITE-RIGHT isn't true.

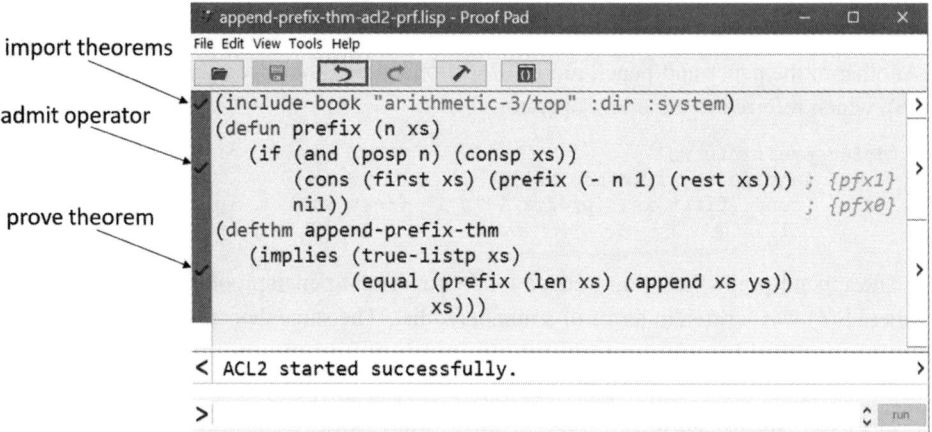

Figure 5.3
Theorem with constraint: append-prefix.

Box 5.3
Using Implication to Constrain the Domain of a Theorem

A theorem that takes the form of an implication, $x \to y$, says that the conclusion, y, will be true when the hypothesis, x, is true, but it says nothing about the status of the conclusion when the hypothesis is false. The ACL2 equivalent of the Boolean formula $x \to y$ is (implies x y). For example, one can conclude that $u - 1 < v - 1$ if one knows that $u < v$. In ACL2, this fact would be stated as an implication.

```
(defthm simple-theorem-about-numbers
  (implies (< u v)
           (< (- u 1) (- v 1))))
```

The theorem is true, however, when the variable *xs* is constrained to the domain of true lists. In the statement of the append-prefix theorem in section 4.4 (page 97), the role of *xs* was played by the numbered list $[x_1 \; x_2 \; \ldots \; x_n]$, which is a true list by definition (page 81). ACL2 does not permit the use of the numbered list syntax, so the true-list constraint must be explicit. An implication formula imposes the necessary constraint.

```
(defthm append-prefix-thm
  (implies (true-listp xs)
           (equal (prefix (len xs) (append xs ys))
                  xs)))
```

Now we have a theorem that we believe is true based on our paper-and-pencil proof. The mechanized logic of ACL2 succeeds in proving this theorem, but, as was the case with the append-suffix theorem (figure 5.2, page 104), it needs to cite some theorems of numeric algebra. Figure 5.3 (page 106) displays a Proof Pad session that imports those theorems, defines the prefix operator, and then states and proves the theorem. The figure indicates that ACL2 "admits" the prefix operator when it encounters its definition. This means that ACL2 allows the definition to join the universe of entities that can participate in the mechanized reasoning process. Box 5.4 explains what this entails.

Box 5.4
ACL2 Must Prove That Operators Terminate

In figure 5.3 (page 106), there is a step that has the label "admit operator." That is the terminology ACL2 uses for the process of accepting an operator into its mechanized logic. A domain in which operators may fail to terminate calls for a more complex reasoning process than a domain that is restricted to operators that are guaranteed to complete their computation in a finite number of steps. To give itself a better shot at succeeding in mechanized proofs, ACL2 does not deal with operators with a potential for nontermination. It will *admit* an operator to its domain of logic only if it can verify termination in all cases. Sometimes, coming up with an operator definition that makes it possible for ACL2 to prove termination is, by itself, a major project, including importing theorems or coming up with new theorems to facilitate the reasoning process. Hardware and software with guaranteed properties are not easy to come by.

Exercises

1. Define the {*rep-len*} theorem (page 99) in ACL2, and use Proof Pad to run it through the mechanized logic. (*Hint*: Use natp to constrain the first operand of rep.)

2. Exercise 5 on page 99 required a paper-and-pencil proof of the following proposition: $\forall xs.((\text{nthcdr} \ (\text{len} \ xs) \ xs) = \text{nil})$. Define this theorem in ACL2 and get the mechanized logic to prove it.

3. The formula (equal (prefix (len *xs*) *xs*) *xs*) is true with a certain constraint on *xs*. Define an ACL2 theorem with this equation as its conclusion and get the mechanized logic to prove it.

5.4 Helping Mechanized Logic Find Its Way

The mechanized logic of ACL2 was able to carry out proofs of the theorems of section 5.1 without assistance. To prove the theorems in section 5.2, ACL2 needed to cite some proven theorems packaged in books developed by experts. These were carefully chosen examples to get started with the mechanized proofs. Usually, the process of using a mechanized logic requires both proven theorems packaged in books and specialized theorems chosen to match the needs of a particular goal. That is, to succeed in proving a complex property of an operator defined in ACL2, it is usually necessary to prove some simpler properties that the mechanized logic can cite in a proof of the more complex property.

$$\text{Axioms } \{Fibonacci\} \quad \begin{aligned} f_0 &= 0 \\ f_1 &= 1 \\ f_{n+2} &= f_{n+1} + f_n \end{aligned} \quad \begin{aligned} &\{f0\} \\ &\{f1\} \\ &\{f2\} \end{aligned}$$

```
(defun fib(n) ; n-th Fibonacci number
  (if (posp (- n 1))
      (+ (fib(- n 1)) (fib(- n 2)))    ; n > 1
                                       ; {fib2}
      n))                              ; {fib1}
```

...	(fib 2)	(fib 3)	(fib 4)	(fib 5)	...	(fib 10)	(fib 11)	(fib 12)	...
...	1	2	3	5	...	55	89	144	...

Figure 5.4
Fibonacci numbers.

Choosing simpler properties that ACL2 can prove and building, finally, to the more complex proof calls for insight and creativity. This is where your experience with paper-and-pencil proofs comes in handy. You can plan a strategy by thinking of major steps in a proof, stating those steps as separate theorems, and proving them one by one to build up a body of helpful theorems that the mechanized logic can cite to move closer to the goal.

It may help to see an example of how this can work. The Fibonacci numbers are a well-studied sequence that scientists have used to study patterns of development in leaves and flowers, growth rates in animal populations, and other natural phenomena. The sequence can be defined inductively with the Fibonacci equations (figure 5.4).

The Fibonacci operator, fib, defined in ACL2, mirrors the algebraic equations. It selects the inductive formula (fib($n - 1$)) + (fib($n - 2$)) for an operand n that is 2 or bigger (that is, when ($n - 1$) is a positive, natural number). For smaller operands, the ACL2 definition observes that the corresponding Fibonacci number is the same as the operand: (fib 0) = f_0 = 0 by axiom $\{f0\}$ and (fib 1) = f_1 = 1 by axiom $\{f1\}$.

We can be confident that $\forall n.((\text{fib } n) = f_n)$ because the ACL2 definition of fib is a direct transliteration of the Fibonacci equations. We can use the operator fib as a calculator to compute some Fibonacci numbers. This works well for small numbers, but it turns out that there are a huge number of computational steps required to derive (fib n) from the Fibonacci axioms when n gets above a few dozen. For example, the computation of (fib 30) proceeds quickly, but you will see a noticeable delay if you ask ACL2 to compute (fib 40), and you would have to wait a long time for (fib 50).

It's not too hard to see the reason for this. The axioms calculate (fib ($n + 2$)) by first calculating (fib ($n + 1$)), then calculating (fib n), and finally adding those two numbers together. However many computational steps it takes to compute (fib n), it will take at least a few more to compute (fib ($n + 1$)). So, there will be more than twice as many steps in the computation of (fib ($n + 2$)) as there are in the computation of (fib n). That is, when n increases by 2, the number of computational steps more than doubles.

5.4 Helping Mechanized Logic Find Its Way

If we let c_n stand for the number of computational steps in the calculation of (fib n), then our observation about doubling amounts to the inductive relationship $c_{n+2} \geq 2c_n$. We think you have had enough experience to prove (by induction, of course) that $\forall n.(c_{2n+1} \geq 2^n)$, assuming that it takes at least one computational step to compute (fib 1). That's what they call *exponential growth*.[29] It is dramatic, to say the least. We estimate that it would take an hour to compute (fib 50) on a typical laptop computer and upwards of a year to compute (fib 75).

The Fibonacci axioms comprise an inductive definition. They conform to the three C's (figure 4.10, page 91), so they guarantee delivery of a Fibonacci number in a finite number of computational steps. Unfortunately, in this case it turns out to be a huge number of computational steps. Fortunately, there are alternatives.

If a particular inductive definition leads to a long computation, there may be another inductive definition that produces the same results with less work, and that is the case with Fibonacci numbers. Figure 5.5 (page 110) displays another definition that uses a method known as *tail recursion*.[30] A tail recursion is a circular reference in an operator definition that occurs at the *top level*, which means that it is not nested inside a formula to produce an operand for another operator. The operator h occurs at the top level in the formula (h (+ x 1)), but it is nested in the formula (+ (h x) 1). Most computing systems, including ACL2, implement tail recursions efficiently. Nested recursions are more problematic. They don't always lead to inefficient computations, but sometimes, as with Fibonacci numbers, they do.

The downside is that reasoning about tail recursive definitions is often more challenging than reasoning about definitions with nested recursions. One way to proceed, however, is to define an operator both ways, prove that the two definitions produce the same results, and then use the efficient definition for computation and the inefficient one when it simplifies the reasoning process.

---

[29] The term "exponential growth" is bandied about a lot, but mostly in ways that do not match the standard mathematical meaning. The most common usage is to describe something that gets big fast, which is certainly true of exponential growth, but it is also true of quadratic growth. Quadratic growth often gets passed off, informally, as exponential growth, but it's not even close. If the number of computational steps in computing (fib n) from the axioms grew quadratically instead of exponentially, it would not take a lot longer to compute (fib 40) than it does to compute (fib 30).

[30] "Recursive definition" is the most commonly used term for what we call "inductive definition." We don't say "recursive" because the term is often conflated with a computation strategy based on a data structure called a stack, and we think fixating on computational detail obscures the meaning of the definition. We want you to think of inductive (aka recursive) definitions as axiomatic equations that can be used to reason about the operators they define. We leave it to the computer system to determine how to carry out the computation. Sometimes, as in the Fibonacci problem where an inductive definition leads to inefficient computation, we will look for more efficient alternatives, but we will not relinquish our view of operator definitions as axioms to support reasoning. If our programming language allowed mutable variables, as most programming languages do, we would delve into a stack-based view of recursion. Since it doesn't, we won't.

```
(defun gib (n b a)   ; b = (fib(n-1)), a=(fib(n-2))
   (if (posp (- n 1))
       (gib (- n 1) (+ b a) b)           ; {gib2}
       (if (= n 1)
           b                              ; {gib1}
           a)))                           ; {gib0}
(defun fib-fast (n)  ; (fib-fast n)=(fib n)
   (gib n 1 0))      ; (fib 1)=1, (fib 0)=0
```

Figure 5.5
Fibonacci numbers: quick delivery.

If you look at the definition of fib-fast (figure 5.5) closely, we think you will agree that, although it may seem plausible that (gib n 1 0) is the nth Fibonacci number, it isn't obvious. Some reasoning is required. Maybe ACL2 can take it on successfully. We would like to know that $\forall n.((\text{fib-fast } n) = (\text{fib } n))$. The theorem in ACL2 terms can be stated as follows:

```
(defthm fib=fib-fast ; (fib-fast n) = (fib n)
   (implies (natp n)
            (= (fib n) (fib-fast n))))
```

When ACL2 tried to prove this theorem it failed, or at least it took such a long time floundering that we interrupted it to try something else. We thought maybe it needed to know some theorems from numeric algebra, so we imported the arithmetic-3/top book, as usual, but that didn't help. It still sat there spinning, so we interrupted it again.

We guessed that it might be having trouble coming up with an effective induction hypothesis, so we decided to ask it to prove that the gib operator satisfies the basic Fibonacci axioms. This is, to prove that the nth value of gib is the sum of the previous two values (as in axiom {f2}). In ACL2, that theorem can be stated as follows:

```
(defthm gib-inductive-equation ; a la {fib2}
   (implies (posp (- n 1))             ; n > 1
            (= (gib n b a)              ; n-th
               (+ (gib (- n 1) b a)     ; (n-1)th
                  (gib (- n 2) b a))))) ; (n-2)th
```

ACL2 failed quickly this time and reported that it was not able to find an induction scheme that worked. So, our attempt to help ACL2 along didn't work, exactly, but at least ACL2 could determine quickly that things were not going well. Next, we guessed that ACL2 might need help with the base case as well as the inductive case, so we stated a base case for the gib operator as an ACL2 theorem.

```
(defthm gib-base-equation ; a la {fib1}
   (= (gib 1 b a) b))
```

This idea worked. ACL2 was able to prove the gib base equation and the gib inductive equation, and then finally it used them to prove that $\forall n.((\text{fib-fast } n) = (\text{fib } n))$, in the form defined in the fib=fib-fast theorem. Figure 5.6 (page 111) displays the Proof Pad session with the check marks indicating completed proofs by the mechanized logic of ACL2. Try

Exercises

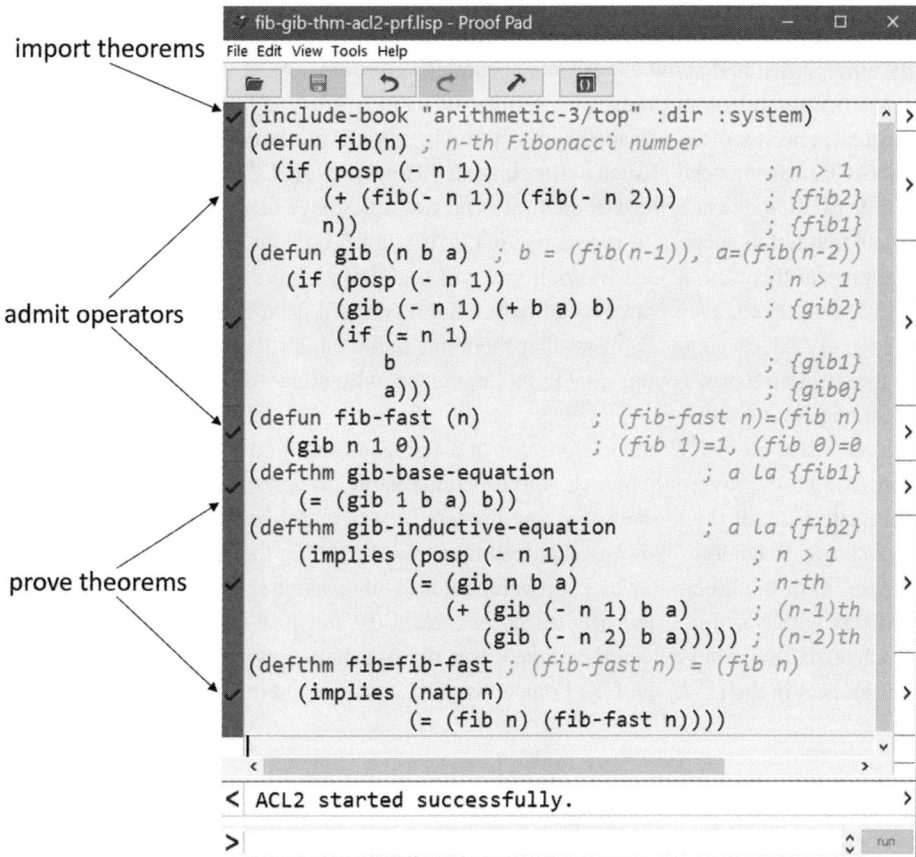

Figure 5.6
Fast Fibonacci delivers Fibonacci numbers.

out the fib-fast operator in a Proof Pad session. The Fibonacci number (fib-fast n) seems to appear instantaneously, even for large values of n.

Exercises

1. Assume that $c_1 = 1$ and $c_{n+2} = 2c_n$. Prove that $\forall n.(c_{2n+1} = 2^n)$.
   Hint: Define $x_n = c_{2n+1}$. First, verify that $x_{n+1} = 2x_n$. Then, use mathematical induction to prove that $x_n = 2^n$, and translate this formula for x_n to a formula for c_{2n+1}.
   Note: $2^0 = 1$, $\forall n.(2^{n+1} = 2 \times 2^n)$ {*Law of Exponents*}

5.5 Proof Automation and Things That Can't Be Done

By now you've had some experience constructing proofs, and if you're like most people, it has been tough going most of the time. It's rarely easy to figure out how to prove a theorem, and it's often extraordinarily difficult. How is it, then, that a mechanized logic like ACL2 can succeed at such a difficult task? How does ACL2 prove theorems?

First of all, it doesn't, most of the time. Our examples have been carefully constructed in a way that led to successful proofs by ACL2. If you try on your own to propose theorems and present them to ACL2 for proof, you will find that it can be extremely difficult to get ACL2 to succeed, even with a lot of help. Often, you will need to sketch a proof yourself, then give ACL2 a sequence of smaller theorems that it can prove, one by one, building up to the theorem that was your goal in the first place, with plenty of roadblocks and changes in strategy along the way.

Even when all ACL2 has to do is fill in a few gaps, we find it remarkable that ACL2 can succeed in proving theorems. And it *is* remarkable. When researchers began trying to automate parts of the theorem-proving process, it was many years before really effective tools began to emerge. Now, mechanized logic plays a role in the verification of important properties of digital circuits and software, not only in research labs but also in engineering projects with deadlines and product cycles. And it's not just ACL2. Many systems of mechanized logic have been developed over the past half-century or so by distinguished researchers in the USA, the UK, France, Sweden, and elsewhere.

Box 5.5
Mechanized Logics: Fifty Years of R&D, Mostly R

- ACL2 (and precursor nqthm): J Moore, Robert Boyer, and Matt Kaufmann, University of Texas, with a long history starting at the University of Edinburgh and continuing at Xerox PARC and SRI
- LCF, HOL, Isabelle: Robin Milner, Mike Gordon, Lawrence Paulson, Stanford University, Cambridge University, University of Edinburgh
- Coq: Thierry Coquand, Gérard Huet, INRIA, University of Gothenburg
- PVS: Sam Owre, Natarajan Shankar, John Rushby, SRI
- Agda: Ulf Norell, Chalmers University
- LF: Robert Harper, Furio Honsell, Gordon Plotkin, Carnegie Mellon University, University of Udine, University of Edinburgh
- Twelf: Frank Pfenning, Carsten Schürmann, Carnegie Mellon University, University of Copenhagen

5.5 Proof Automation and Things That Can't Be Done

Proofs are not only hard. They are sometimes impossible, as Gödel famously proved in 1931, to the great surprise of many mathematicians of the day. In the same vein, there are things computers cannot do. A well-known example is the *halting problem*, which was proved to be outside the realm of computation by Alan Turing in 1936 and, independently, by Alonzo Church. There were no computers at the time, but there were mathematical models of computation that are used still today to study the capabilities of computers.

A program solving the halting problem would be able to determine, given a computer program, whether or not the program would terminate in a finite amount of time given a particular input. There are programs that can solve the halting problem for a limited set of programs, but no program can solve the halting problem for all programs. Proving that no computer program can solve the halting problem is tricky, but the ideas are not too difficult to follow, and they provide an example of reasoning that, in a perverse sense, fits into a discussion of mechanized logic because it exhibits something that neither computers nor people can do. There are many such things, as a search on the term "uncomputable" shows.

To be specific, the following discussion of the halting problem limits itself to a single programming language, but that restriction turns out to be immaterial.[31] Because discussion of the halting problem is clumsy in ACL2, we'll present the ideas in terms of a different language that has the same syntax and a similar interpretation. That language is Lisp. For the purposes of this discussion, you can think of Lisp as ACL2 with the added feature of allowing operators to act as operands. That is, a formula invoking an operator can supply another operator as input. Most programming languages allow this. ACL2 doesn't because the extra facility interferes with some of the strategies that ACL2 uses to mechanize the process of proving theorems.

The discussion will include some logic formulas with quantification, so we need to specify the universe of discourse (page 51). When the bound variable in the formula is h, the universe of discourse will be the set of operators that have two operands and can be defined in Lisp: (defun h (f x) ... *Lisp formula* ...). When the bound variable is f, the universe of discourse will be the set of operators that have one operand and can be defined in Lisp: (defun f (x) ... *Lisp formula* ...).[32] When the bound variable is x, the universe of discourse is the set of values that Lisp operators can deliver.

We define a predicate, H, that tells us whether or not a particular formula in Lisp represents a computation that terminates, that is, a computation that would be completed in a finite amount of time. H is not a Lisp operator and does not itself represent a computation.

---

[31] The halting problem cannot be solved for any "general purpose" programming language, by which we mean a Turing complete language, which itself calls for a long explanation even before discussing how Turing completeness interacts with the halting problem. You can track that down if you're interested. All widely used programming languages (ACL2, Java, and C++, for example) are Turing complete. We'll leave it at that.

[32] The restriction to one operand simplifies the presentation, but it is not really a restriction because the operand could be a list containing any number of values.

It is a mathematical entity outside the realm of computation that gives us a way to use symbols and formulas to talk about whether or not a computation terminates.

The definition of the predicate H uses the Lisp formula $(f\ x)$ to designate a computation that may or may not be completed in a finite amount of time. Since H is a logic predicate, not a computation, it has a value whether or not the formula $(f\ x)$ terminates.

Definition of predicate H:
$H(f, x) = True$, if $(f\ x)$ terminates
$H(f, x) = False$, if $(f\ x)$ does not terminate

The theorem that Turing and Church proved is that no operator h can be defined in Lisp that accepts a Lisp operator f as its first operand and a Lisp value x as its second operand such that, regardless of the definition of the operator f, $(h\ f\ x) = 0$ if the computation $(f\ x)$ terminates and $(h\ f\ x) = 1$ if $(f\ x)$ does not terminate. This means not just that it would be hard to define the operator h. It means that nobody can define h with any amount of effort or cleverness. There is no such definition. There are some things that cannot be done, and this is one of them.

A conjecture of Church and Turing concerning the effectiveness of general-purpose programming languages asserts that for any computation that can be carried out, there is a program that can be written in Lisp (or any other general-purpose programming language) that specifies a way to carry out the computation. If one believes Church–Turing conjecture, and most computer scientists do, then the theorem that Church and Turing proved about the prospects of automating the prediction of program termination says that no program can be written that solves the halting problem. Neither Turing nor Church stated the theorem in terms of Lisp. Turing used a model of computation known as the Turing machine, and Church used the lambda calculus, another model of computation. Both models are still widely studied in computer science theory, and the programming language Lisp was based on the lambda calculus.

We express the halting problem theorem as a logic formula that we will prove using natural deduction (section 2.5). The theorem has no hypotheses.

Theorem {halting problem}:
$\vdash \neg(\exists h.\forall f.\forall x.((H(f, x) \rightarrow ((h\ f\ x) = 1)) \wedge ((\neg H(f, x)) \rightarrow ((h\ f\ x) = 0))))$

There are two Lisp formulas embedded in the logic formula of the theorem, both the same: $(h\ f\ x)$. Those formulas invoke an operator h, which is defined in Lisp because that is the universe of discourse for h, the bound variable in the $\exists$ quantification. The logic formula then compares the value that the formula $(h\ f\ x)$ computes to a natural number.

To make the proof more compact and, we hope, more readable, we will use E as an abbreviation for the long $\exists$ formula whose negation is the conclusion of the theorem.

$E \equiv \exists h.\forall f.\forall x.((H(f, x) \rightarrow ((h\ f\ x) = 1)) \wedge ((\neg H(f, x)) \rightarrow ((h\ f\ x) = 0)))$

5.5 Proof Automation and Things That Can't Be Done

Theorem {halting problem}: ⊢ ¬E   Note: This theorem has no hypotheses.
  where $E \equiv \exists h.\forall f.\forall x.((H(f,x) \to ((h\ f\ x) = 1)) \land ((\neg H(f,x)) \to ((h\ f\ x) = 0)))$
proof

 Assume $(\neg(\neg E))$ *this assumption will be discharged*
 ———————————{¬¬ forward}   *proved in figure 2.22 (page 48)*
   E
 ———————————{paradox}   *proved in figure 5.10 (page 118)*
  False
 ———————————{→ introduction}   *assumed $(\neg(\neg E))$, proved False,*
 $(\neg(\neg E)) \to False$ *conclude $(\neg(\neg E)) \to False$*
 Discharge $(\neg(\neg E))$ *this discharge leaves no hypotheses*
 ———————————{reductio ad absurdum}   *figure 2.15 (page 39)*
  $\neg E$

Figure 5.7
Theorem {halting problem}: a proof by contradiction.

With the abbreviation, the {halting problem} theorem is: ⊢ ¬E. Figure 5.7 displays the proof, which uses the natural deduction formalism. It cites a theorem we already proved, theorem {¬¬ forward}, and a theorem, {paradox}, which we will prove after showing how our goal, the {halting problem} theorem, can be derived from it. To clarify the derivation, we need to state the {paradox} theorem, which we will prove later.

 Theorem {paradox}: $E \vdash$ *False*

In addition to citing the {paradox} theorem, the proof of the {halting problem} theorem cites the {reductio ad absurdum} inference rule (figure 2.15, page 39), so our proof of the {halting problem} theorem is a proof by contradiction. It begins by assuming the negation of the formula in its conclusion. The assumption, as always, stands in lieu of a proof. As it happens, the assumption is discharged later in the proof, leaving the {halting problem} theorem with no hypotheses, as it is stated. The conclusion of the theorem is the negation of a long there-exists formula. What this means is that there is no program that, given an arbitrary program p, can determine in a finite number of computation steps whether the program p will terminate.

We are not quite ready to tackle the {paradox} theorem. The proof will be easier to follow if we break it into parts, then connect the parts in the final analysis.

Recall that E, the hypothesis of the {paradox} theorem, is a ∃ formula. Since E is a hypothesis of the {paradox} theorem, we will assume at the beginning of the proof of the {paradox} theorem that E is true. E asserts that there is at least one definition of an operator h that provides a universal, computational solution to the halting problem. Since

Axioms p		
(p x) = 0	if (h p x) = 0	{p0}
(p x) = (loop x)	if (h p x) ≠ 0	{p1}

```
(defun loop (x)   ; a nonterminating computation
  (loop (loop x)))
(defun p (x)      ; definition derived from axioms of p
  (if (equal (h p x) 0)
      0
      (loop x)))
```

Figure 5.8
Definitions of operators p and loop.

there is at least one such operator, let's assume someone has handed us a definition of the operator and that the operator has the name h: (defun h (f x) ... *Lisp formula* ...).

We don't know the definition of h, but the formula E tells us some of its properties. In particular, E says that for any operator f and any value x, the formula (h f x) = 0 if $H(f, x)$ is *False* and (h f x) = 1 if $H(f, x)$ is *True*. We specify those properties in the following two theorems:

Theorem {hF}: $E \vdash \forall f. \forall x.(\neg H(f, x)) \rightarrow ((h \ f \ x) = 0)$
Theorem {hT}: $E \vdash \forall f. \forall x. H(f, x) \rightarrow ((h \ f \ x) = 1)$

Since h is a Lisp operator, we can refer to it in the definition of another Lisp operator. Figure 5.8 defines the operator p. It invokes h to decide whether to deliver the value 0 or the computation represented by an invocation of an operator called loop, which is also defined in figure 5.8. The loop operator doesn't do anything. Or, rather, it does way too much by doing nothing over and over, forever. We are going to take it on faith that (loop x) does not terminate, regardless of the value of its operand x. We think you can probably convince yourself of that,[33] so we assert that $\forall x.(H(\text{loop}, x) = False)$.

We can inquire about p using the predicate H. In particular, we would like to know, given a value x, whether $H(\text{p}, x) = True$ or *False*. It has to be one or the other because H is a predicate. Since p is an operator, x is a value, and 1 is not 0, the following theorems are special cases of theorem {hF} and theorem {hT}:

Theorem {pF}: $E \vdash (\neg H(\text{p}, x)) \rightarrow ((h \ p \ x) = 0)$
Theorem {pT}: $E \vdash H(\text{p}, x) \rightarrow (\neg((h \ p \ x) = 0))$

Now, let's reason from the definition of p (figure 5.8). In the definition, we find that if (h p x) = 0, then (p x) = 0. We also find that if (h p x) ≠ 0, then (p x) represents the

---
[33] Some of the exercises will help you reason why.

5.5 Proof Automation and Things That Can't Be Done

computation (loop x). We will use the notation ($p\ x$) = (loop x) to indicate this relationship between ($p\ x$) and (loop x).[34] This reasoning verifies two more theorems.

Theorem {h0}: ⊢ (h p x) = 0 → (p x) = 0
Theorem {h1}: ⊢ ¬((h p x) = 0) → (p x) = (loop x)

Furthermore, to compute the value of the formula (p x), the if operator in the definition of p (figure 5.8, page 116) selects one of two formulas: 0 or (loop x). If it selects the formula 0, then (p x) terminates. Therefore, from the definition of the predicate H (page 114), we conclude that $H(p, x) = True$. That is, ((p x)=0) → H(p, x). Label this implication {p0}.

On the other hand, if the if operator in the definition of p selects the formula (loop x), then (p x) does not terminate.[35] Therefore, from the definition of the predicate H, we conclude that $H(p, x) = False$. That is, ((p x)=(loop x)) → (¬H(p, x)). Label this implication {p1}. Theorems {p0} and {p1} restate these implications.

Theorem {p0}: ⊢ ((p x) = 0) → H(p, x)
Theorem {p1}: ⊢ ((p x) = (loop x)) → (¬H(p, x))

From the three theorems, {pF}, {h0}, and {p0}, we can derive theorem {H+}, and from the other three theorems, {pT}, {h1}, and {p1}, we can derive theorem {H−}. Figure 5.9 (page 118) displays a proof of theorem {H−}. The proof of theorem {H+} is similar, and working through it would be good practice (exercise 1, page 119).

Theorem {H+}: E ⊢ (¬H(p, x)) → H(p, x)
Theorem {H−}: E ⊢ H(p, x) → (¬H(p, x))

The {contradiction} equation (figure 2.12, page 30), together with the {double negation} equation (figure 2.11, page 29), prove that ((¬H(p, x)) → H(p, x)) = H(p, x) and (H(p, x) → (¬H(p, x))) = (¬H(p, x)). So, we can rewrite the {H+} and {H−} theorems as follows:

Theorem {H+ version 2}: E ⊢ H(p, x)
Theorem {H− version 2}: E ⊢ ¬H(p, x)

Now, at last, we're ready to take on the proof of theorem {paradox}. It derives *False* from the contradiction that is apparent in theorems {H+} and {H−}. Figure 5.10 (page 118) displays this final step in our proof of the theorem of Turing and Church, confirming that no computer program can solve the halting problem.

---

[34] This interpretation of the equation (p x) = (loop x) conforms with our usual practice of substituting the formula designated in the definition of f for an invocation ($f\ x$). That is, we interpret the substitution of a new formula in place of an old one with the same meaning as an equation between the old formula and the new one. It is an odd sort of equality in the case of (p x) = (loop x) because the formula (loop x) represents a computation, not a value. The loop operator at every stage delivers a new formula, but it never manages to produce a value.

[35] The formula (loop x) does not complete its computation in a finite amount of time (exercise 5, page 119).

Theorem {H−}: $E \vdash H(p, x) \to (\neg H(p, x))$
proof

        Assume E *hypothesis of theorem*
  ─────────────────────{pT}
$H(p, x) \to (\neg((h\ p\ x) = 0))$
  -      *separates proofs of {→ chain} hyps*
                                              no proofs above the line for {h1}
  ─────────────────────{h1}   *because thm {h1} has no hypotheses*
$\neg((h\ p\ x) = 0) \to (p\ x) = (\text{loop}\ x)$
  ─────────────────────{→ chain}   {→ chain} *thm (figure 2.20, page 47)*
$H(p, x) \to (p\ x) = (\text{loop}\ x)$
  -      *separates proofs of {→ chain} hyps*
                                              no proofs above the line for {p1}
  ─────────────────────{p1}   *because thm {p1} has no hypotheses*
$(p\ x) = (\text{loop}\ x) \to (\neg H(p, x))$
  ─────────────────────{→ chain}   *another citation of {→ chain} theorem*
$H(p, x) \to (\neg H(p, x))$

Figure 5.9
Proof of theorem {H−}.

Theorem {paradox}: $E \vdash \textit{False}$
proof

       Assume E *hypothesis of theorem*
  ──────────{H+ version 2}   *proof of {H+} left as exercise*
       $H(p, x)$
  -    *separates proofs for {∧ complement} theorem*
       Assume E *hypothesis of theorem*
  ──────────{H− version 2}   {H−} *proved in figure 5.9 (page 118)*
       $\neg H(p, x)$
  ──────────{∧ complement}   *exercise 1, page 49*
       False

Figure 5.10
Theorem {halting problem}: a proof by contradiction.

Exercises

1. Prove theorem {H+} (page 117).

2. Suppose x is an operand for the operator loop (figure 5.8, page 116). Prove that the formula (loop x) represents the same computation as the formula (loop (loop x)).

3. Suppose f is an operator, x is an operand, and n is a natural number. The following equations define the notation $(f^n\ x)$:

   $(f^0\ x) = x$ {iter0}
   $(f^{n+1}\ x) = (f\ (f^n\ x))$ {iter1}   Example: $(f^3\ x) = (f\ (f\ (f\ x)))$

   Interpreting the relationship between (loop x) and (loop (loop x)) stated in exercise 2 as an equation (see footnote, page 117), use mathematical induction to verify the following formula:
   $$\forall n.\forall x.((\text{loop}\ x) = (\text{loop}^{n+1}\ x))$$

4. Assume that if f is an operator and x is an operand, then it takes at least one computational step to compute $(f\ x)$. Prove by induction that it takes at least n computational steps to compute $(f^n\ x)$, where $(f^n\ x)$ satisfies the axioms {iter0} and {iter1} in exercise 3.

5. Suppose x is an operand for the operator loop (figure 5.8, page 116). Let T stand for the number of computation steps required to compute (loop x). Using the definitions and theorems of exercise 2, exercise 3, and exercise 4, prove $(\forall n.(T > n))$.

II Computer Arithmetic

6 Binary Numerals

6.1 Numbers and Numerals

Numbers are mathematical objects with certain properties, and they come with operators, such as addition and multiplication, that produce new numbers from numeric operands. Because numbers are mathematical objects, they are ephemeral. You can't really get your hands on them. They are figments of the imagination.

However, numbers are useful, and to use them we need to be able to write them down. Decimal numerals are one way to do this. The numeral 144 stands for the number of eggs in a dozen cartons of eggs. The numeral 1215 stands for the number of years between the twenty-seventh year of the reign of Caesar Augustus and the signing of the Magna Carta.

However, "144" and "1215" are numerals. They are not numbers. They are symbols that stand for numbers, and they are not the only symbols in use for that purpose. The symbols CXLIV and MCCXV (Roman numerals) stand for the same two numbers. So do the symbols 90_{16} and $4BF_{16}$ (hexadecimal numerals), but most people use decimal numerals like 144 and 1215 when they do arithmetic.

The decimal representation is so embedded in our experience and practice that we often conflate numerals and numbers. Some dictionaries treat the terms as synonyms. Usually there is no harm in considering numbers and numerals to be the same thing, but we are going to use numerals to do arithmetic in a mechanized way, so we will do well to distinguish between numbers (mathematical objects) and numerals (symbols for numbers).

Let's think about how we interpret a decimal numeral as a number. Consider the numeral 1215, for example. Each digit in the numeral has a different interpretation. The first digit is the number of thousands in the number that 1215 stands for. The second tells us the number of hundreds, then the tens, and finally the units. The following formula is a way to express this interpretation:

$$\mathbf{1} \times 10^3 + \mathbf{2} \times 10^2 + \mathbf{1} \times 10^1 + \mathbf{5} \times 10^0$$

This formula computes a number from the individual digits in the numeral using standard arithmetic operations (addition, multiplication, and exponentiation). It shows us what the individual digits in the numeral stand for, and it gives us a leg up on figuring out other kinds of numerals. The digits in the hexadecimal numeral have a similar meaning but with a different base. Decimal numerals are based on powers of ten and hexadecimal numerals are based on powers of sixteen.

Box 6.1
Digits as Numbers

Perhaps you noticed a subtle confusion in the formulas we use to explain the meaning of numerals. At first, we claim that 1215 is merely a symbol standing for a mathematical object. And we claim that the digit 2 is merely a symbol standing for the number of items in a pair, along with similar claims for the digits 1 and 5. Then, we use those symbols in the formula $1 \times 10^3 + 2 \times 10^2 + 1 \times 10^1 + 5 \times 10^0$ as if they were numbers.

There is some sleight of hand going on here. Numbers as mathematical objects are figments of our imagination, but when we write formulas, we have to choose some symbols to represent them. So, in the formula $1 \times 10^3 + 2 \times 10^2 + 1 \times 10^1 + 5 \times 10^0$, we use the symbols 1, 10, 3, 2, 5, and 0 as if they were numbers. But, in the numeral 1215, the symbols 1, 2, and 5 are not numbers. They are symbols standing for numbers.

It's even worse with the hexadecimal numeral $4BF_{16}$ and the formula $4 \times 16^2 + 11 \times 16^1 + 15 \times 16^0$. In the formula, we have rewritten the symbol B as the hexadecimal numeral for 11 and the symbol F as the hexadecimal numeral for 15. And we've had the temerity to pretend that symbols in the formula are numbers when they are really hexadecimal numerals, as they were in the formula for the meaning of the decimal numeral 1215.

Furthermore, we've really mixed things up in the numeral $4BF_{16}$ because the "4BF" part is in hexadecimal notation and the "16" part is a decimal numeral, indicating that we are to interpret the digits in base sixteen rather than base ten. Try to get your head around this. We're more or less stuck with it. Figments of our imagination have to be materialized somehow if we are going to talk about them.

The system of decimal numerals calls on ten different symbols to represent digits: 0, 1, 2, ...9. The hexadecimal system requires sixteen different symbols, conventionally 0, 1, 2, ...9, A, B, C, D, E, F. The digits stand for the customary numbers (0 for zero, 1 for one, 2 for two, and so on) and the letters A through F stand for the numbers ten through fifteen.[36] That leads to the following formula to express the meaning of the hexadecimal numeral $4BF_{16}$. (Remember, B stands for eleven, F for fifteen.)

$$4 \times 16^2 + 11 \times 16^1 + 15 \times 16^0$$

---

[36] There are no conventional squiggles for digits beyond fifteen, presumably because no numeral system with a base exceeding sixteen is in common use. A few thousand years ago, the Mayan civilization used a base twenty system with twenty different symbols for digits. The ancient Sumerians used a system with base sixty but with special arrangements to deal with the lack of a symbol for zero.

6.1 Numbers and Numerals

Formulas like this convert numerals to numbers. No doubt you could construct the appropriate formula for any given numeral: decimal, hexadecimal, or any other base.

We'll say more about converting numerals to numbers later, but what about going the other direction, converting numbers to numerals? Suppose someone gives you an operator called dgts that converts a number to a decimal numeral. Let's say that (dgts 1215) would deliver the list [5 1 2 1]. That is, dgts delivers a list of the decimal digits of its operand, starting with the ones digit (a 5 in this case), then the tens digit, and so on, reading right to left from the customary way of writing the numeral. If you had a definition of dgts, you might test it on a few numbers to see if it does what you expect.

```
(check-expect (dgts 1215) (list 5 1 2 1))
(check-expect (dgts 1964) (list 4 6 9 1))
(check-expect (dgts 12345) (list 5 4 3 2 1))
(check-expect (dgts 0) nil)
```

Box 6.2
Numerals as Lists ... Backwards

What! The dgts operator delivers the digits backwards! Why is that?

Of course dgts could have delivered the digits in the customary order, but reverse order simplifies some of the equations we will use to interpret numerals. We will write numerals like 1215 in the usual way, but the dgts operator delivers them in the form of a list with the digits in reverse order: [5 1 2 1].

Besides being backwards, the elements in the list are numbers, not symbolic digits. We could use pure symbols, but we indulge in this sleight of hand to simplify parts of the discussion. In a similar vein, the list notation [5 1 2 1] is the symbol we use to describe the list, but the list itself is a mathematical object. It's another figment of our imagination, ephemeral in the same sense as a number.

Wait a minute! Why does (dgts 0) deliver the empty list instead of the one-element list [0]? That's another little trick. Besides delivering the digits in reverse order, leading zeros are omitted. We could write the numeral 1964 with as many leading zeros as we like. The numerals 1964, 01964, and 000001964 all stand for the same number. Those numerals correspond to [4 6 9 1], [4 6 9 1 0], and [4 6 9 1 0 0 0 0 0] in our list format. "Leading zeros" come at the end when the digits are in reverse order.

However, dgts doesn't include any leading zeros in the numerals it delivers. It leaves them all off, even for the number zero. That's why (dgts 0) is nil. The numeral (dgts 012345) is the same as the numeral (dgts 12345) too because dgts interprets its operand as a number. Since 012345 and 12345 stand for the same number, the formula (dgts 012345) and the formula (dgts 12345) both deliver the list [5 4 3 2 1]. No leading zeros.

The computer interprets the decimal numeral 012345 in the formula (dgts 012345) as a mathematical object. How does the computer represent the object? None of your business. That's the computer's business. It has its own way of dealing with numbers. Later, we'll study the way most computers do this, but for now we will assume that the computer has some way of turning numerals into whatever form it uses to represent numbers.

After running a few sanity checks on the dgts operator, you might want to do some serious testing. Big batches of automated tests using random data, perhaps. Coming up with automated tests calls for some thought. Let's start small. How about the units digit in a decimal numeral? What mathematical formula would deliver the units digit in a decimal numeral given an arbitrary positive integer n?

The units digit in a decimal numeral is the remainder when you divide the number by ten. The formula that converts a numeral to a number makes that clear.

$$1 \times 10^3 + 2 \times 10^2 + 1 \times 10^1 + 5 \times 10^0$$

Each of the terms in the formula is a product of a power of ten with another number. A power of ten is, of course, a multiple of ten, so none of those terms contribute to the remainder when dividing by ten. None of them, that is, except the units digit. It does not have a factor of ten in it because 10^0 is one, which is not a multiple of ten. So, to get the units digit in the numeral, compute the remainder in the division of the number by ten.

Box 6.3
mod and floor: Think Third-Grade Division

Transport yourself back to the third grade, or whenever you learned long division. There were four parts to the problem, and they all had names.

divisor	number you divide by
dividend	number you divide by the divisor
quotient	what you get when you do the division
remainder	what's left over to make up the difference

$$q = (\text{floor } n\ d) \quad \textit{quotient}$$
$$r = (\text{mod } n\ d) \quad \textit{remainder}$$
$$n = qd + r \quad \{check \div\}$$

The remainder is what the mod operator delivers (box 6.3). The following test uses mod to make sure the units digit in the numeral that the dgts operator delivers for the number n is correct.

```
(= (first (dgts n)) (mod n 10))
```

6.1 Numbers and Numerals

Since **dgts** delivers the digits backwards, (first (dgts *n*)), the first digit in the list, is the units digit in the numeral, which is the last digit when the numeral is written in the conventional format. The formula checks to make sure that the units digit of the numeral that **dgts** delivers is (mod *n* 10), the remainder when dividing *n* by 10.

We can use the DoubleCheck facility of Proof Pad to run this test on a batch of random numbers. We need to be careful not to allow zero to pop up in the testing because (**dgts** 0) is nil, so there is no first digit to check. Besides, we've already completed the testing of (**dgts** 0) in our sanity checks. We can avoid retesting zero by adding one to a random natural number. That produces a random, nonzero, positive integer.

```
(defproperty dgts-last-digit-tst
  (n-1   :value (random-natural))
  (let* ((n (+ n-1 1)))  ; avoid n=0
    (= (first (dgts n))
       (mod n 10))))
```

That takes care of testing the units digit, but what about the others? We can do something about those by observing that the quotient when *n* is divided by ten is a number with the same digits as *n* except that the units digit is missing. Remember, we're doing third-grade arithmetic here. The quotient is the main result of the division. No fraction, no decimal point, no remainder. Just the whole-number quotient. Since we've already tested to make sure the units digit is correct, we don't need to worry about that. We only need to worry about the other digits. The intrinsic operator **floor** (box 6.3, page 126) produces the quotient, discarding the remainder. We can get those other digits by applying **dgts** to the operand (floor *n* 10).

The following formula implements the test we have in mind. It checks to make sure the digits other than the units digit in the list that (**dgts** *n*) delivers are the same as the digits in the list that **dgts** delivers when its operand is the quotient in the division (*n* ÷ 10).

```
(equal (rest (dgts n))      ; all digits except the units digit
       (dgts (floor n 10))) ; digits of the quotient
```

As with the test of the units digit, we can run a batch of tests based on our rest-of-the-digits observation by defining a DoubleCheck property.
observation.

```
(defproperty dgts-other-digits-tst
  (n-1   :value (random-natural))
  (let* ((n (+ n-1 1)))           ; avoid n=0
    (equal (rest (dgts n))
           (dgts (floor n 10)))))
```

It would be nice to run these tests, but **dgts** is not an intrinsic operator. We have to provide a definition for it. To do that we use the **defun** command, which is similar to **defproperty** but without any value specifications.

The definition of the **dgts** operator will be inductive, will use some ideas we discussed in putting together tests, and will conform to the requirements of the three C's (figure 4.10, page 91), repeated here and customized for **dgts**.

Complete   Two cases: the number is zero or it isn't. So, two formulas.
Consistent   The cases do not overlap—no chance for inconsistency.
Computational   Inductive case ($n > 0$): operand is divided by ten, making it closer to zero (the noninductive case).

```
(defun dgts (n)
  (if (zp n)
      nil                                       ; {dgts0}
      (cons (mod n 10) (dgts (floor n 10)))))   ; {dgts1}
```

This definition uses the predicate **zp** (box 4.6, page 94) to detect the value zero within the domain of natural numbers. If you put the definition of **dgts** at the beginning of a program, import the "testing" and "DoubleCheck" facilities (include-book, chapter 3, pages 71–72), and import some theorems about modular arithmetic (box 6.4), you can enter the tests and run them using Proof Pad. You can also enter formulas in the command panel to compute decimal numerals for any natural numbers you choose.

Box 6.4
Termination, ACL2 Admit, and floor/mod Equations

ACL2 will not accept the definition of an operator unless it can prove that the operator always delivers a value in a finite number of computation steps. Nonterminating operators complicate the reasoning process. Proving that **dgts** terminates requires applying some theorems of modular arithmetic. Fortunately, experts have put together some theorems on that topic, and the following include-book directive will import them to make it possible for the mechanized logic to prove termination and admit the **dgts** operator to the ACL2 logic.

    (include-book "arithmetic-3/floor-mod/floor-mod" :dir :system)

Exercises

1. Let *y* stand for the number of years between the signing of the Magna Carta and the signing of the United States Declaration of Independence. Find the numeral for *y* and use **dgts** (page 128) to verify that you got it right.

2. Prove theorem {*mod-div*}: (mod (∗ a x) (∗ a b)) = (∗ a (mod x b))
   Hint: You won't need induction, but the following facts will help. Suppose *x* is the dividend and *d* the divisor in a third-grade division problem (box 6.3, page 126). Then, r = (mod x d) is the remainder and q = (floor x d) is the quotient. Third-graders use the equation $(qd + r) = x$ to make sure they have done the division correctly. They also know that $0 \leq r < d$: (mod x d) is the number *r* in the range $0 \leq r < d$ such that $qd + r = x$.

6.2 Numbers from Numerals

3. Define the {*mod-div*} theorem in ACL2 notation, and use ACL2 to verify that it is a theorem. Since the theorem does not hold for all numbers *a*, *b*, and *x*, you will need to ask ACL2 to prove an implication with hypotheses that constrain the theorem to the domain in which it is true. If you state it correctly and import the theorems about modular arithmetic contained in the floor-mod book (box 6.4, page 128), ACL2 will succeed.

Hint: (posp *n*) is true if *n* is a nonzero natural number and false otherwise.

6.2 Numbers from Numerals

The **dgts** operator (page 128) provides a way to compute a decimal numeral given a number. How about going in the other direction? Given a decimal numeral, compute the corresponding number. You already know the formula.

$$1 \times 10^3 + 2 \times 10^2 + 1 \times 10^1 + 5 \times 10^0 = 1215$$
$$1 \times 10^2 + 4 \times 10^1 + 4 \times 10^0 = 144$$

What properties would an operator converting decimal numerals to numbers have? Let's assume that numerals are represented in the manner of the **dgts** operator: units digit first, then the tens digit, then hundreds, and so on. We want to define an operator, **nmb10**, that converts decimal numerals in that form to numbers. We know that (nmb10 nil) must be zero because (dgts 0) = nil, and we are trying to convert numerals produced by **dgts** back to the numbers they came from. How about a one-digit numeral $[x_0]$? The equation in that case would be (nmb10 $[x_0]$) = x_0. If there are two or more digits, $[x_0\ x_1\ \ldots x_{n+1}]$, then the equation would take the following form:

$$\text{(nmb10 } [x_0\ x_1\ x_2 \ldots x_{n+1}]) = x_0 + x_1 \times 10^1 + x_2 \times 10^2 + \ldots x_{n+1} \times 10^{n+1} \quad \{a\}$$

All of the terms in the sum include a factor of ten except the first term, so we can factor ten out of those terms. Factoring the formula in this way produces a new equation.

$$\text{(nmb10 } [x_0\ x_1\ x_2 \ldots x_{n+1}]) = x_0 + 10 \times (x_1 \times 10^0 + x_2 \times 10^1 + \ldots x_{n+1} \times 10^n) \quad \{b\}$$

The list $[x_1\ x_2\ \ldots x_{n+1}]$ is also a decimal numeral, albeit for a different number. The number it denotes is $(x_1 \times 10^0 + x_2 \times 10^1 + \ldots x_{n+1} \times 10^n)$, which is the value **nmb10** should deliver given the numeral $[x_1\ x_2\ \ldots x_{n+1}]$.

$$\text{(nmb10 } [x_1\ x_2 \ldots x_{n+1}]) = x_1 + x_2 \times 10^1 + \ldots x_{n+1} \times 10^n \quad \{c\}$$

Observe that the right-hand side of equation $\{c\}$ is equal to the part of the formula in equation $\{b\}$ that is multiplied by ten. Therefore, we can rewrite equation $\{b\}$ as follows:

$$\text{(nmb10 } [x_0\ x_1\ x_2 \ldots x_{n+1}]) = x_0 + 10 \times \text{(nmb10 } [x_1\ x_2 \ldots x_{n+1}]) \quad \{d\}$$

Equation $\{d\}$ is an inductive equation that delivers the right value for numerals with two or more digits. It also works for one-digit numerals because (nmb10 nil) is zero.

$$\text{(nmb10 } [x_0]) = x_0 + 10 \times \text{(nmb10 nil)} = (x_0 + 10 \times 0) = x_0 \quad \{d*\}$$

Together, equation $\{d\}$ for numerals with one or more digits and the equation for empty numerals, (nmb10 nil) = 0, conform to the rule of the three C's (page 91), so we have the makings of an inductive definition of nmb10. The following equations summarize our analysis and show how the first and rest operators (page 82) extract the required digits from the numeral. From that point, constructing the ACL2 definition is a straightforward translation of the equations to prefix notation.

$$\begin{aligned}
(\text{nmb10 } [x_0 \; x_1 \; x_2 \ldots x_{n+1}]) &= x_0 + 10 \times (\text{nmb10 } [x_1 \; x_2 \ldots x_{n+1}]) & \{n10.1\} \text{ (below)} \\
(\text{nmb10 nil}) &= 0 & \{n10.0\} \text{ (below)} \\
x_0 &= (\text{first } [x_0 \; x_1 \ldots x_n]) & \{cons\} \text{ (page 82)} \\
[x_1 \ldots x_n] &= (\text{rest } [x_0 \; x_1 \ldots x_n]) & \{rest\} \text{ (page 82)}
\end{aligned}$$

```
(defun nmb10 (xs)
  (if (consp xs)
      (+ (first xs) (* 10 (nmb10 (rest xs))))   ; {n10.1}
      0))                                        ; {n10.0}
```

We have derived this definition carefully from things we know about numbers, and we can use logic to be sure we got it right. We want to prove that (nmb10 $[x_0 \; x_1 \ldots x_n]$) delivers the same number as the formula $(x_0 + x_1 \times 10^1 + x_2 \times 10^2 + \ldots x_n \times 10^n)$.

Theorem $\{Horner \; 10\}$: (nmb10 $[x_0 \; x_1 \ldots x_n]$) = $x_0 + x_1 \times 10^1 + x_2 \times 10^2 + \ldots x_n \times 10^n$

The theorem asserts that the operator nmb10 computes a sum of multiples of successive powers of ten. The multipliers (known as polynomial coefficients) of the powers of ten are the digits in a decimal numeral. We call the theorem *Horner 10* because the scheme that the operator nmb10 uses to carry out the computation is known as Horner's rule. Proving theorem $\{Horner \; 10\}$ amounts to verifying that $(\forall n.H(n))$ is true, where the predicate H is defined for each natural number n as follows:

$$H(n) \equiv ((\text{nmb10 } [x_0 \; x_1 \ldots x_n]) = x_0 + x_1 \times 10^1 + x_2 \times 10^2 + \ldots x_n \times 10^n)$$

Figures 6.1 (base case) and 6.2 (inductive case) provide a proof by induction of theorem $\{Horner \; 10\}$. As usual, the inductive case, $\forall n.(H(n) \rightarrow H(n+1))$, derives $H(n+1)$ from the induction hypothesis $H(n)$ and other known equations. We conclude that $\forall n.H(n)$ is true, citing mathematical induction (figure 4.9, page 89).

Theorem $\{Horner \; 10\}$ confirms that nmb10 delivers the number that its operand, which is a base 10 numeral, denotes. We also expect the inverse to be true. That is, we expect the formula (dgts n) to deliver the base 10 numeral for the natural number n as stated in the following theorem:

Theorem $\{dgts\text{-}ok\}$: $\forall n.((\text{nmb10 (dgts } n)) = n)$

6.2 Numbers from Numerals

$$H(0) \equiv ((\text{nmb10 } [x_0]) = x_0)$$

	(nmb10 $[x_0]$)	
=	(nmb10 (cons x_0 nil))	{*cons*} (page 82)
=	(+ (first (cons x_0 nil))	{n10.1} (page 130)
	($*$ 10 (nmb10 (rest (cons x_0 nil)))))	
=	(+ x_0 ($*$ 10 (nmb10 (rest (cons x_0 nil)))))	{*first*} (page 82)
=	(+ x_0 ($*$ 10 (nmb10 nil)))	{*rest*} (page 82)
=	(+ x_0 ($*$ 10 0))	{n10.0} (page 130)
=	x_0	{*algebra*}

Figure 6.1
{*Horner 10*}: proof of base case.

$$H(n+1) \equiv ((\text{nmb10 } [x_0\ x_1\ \ldots x_{n+1}]) = x_0 + x_1 \times 10^1 + \ldots x_{n+1} \times 10^{n+1})$$

	(nmb10 $[x_0\ x_1\ \ldots\ x_{n+1}]$)	
=	(nmb10 (cons x_0 $[x_1\ \ldots\ x_{n+1}]$))	{*cons*}
=	(+ (first (cons x_0 $[x_1\ \ldots\ x_{n+1}]$))	{n10.1}
	($*$ 10 (nmb10 (rest (cons x_0 $[x_1\ \ldots\ x_{n+1}]$)))))	
=	(+ x_0 ($*$ 10 (nmb10 (rest (cons x_0 $[x_1\ \ldots\ x_{n+1}]$)))))	{*first*}
=	(+ x_0 ($*$ 10 (nmb10 $[x_1\ \ldots\ x_{n+1}]$)))	{*rest*}
=	(+ x_0 ($*$ 10 ($x_1 + x_2 \times 10^1 + \ldots x_{n+1} \times 10^n$)))	{$H(n)$}
=	$x_0 + x_1 \times 10^1 + x_2 \times 10^2 + \ldots x_{n+1} \times 10^{n+1}$	{*algebra*}

Figure 6.2
{*Horner 10*}: proof of inductive case.

$$\frac{\text{Base Case: } D(0) \equiv ((\textsf{nmb10 (dgts 0)}) = 0)}{\begin{array}{ll} & (\textsf{nmb10 (dgts 0)}) \\ = & (\textsf{nmb10 nil}) \qquad \{\textit{dgts0}\} \text{ (page 128)} \\ = & 0 \qquad\qquad\qquad \{\textit{n10.0}\} \text{ (page 130)} \end{array}}$$

$$\frac{\text{Inductive Case: } D(n+1) \equiv ((\textsf{nmb10 (dgts }(n+1))) = (n+1))}{\begin{array}{ll} & (\textsf{nmb10 (dgts }(n+1))) \\ = & (\textsf{nmb10 (cons (mod }(n+1)\text{ 10) (dgts (floor }(n+1)\text{ 10)))}) \qquad \{\textit{dgts1}\} \text{ (page 128)} \\ = & (+ \text{ (mod }(n+1)\text{ 10)} \\ & \quad (* \text{ 10 (nmb10 (dgts (floor }(n+1)\text{ 10)))))} \qquad\qquad\qquad \{\textit{n10.1}\} \text{ (page 130)} \\ = & (+ \text{ (mod }(n+1))\text{ 10) (* 10 (floor }(n+1)\text{ 10)))} \qquad\qquad\;\; \{D(\textsf{floor }(n+1)\text{ 10})\} \\ = & (n+1) \qquad\qquad\qquad\qquad\qquad\qquad\qquad\qquad\qquad\qquad\;\; \{\textit{check}\div\} \text{ (page 126)} \end{array}}$$

Figure 6.3
Proof by induction of theorem {*dgts-ok*}.

Define $D(n) \equiv ((\textsf{nmb10 (dgts } n)) = n)$. We want to prove that $(\forall n.D(n))$ is true. The universe of discourse of the predicate D is the natural numbers, so a proof of $D(0)$ together with a proof of $(\forall n.(D(n) \to D(n+1)))$ leads by natural deduction to the desired conclusion. Both of the required proofs appear in figure 6.3.

If you look carefully at the proof of $D(n) \to D(n+1)$, you may notice that it is not according to Hoyle. In a proof by mathematical induction, we can cite $D(n)$ to justify any step in a proof of $D(n+1)$, but figure 6.3 cites a different proposition, $D(\textsf{floor }(n+1)\text{ 10})$, in its proof of $D(n+1)$. However, $(\textsf{floor }(n+1)\text{ 10})$ is strictly smaller than $(n+1)$, and the proof relies on an inference rule known as *strong induction*. Strong induction is equivalent to ordinary mathematical induction, even though it looks more powerful because any or all of the propositions $D(0), D(1), \ldots D(n)$ can be cited in the proof of $D(n+1)$. Ordinary mathematical induction allows citing $D(n)$, but not the other propositions. Nevertheless, it is possible to verify that if the rule of ordinary mathematical induction is a valid rule of inference, then so is strong induction, and vice versa. The proof is not difficult, but it's a distraction, so we will present a rationale and leave it at that.

The rationale for strong induction is similar to the rationale for ordinary mathematical induction (page 88). Suppose you are proving the propositions $P(0), P(1), P(2), \ldots$ and so on, one by one, in sequence. When you get to the point where you want to prove $P(n+1)$, you will have already proven all of the propositions with smaller indices: $P(0)$, $P(1), P(2), \ldots P(n)$. So, in the proof of $P(n+1)$, you would be able to cite any of the previous propositions, not just $P(n)$. When you cite $P(n)$, but not propositions with smaller indices, in the proof of $P(n+1)$, you are using ordinary mathematical induction. When you cite one or more propositions with indices smaller than n, you are using strong induction.

Exercises

$$\frac{\text{Prove } (\forall m < n.P(m)) \to P(n)}{\text{Infer } (\forall n.P(n))} \text{\{strong induction\}}$$

Figure 6.4
Mathematical induction (strong induction version).

To put it another way, the formal statement of the strong induction rule (figure 6.4) is different from the formal statement of ordinary induction (figure 4.9, page 89), but in practice strong induction encompasses the ordinary induction rule as a special case. A proof of $P(n + 1)$ that cites only $P(n)$ and not propositions in the predicate P with smaller indices could cite strong induction because $P(n)$ is one of the propositions that the strong-induction hypothesis, $(\forall m < n + 1.P(m))$, assumes are true. So, we may as well cite strong induction, even when we're only relying on the ordinary induction rule.

Exercises

1. Let d stand for the number of furlongs in the Boston Marathon, not counting the last 165 yards. Prove that (nmb10 (dgts d)) = d using the definitions of dgts (page 128) and nmb10 (page 130) but without citing any of the theorems from this section.

2. Define a DoubleCheck property to test the equation (nmb10 (dgts n)) = n for random natural numbers n. Of course, all of the tests should succeed because we proved that the formula always delivers true. If a test fails, something is wrong with the definition of the property or the operators it refers to.

3. Define a DoubleCheck property to test (equal (dgts (nmb10 xs)) xs) for random decimal numerals xs.
   Note: (random-list-of (random-between 0 9)) generates random decimal numerals.
   Note: The test must use the operator equal because (= x y) requires x and y to be numbers.
   Note: This test can fail. If it does, check out the data that causes the failure.

4. We proved that $\forall n.((\text{nmb10 (dgts } n)) = n)$ (page 130). That is, the operator nmb10 inverts the operator dgts. However, it is not quite true that dgts inverts nmb10. Why not? Give an example of a decimal numeral xs for which (dgts (nmb10 xs)) ≠ xs.

5. Describe a constraint on xs such that (dgts (nmb10 xs)) = xs.

6. Prove that (dgts (nmb10 xs)) = xs if xs satisfies the constraint of exercise 5.

7. Prove that $\forall n.((\text{len (dgts } (n + 1))) = \lfloor log(n + 1) \rfloor + 1)$ {len-bits 10}
   Note: $\lfloor log(n + 1) \rfloor$ = an integer such that $10^{\lfloor log(n+1) \rfloor} \leq (n + 1) < 10^{\lfloor log(n+1) \rfloor + 1}$
   Note: The operators len and dgts are defined on pages 84 and 128.

> **Box 6.5**
> Strong Induction Requires Two Proofs or One?
>
> The inference rule for ordinary mathematical induction (figure 4.9, page 89) requires two proofs above the line: (1) Prove $P(0)$ and (2) Prove $\forall n.(P(n) \to P(n+1))$. The strong induction rule (figure 6.4) calls for only one proof: Prove $(\forall m < n.P(m)) \to P(n)$. However, when $n = 0$, this is $(\forall m < 0.P(m)) \to P(0)$. Since there are no natural numbers less than zero, $(\forall m < 0.P(m))$ is true because a $\forall$ quantification with an empty universe of discourse is true by default (page 52).
>
> So, proving $(\forall m < 0.P(m)) \to P(0)$ is the same as proving $True \to P(0)$, which is equivalent to proving $P(0)$, just as in ordinary induction. In other words, in strong induction there are really two proofs to do, one for $P(0)$ and one for $(\forall m < n.P(m)) \to P(n)$ when n is not zero, which we usually state in the equivalent form $(\forall m < n + 1.P(m)) \to P(n + 1)$.
>
> Looking at it this way makes a proof citing strong induction look like a proof citing ordinary mathematical induction except that in the proof of $P(n + 1)$, the induction hypothesis includes all of the propositions $P(0), P(1), P(2) \ldots P(n)$, not just $P(n)$. The key is that a proof by strong induction can cite any (or some, or even all) of those propositions to justify steps in the proof of $P(n + 1)$.

6.3 Binary Numerals

Digital circuits, since they are materializations of formulas in mathematical logic, have components that can represent two different values. We call them zero and one, and it happens that those names make it convenient to discuss circuits that deal with binary numerals, which use 0 and 1 to denote binary digits (that is, *bits*). Before delving into circuits, let's talk about binary numerals.

Decimal numerals represent numbers as sums of multiples of powers of ten. Binary numerals are similar, but they use two as a base instead of ten. So, a binary numeral with bits $x_n x_{n-1} \ldots x_2 x_1 x_0$, where each x_i is either a zero or a one, stands for the number $(x_0 + x_1 \times 2^1 + x_2 \times 2^2 + \cdots + x_n \times 2^n)$. The only differences between this formula and the one that interprets decimal numerals is that it uses powers of two instead of powers of ten and the multipliers are bits (0, 1) instead of digits (0, 1, 2, ... 9).

Therefore, we can convert the operators for decimal numerals to binary by changing the base from ten to two. The operators **bits** and **numb**, which are used to construct and interpret binary numerals, are defined in figure 6.5 (page 135). Like the corresponding operators for decimal numerals (**dgts**, page 128, and **nmb10**, page 130), **bits** and **numb** use the predicate **zp** (page 94) to choose between the base case ($n = 0$) and the inductive case ($n > 0$). ACL2 needs access to the theorems in the floor-mod book (box 6.4, page 128) to admit the **bits** operator to its logic, just as it did for the definition of **dgts**.

A theorem about **numb** concerning the interpretation of binary numerals as numbers and a theorem about **numb** being the inverse of the **bits** operator are true, and the proofs

Exercises   135

```
(defun bits (n)
  (if (zp n)
      nil                                    ; {bits0}
      (cons (mod n 2) (bits (floor n 2))))) ; {bits1}

(defun numb (xs)
  (if (consp xs)
      (if (= (first xs) 1)
          (+ 1 (* 2 (numb (rest xs))))   ; {2numb+1}
          (* 2 (numb (rest xs))))        ; {2numb}
      0))                                ; {numb0}
```

Figure 6.5
Definitions of operators bits and numb.

are similar to the proofs of the corresponding theorems about decimal numerals presented earlier in this chapter. Constructing those proofs can clarify your understanding of both decimal numerals and binary numerals.

Box 6.6
Representation Trick: Any List Is a Binary Numeral

A difference between the definition of numb (figure 6.5) and the definition of nmb10 (page 130) is that numb treats bits as symbols, whereas nmb10 assumes that digits are numbers. A 1 in our representation of binary numerals stands for a one-bit. Anything else stands for a zero-bit. That's why the definition of numb has two equations ({2numb+1} and {2numb}) for nonempty numerals, whereas the definition of nmb10 has only one equation ({n10.1}).

There are two motivations for this design decision. One is that it makes binary numerals entirely symbolic, with no ephemeral entities like mathematical numbers. Circuits represent bits symbolically by electronic signals, and so the definition of numb is more closely related to a circuit than it would be if it treated bits as numbers.

The other motivation is to reduce the number of constraints in theorems about ACL2 models of circuits. Bits don't have to be lists of zeros and ones. Any list is a numeral. That numeral has one-bits where the elements of the list are ones and zero-bits where its elements are anything other than ones. In fact, numerals are entirely unconstrained. The empty list represents the number zero, but so does any ACL2 entity x such that (consp x) is false. Many picayune details in theorems about binary numerals are avoided by this elimination of constraints.

Exercises

1. Adapt the proof of *{Horner 10}* (page 130) to prove theorem *{Horner 2}*.
   $$\forall n.((\text{numb } [x_0\ x_1\ \ldots\ x_n]) = x_0 + x_1 \times 2^1 + x_2 \times 2^2 + \ldots\ x_n \times 2^n)\quad \{\textit{Horner 2}\}$$

2. Prove theorem *{bits ok}*. $\forall n.(((\text{numb (bits } n)) = n))\quad \{\textit{bits ok}\}$

3. Prove theorem {nmb1}.
   $\forall n.((\text{numb } [x_0\ x_1\ \ldots\ x_n]) = (\text{numb } [x_0]) + 2\times(\text{numb } [x_1\ x_2\ \ldots\ x_n]))$ {nmb1}

4. ACL2 succeeds in proving theorem {nmb1} (exercise 3). Confirm that assertion by running the following formalization of the theorem through the mechanized logic:

   ```
   (defthm nmb1
     (implies (consp xs)
              (= (+ (numb (list (first xs)))
                    (* 2 (numb (rest xs))))
                 (numb xs))))
   ```

5. Given the following definition of operator pad, prove theorem {*len-pad*}:
   $\forall n.((\text{len }(\text{pad } n\ x\ xs)) = n)$ {*len-pad*}

   ```
   (defun pad (n x xs)
     (let* ((padding (- n (len xs))))
       (if (natp padding)
           (append xs (rep padding x))  ; {pad+}
           (prefix n xs))))              ; {pad-}
   ```

   Note: The definition of pad refers to the operators rep (page 99) and prefix (page 98). It uses a let∗ formula (box 3.4, page 75) to associate the name "padding" with the number of elements to append to *xs* to form a list of the desired length. If *n* is less than (len *xs*), pad delivers the first *n* elements of *xs*. In any case, the length of (pad *n x xs*) is *n*.

6. Given the following definition of operator fin, prove theorem {*hi-1*}:
   $\forall n.((\text{fin }(\text{bits } n)) = 1)$ if $n > 0$ {*hi-1*}

   ```
   (defun fin (xs)
     (if (consp (rest xs))
         (fin (rest xs))   ; {fin2}
         (first xs)))      ; {fin1}
   ```

7. Define theorem {*hi-1*} of exercise 6 in ACL2 and run it through the mechanized logic. ACL2 will succeed if you use implies and posp (box 4.6, page 94) to constrain the theorem to nonzero, natural numbers. Of course, the definition of the operator bits (page 135) will need to be admitted to the ACL2 logic (box 6.4, page 128) before it can attempt the proof of this theorem.

8. Prove theorem {*len-bits*}. $\forall n.((\lfloor 2^{n-1}\rfloor \leq m < 2^n) \rightarrow ((\text{len }(\text{bits } m)) = n))$ {*len-bits*}

9. Prove theorem {*len-bits≤*}. $\forall n.((0 \leq m < 2^n) \rightarrow ((\text{len }(\text{bits } m)) \leq n))$ {*len-bits≤*}

10. Prove theorem {*log-bits*}. $\forall n.((\text{len }(\text{bits }(n+1))) = \lfloor \log_2(n+1)\rfloor + 1)$ {*log-bits*}
    Note: $\lfloor x \rfloor$ is the greatest integer not exceeding *x*.

11. Prove theorem {*leading-0*}.
    $\forall n.((\text{numb }(\text{append }(\text{bits } n)\ (\text{list } 0))) = (\text{numb }(\text{bits } n)))$ {*leading-0*}

12. Assume theorem {*leading-0*} (exercise 11) is true. Prove theorem {*leading-0s*}.
    $\forall m.(\forall n.((\text{numb }(\text{append }(\text{bits } n)\ (\text{rep } m\ 0))) = (\text{numb }(\text{bits } n))))$ {*leading-0s*}
    Note: The operator rep is defined on page 99.

Exercises 137

13. Express theorem {*leading-0s*} of the previous exercise in ACL2 notation and run it through the ACL2 mechanized logic. ACL2 will succeed if you state the theorem correctly as an implication, referring to the predicate natp to constrain n and m to be natural numbers. Of course, the definitions of the operators bits and numb (page 135) will need to be admitted to the ACL2 logic (box 6.4, page 128) before it can attempt the proof of this theorem.

14. Prove theorem {*pfx-mod*}:

$$\forall w.(\forall n.((\text{numb } (\text{prefix } w \text{ (bits } n))) = (\text{mod } n \ 2^w))) \quad \{pfx\text{-}mod\}$$

Hint: The universe of discourse of $\forall w$ is the natural numbers. Induct on w. Split the inductive case into two cases: $n = 0$ and $n > 0$. Do a separate proof of the theorem for each case. In the $n > 0$ case, the {*mod-div*} theorem (page 128) will be helpful.

Note: The {∨ elimination} rule (figure 2.15, page 39) formally justifies case-by-case proofs.

7 Adders

7.1 Adding Numerals

When people do paper-and-pencil arithmetic, they use decimal numerals. For example, to add two numbers, a person writes down the decimal numerals for the numbers, one under the other with the digits lined up so that the units digit of one number is directly under the units digit of the other, and similarly for the tens digits, hundreds digits, and so on. Then, the units digits are added together, and the low-order digit of that sum is written below the units-digit column.

If the sum of the units digits is ten or more, a carry is marked above the tens-digit column. Then, the carry (if there is one) and the tens digits of the two numerals are added together. As before, the low-order digit goes into the numeral for the sum, this time in the tens-digit column, and the carry, if there is one, is marked above the hundreds-digit column. This process moves across the digits of the addends until all the digits are accounted for (figure 7.1, page 140).

To add decimal numerals in this way, a person needs to know the table for one-digit sums ($0 + 0 = 0, 0 + 1 = 1, \ldots 2 + 2 = 4 \ldots 9 + 8 = 17, 9 + 9 = 18$). To add binary numerals, a similar table of one-bit sums is needed, but the table is much smaller for binary numerals than for decimal numerals because it only has to account for two kinds of bits (0 and 1), not ten digits (0, 1, 2, ... 9). The process of adding numerals is in other respects the same for binary numerals as for decimal numerals (figure 7.2, page 140). The small size of the table for one-bit addition simplifies both the paper-and-pencil process and the design of digital circuits for addition of binary numerals compared to designing circuits for decimal numerals.[37]

---

[37] The table in figure 7.2 relies on commutativity and associativity (figure 2.2, page 18) for completeness.

```
  11 1      carries
  ----
  9542      first addend
+  638      second addend
  ----
 10180      sum
```

Figure 7.1
Adding decimal numerals.

```
  11 111 1    carries
  --------
  01011101    first addend
+ 11010101    second addend
  --------
 100110010    sum
```

one-bit addition		
+	c	s
0 + 0	0	0
0 + 1	0	1
1 + 1	1	0
1 + 1 + 1	1	1

Figure 7.2
Adding binary numerals.

Exercises

1. Describe by example the process of multiplying a pair of decimal numerals.

2. Describe by example the process of multiplying a pair of binary numerals.

7.2 Circuits for Adding One-Bit Binary Numerals

The addition table for one-bit binary numerals in figure 7.3 (page 141) displays the sum as two separate bits: a carry-bit c and a sum-bit s. A close look at the table shows that the carry-bit matches the table of values of the digital gate for logical-and (figure 2.13, page 34). That is, the carry-bit is 1 only if both inputs are 1s. Otherwise, the carry-bit is 0. So, a logical-and gate can serve as a digital circuit to compute the carry-bit in the addition of two one-bit binary numerals. Feed the signals for the one-bit numerals into a logical-and gate and the output signal will represent the carry-bit correctly.

Another close look reveals that the sum-bit matches the table of values of the digital gate for exclusive-or (XOR, figure 2.13). That is, the sum-bit is 0 if the two inputs are the same and 1 if they are different. So, constructing a digital circuit to compute the sum-bit amounts to feeding the signals for the one-bit numerals into an exclusive-or gate. Combining these ideas for carry-bit and sum-bit circuits leads to a two-input, two-output circuit known as a half-adder (figure 7.3, page 141).

7.2 Circuits for Adding One-Bit Binary Numerals

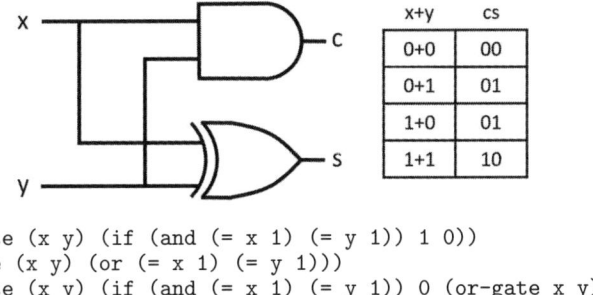

```
(defun and-gate (x y) (if (and (= x 1) (= y 1)) 1 0))
(defun or-gate (x y) (or (= x 1) (= y 1)))
(defun xor-gate (x y) (if (and (= x 1) (= y 1)) 0 (or-gate x y)))
(defun half-adder (x y) (list (xor-gate x y) (and-gate x y)))
```

Figure 7.3
Half-adder circuit and ACL2 model.

Since we reason about digital circuits using the methods of Boolean algebra, we need an algebraic representation of the circuit diagram for the half-adder circuit. Remember, digital circuits are only one of four equivalent representations of Boolean formulas that we have studied: circuit diagrams, well-formed formulas in the notation of mathematical logic (for example, $x \wedge y$), Boolean formulas in engineering notation (juxtaposition for $\wedge$, + for $\vee$, and over-bar for $\neg$), and ACL2 notation. The ACL2 formalization allows us to mechanize some aspects of the reasoning process. So, figure 7.3 also specifies the half-adder circuit in ACL2 terms. We refer to this specification as an ACL2 model of the half-adder circuit. We use the same name for the model as the circuit: the operator **half-adder** delivers the two output signals as a list of two elements, the first element being the sum-bit and the second, the carry-bit.

In the end, we would like to have a circuit that adds binary numerals, and we saw in an example (figure 7.2, page 140) that this would require us to deal with three input bits in each column: the corresponding bits in the two addends and the carry-bit brought from adding the bits in the previous column. The half-adder circuit is not up to this task because it has only two input signals. However, we can put together a full-adder circuit by combining two half-adders and a logical-or gate, as shown in figure 7.4 (page 142). Since the full-adder circuit has three inputs, each of which is either 0 or 1, there are eight possible input configurations, as shown in the full-adder table.

The **full-adder** operator defined in the figure is a formal model in ACL2 of the circuit diagram, and the following tests, one for each line in the table, comprise a comprehensive, mechanized verification of the model:

```
(check-expect (full-adder 0 0 0) (list 0 0))
(check-expect (full-adder 0 0 1) (list 1 0))
(check-expect (full-adder 0 1 0) (list 1 0))
(check-expect (full-adder 0 1 1) (list 0 1))
(check-expect (full-adder 1 0 0) (list 1 0))
```

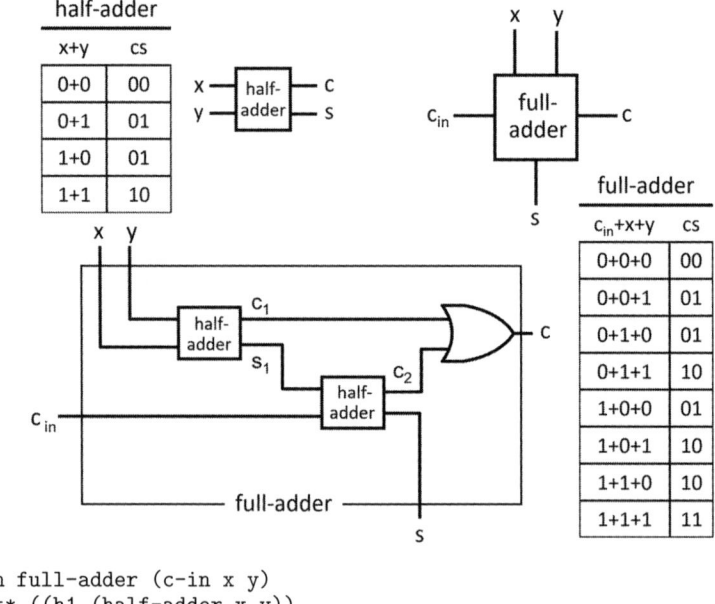

```
(defun full-adder (c-in x y)
  (let* ((h1 (half-adder x y))
         (s1 (first h1)) (c1 (second h1))
         (h2 (half-adder s1 c-in))
         (s  (first h2)) (c2 (second h2))
         (c  (or-gate c1 c2)))
    (list s c)))
```

Axiom {*snd*}: (second xs) = (first (rest xs))

Figure 7.4
Full-adder circuit and ACL2 model.

7.3 Circuit for Adding Two-Bit Binary Numerals

Theorem {full-adder-ok}:   (numb [s c]) = (numb [x]) + (numb [y]) + (numb [c_{in}])
where [s c] = (full-adder c_{in} x y)

```
(defthm full-adder-ok
  (= (numb (full-adder c-in x y))
     (+ (numb (list c-in)) (numb (list x)) (numb (list y)))))
```

Figure 7.5
Full-adder theorem.

```
(check-expect (full-adder 1 0 1) (list 0 1))
(check-expect (full-adder 1 1 0) (list 0 1))
(check-expect (full-adder 1 1 1) (list 1 1))
```

The full-adder operands are symbols for bits, but we can use the numb operator (page 135) to interpret them as numbers. If x is a bit, then [x] is a numeral for the number it denotes, and the {*Horner 2*} theorem (exercise 1, page 135) asserts that (numb [x]) computes that number. The same theorem asserts that if s and c are bits, then (numb [s c]) is the number that the numeral [s c] denotes. Combining these observations with the full-adder table (figure 7.4, page 142) verifies the theorem {full-adder-ok} shown in figure 7.5.[38]

7.3 Circuit for Adding Two-Bit Binary Numerals

A circuit that adds two-bit binary numerals comes from combining two full-adder circuits (figure 7.6, page 144). The first full-adder circuit gets as input the low-order bits, x_0 and y_0, of the two addends. The second full-adder circuit gets the high-order bits, x_1 and y_1. The circuit directs the output carry, c_1, from the first full-adder to the input carry of the second full-adder.

The circuit produces three output signals: one sum-bit from each full-adder (s_0 and s_1) and the carry-out from the second full-adder (c_2). With a carry-in of zero for the first full-adder ($c_0 = 0$), the output signals form a two-bit numeral [s_0 s_1] and a carry-bit c_2 that together represent the sum of the two input numerals. More generally, the output signals represent the sum of the two input numerals and the carry-in bit ($c_0 = 0$ or $c_0 = 1$). The following equation {★} shows how to interpret the input and output signals as numbers:

(numb [s_0 s_1]) + (numb [c_2]) = (numb [c_0]) + (numb [x_0 x_1]) + (numb [y_0 y_1])   {★}

---

[38] The full-adder operator delivers a list [s c] whose first element is the sum-bit and whose second element is the carry-bit. The order of elements in the result delivered by full-adder was designed to form a numeral for the sum of its three, one-bit operands. The list in reverse order, [c s], would contain the same information, but it would not conform to our representation of binary numerals because the low-order bit would no longer come first.

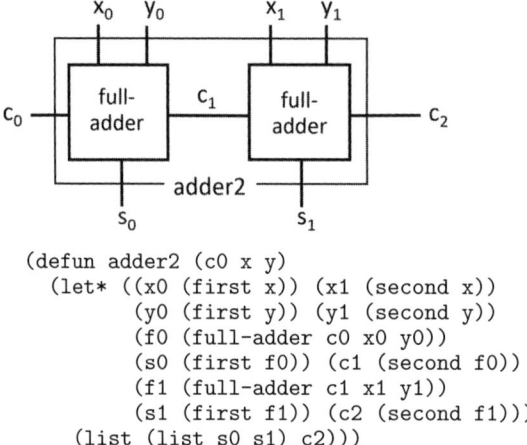

```
(defun adder2 (c0 x y)
  (let* ((x0 (first x)) (x1 (second x))
         (y0 (first y)) (y1 (second y))
         (f0 (full-adder c0 x0 y0))
         (s0 (first f0)) (c1 (second f0))
         (f1 (full-adder c1 x1 y1))
         (s1 (first f1)) (c2 (second f1)))
    (list (list s0 s1) c2)))
```

Figure 7.6
Two-bit adder and ACL2 model.

We refer to the output two-bit numeral, without the carry-bit, as the sum-bits. Ignoring the carry-bit amounts to doing modular arithmetic, mod 2^2 (theorem {*pfx-mod*}, exercise 14, page 137).

A mechanized verification of the arithmetic properties of adder2 could be constructed as a comprehensive sequence of check-expect tests, as we did with the full-adder for one-bit addition (page 141). There would be 32 cases to check because there are five bits of input (a carry-in and two bits in each numeral, 2^5 combinations in all). That makes it tedious to construct comprehensive check-expect tests. It would be easy to make a mistake.

A better approach is to write equation {⋆} formally in ACL2. The equation states our expectations of the ACL2 model of the two-bit adder circuit (adder2, figure 7.6) in the same way that the {full-adder-ok} theorem (figure 7.5, page 143) verified the formal model of the full-adder circuit.

The theorem is an equation that, on one side, is the sum of the numeric interpretation of the input numerals and input carry. On the other side, the equation is the numeric interpretation of the output numeral and output carry. As in the {full-adder-ok} theorem, the {adder2-ok} theorem that specifies the crucial arithmetic property of the two-bit adder uses the numb operator (page 135) to interpret binary numerals as numbers.

```
(defthm adder2-ok
  (let* ((a (adder2 c0 (list x0 x1) (list y0 y1)))
         (s (first a)) (c (second a)))
    (= (numb (append s (list c)))
       (+ (numb (list c0))
          (numb (list x0 x1))
          (numb (list y0 y1))))))
```

Exercises

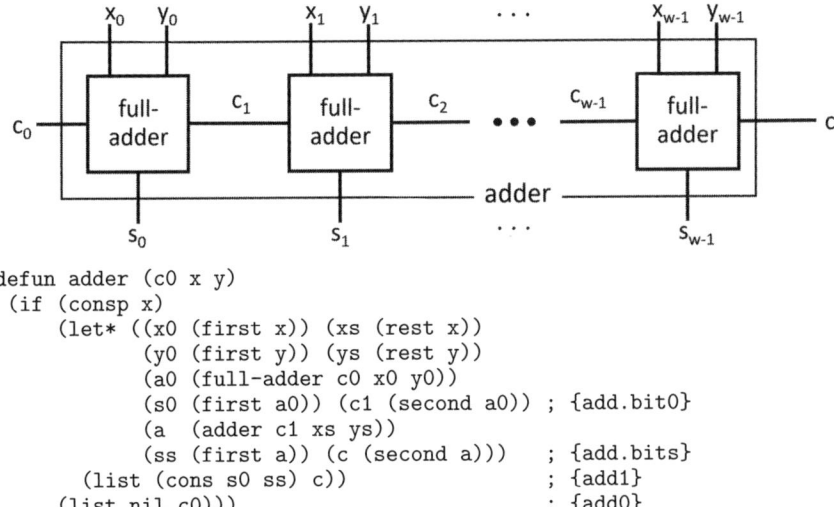

```
(defun adder (c0 x y)
  (if (consp x)
      (let* ((x0 (first x)) (xs (rest x))
             (y0 (first y)) (ys (rest y))
             (a0 (full-adder c0 x0 y0))
             (s0 (first a0)) (c1 (second a0)) ; {add.bit0}
             (a  (adder c1 xs ys))
             (ss (first a)) (c (second a)))   ; {add.bits}
        (list (cons s0 ss) c))                ; {add1}
      (list nil c0)))                         ; {add0}
```

Figure 7.7
Ripple-carry adder and ACL2 model.

Exercises

1. Define a DoubleCheck property that checks the output of the two-bit adder model (figure 7.6, page 144) against expectations. Run the test using Proof Pad.
   Note: The data generator (random-between 0 1) delivers a 0 or a 1 at random.

2. By default, Proof Pad repeats the test fifty times with random data when it runs a DoubleCheck test. How many random tests do you think would be needed to be reasonably confident that all 32 different cases for the two-bit adder have been tested? Make a ballpark estimate if you can.

7.4 Adding w-Bit Binary Numerals

By now, you can predict what a circuit for adding three-bit binary numerals would look like. Put another full-adder in the circuit and feed the high-order bits from the two numerals into the new full-adder. In addition, connect the carry-out from the two-bit circuit to the carry-in of the new full-adder. A two-bit adder circuit augmented in this way becomes a circuit for adding three-bit numerals. A circuit diagram for adding numerals with any number of bits combines the appropriate number of full-adder components in this manner. Figure 7.7 presents a schematic for a circuit that adds w-bit numerals. The circuit is known as a *ripple-carry adder* because of the way the carry-bit propagates across the line of full-adder components.

The ACL2 model in the figure relies on inductive definition. It feeds the carry-in and the low-order bits from the two numerals into the **full-adder** operator. (The low-order bit in a

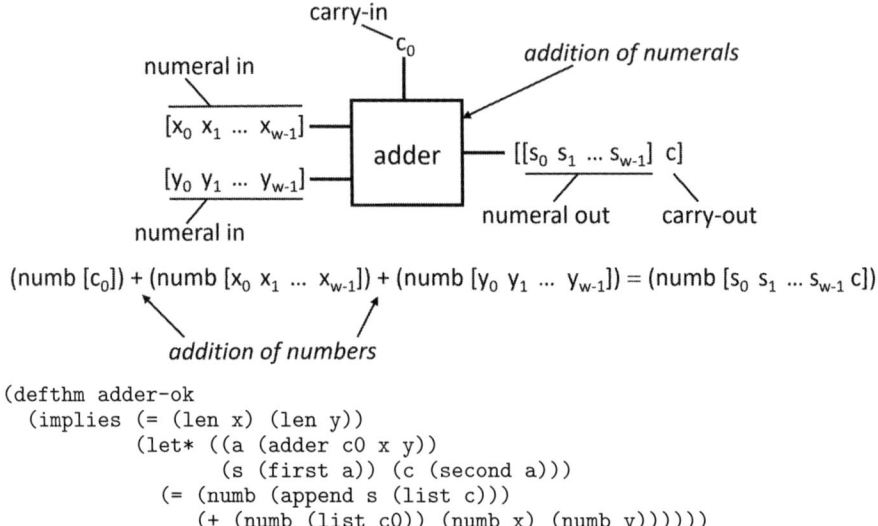

Figure 7.8
Adding *w*-bit binary numerals.

numeral is the "ones bit," which is the first element of the list that we use to represent the numeral.) The sum-bit from the list that the full-adder operator delivers is the low-order bit of the numeral representing the sum of the numbers that the input numerals represent. The remaining sum bits in the list are those delivered by the adder operating on the other bits in the input numerals (that is, on all the bits in the input numerals except the low-order bits).

Because the model defines an operator in ACL2, you can run the operator to see that it works in specific cases. To add two binary numerals, supply lists of 0s and 1s representing those numerals in an invocation of the adder operator and specify zero as the input carry-bit. The output will be the binary numeral for the sum.

The theorem in figure 7.8 explains how input and output signals are interpreted as numbers. As with the two-bit adder (figure 7.3, page 144), the three numbers represented by the two input numerals and the input carry, when added together, equal the number represented by a numeral formed from the output sum-bits and the output carry-bit. However, the theorem for the *w*-bit adder is stated as an implication to constrain inputs to be numerals with the same number of bits. This was not necessary with the theorem for the two-bit adder because the lengths of the numerals were explicit in the ACL2 model.

Theorem {adder-ok} (figure 7.8) states the arithmetic property that we expect the adder circuit to have. The mechanized logic of ACL2 succeeds in verifying the theorem without assistance, but because reasoning about circuits is such an important idea, we think going

7.4 Adding w-Bit Binary Numerals

through a paper-and-pencil proof will be worthwhile. Our proof will, of course, work from the model of the adder circuit (figure 7.7, page 145) rather than from the circuit diagram. Box 7.1 (page 148) discusses some of the ramifications of this approach, which has been our basis for reasoning about circuits.

Figure 7.9 (page 149) displays the {adder-ok} theorem in algebraic notation. Proving the theorem amounts to verifying that $(\forall n.R(n))$ is true, where the predicate R, which has the natural numbers as its universe of discourse, is defined as follows:

$R(n) \equiv ((\text{numb } [s_0 \; s_1 \ldots s_n \; c]) =$
$\quad (\text{numb } [c_0]) + (\text{numb } [x_0 \; x_1 \ldots x_n]) + (\text{numb } [y_0 \; y_1 \ldots y_n]))$
$\quad \text{where } [[s_0 \; s_1 \ldots s_n] \; c] = (\text{adder } c_0 \; [x_0 \; x_1 \ldots x_n] \; [y_0 \; y_1 \ldots y_n])$

The proof will use mathematical induction. Figure 7.9 (page 149) displays the equation of the base case, $R(0)$, and sketches its proof. Here, we elaborate some of the details that were omitted in the sketch. The first step is to compute the value of $(\text{adder } c_0 \; [x_0] \; [y_0])$ by working through the definition of the adder operator (figure 7.7, page 145). With those operands, the if operator in the definition of adder selects the {add1} equation.

$\quad (\text{adder } c_0 \; [x_0] \; [y_0])$
$= [(\text{cons } s_0 \; ss) \; c] \qquad \{\text{add1}\} \text{ (page 145)}$

where

$\quad [ss \; c]$
$= (\text{adder } c_1 \; (\text{rest } [x_0]) \; (\text{rest } [y_0])) \qquad \{\text{add.bits}\} \text{ (page 145)}$
$= (\text{adder } c_1 \; \text{nil nil}) \qquad \{rst1\} \text{ (exercise 1, page 84)}$
$= [\text{nil } c_1] \qquad \{\text{add0}\} \text{ (page 145)}$

Therefore, $ss = \text{nil}$
$\qquad c = c_1 \qquad\qquad\qquad \{\dagger\}$
and $(\text{cons } s_0 \; ss) = (\text{cons } s_0 \; \text{nil}) = [s_0] \quad \{\ddagger\}$

The equation $R(0)$ (figure 7.9, page 149) makes the following requirement:

$\quad [[s_0] \; c] = (\text{adder } c_0 \; [x_0] \; [y_0])$

The following argument, which proceeds from the right-hand side to the left-hand side of the requirement, confirms that it is consistent with the definition of the adder operator:

$\quad (\text{adder } c_0 \; [x_0] \; [y_0])$
$= [(\text{cons } s_0 \; ss) \; c] \qquad \{\text{add1}\}$
$= [[s_0] \; c] \qquad\qquad \{\ddagger\}$

The proof of the base case can be completed as follows:

$$\begin{aligned}
&\quad (\text{numb } [s_0 \ c]) \\
&= (\text{numb } [s_0 \ c_1]) & &\{\dagger\} \\
&= (\text{numb } (\text{full-adder } c_0 \ x_0 \ y_0)) & &\{\text{add.bit0}\} \text{ (page 145)} \\
&= (\text{numb } [c_0]) + (\text{numb } [x_0]) + (\text{numb } [y_0]) & &\{\text{full-adder-ok}\} \text{ (page 143)}
\end{aligned}$$

Box 7.1
Models and Circuit Fabrication

We expect that you could, given a basket of logic gates, wires, and enough time, use the diagram of the adder circuit to build one for any specified word size, and we think you can convince yourself that the model matches the diagram. If it does, then properties of the model that we verify guarantee that the circuits also have those properties.

In a complete formalization, we would need a way to convert models into instructions for fabricating circuits so that the fabricated circuit would have the properties of the model. Such a formalization would use the methods we have employed to formalize other operations. We leave that step to the imagination.

So much for the base case. Figure 7.9 (page 149) also presents a proof of the inductive case: $\forall n.(R(n) \rightarrow R(n+1))$. When that proof arrives at the sum $(\text{numb } [c_1]) + (\text{numb } [x_1 \ldots x_{n+1}]) + (\text{numb } [y_1 \ldots y_{n+1}])$, it recognizes that the induction hypothesis, $R(n)$, applies because the numerals $[x_1 \ldots x_{n+1}]$ and $[y_1 \ldots y_{n+1}]$ have $n+1$ elements, like the numerals in the equation $R(n)$. The subscripts run from 1 to $n+1$ instead of from 0 to n, but it's the number of bits that counts, not the subscripts.

The induction hypothesis says that this sum equals $(\text{numb } [s_1 \ldots s_{n+1} \ c])$, and the {nmb1} theorem (exercise 3, page 136) adds $(\text{numb } [s_0]) + 2\times(\text{numb } [s_1 \ldots s_{n+1} \ c])$ to arrive at $(\text{numb } [s_0 \ s_1 \ldots s_n \ c])$. We have derived the left-hand side of equation $R(n+1)$, having started from the right-hand side. That completes the proof of the {adder-ok} theorem by mathematical induction.

The proof is tedious, to say the least. It requires working out the details of numerous operator invocations from symbolic representations of the operands. Fortunately, ACL2 is on hand to work through the muck and mire and arrive at the same conclusion. That gives us confidence that our ripple-carry circuit for adding binary numerals delivers the expected results.

Exercises

1. Define in ACL2 an operator **add-bin** that adds any two binary numerals, even if the numerals contain a different number of bits. That is, the value (add-bin $c\ x\ y$) should be a binary numeral for the number (numb $[c]$) + (numb x) + (numb y), as long as x and y are binary numerals and c is 0 or 1, regardless of (len x) or (len y). Design and run some sanity checks on your operator.

Exercises

Theorem {adder-ok}:
$\forall n.((\text{numb } [s_0 \ s_1 \ \ldots \ s_n \ c]) = (\text{numb } [c_0]) + (\text{numb } [x_0 \ x_1 \ \ldots \ x_n]) + (\text{numb } [y_0 \ y_1 \ \ldots \ y_n]))$
where $[[s_0 \ s_1 \ \ldots \ s_n] \ c] = (\text{adder } c_0 \ [x_0 \ x_1 \ \ldots \ x_{n+1}] \ [y_0 \ y_1 \ \ldots \ y_n])$

Base Case

$R(0) \equiv ((\text{numb } [s_0 \ c]) = (\text{numb } [c_0]) + (\text{numb } [x_0]) + (\text{numb } [y_0]))$
where $[[s_0] \ c] = (\text{adder } c_0 \ [x_0] \ [y_0])$

$(\text{adder } c_0 \ [x_0] \ [y_0]) = [(\text{cons } s_0 \ \text{nil}) \ c] = [[s_0] \ c]$	{*add1*} ({*adder*}, *fig. 7.7, p.145*)
where $[s_0 \ c] = (\text{full-adder } c_0 \ x_0 \ y_0)$	*note:* $(\text{adder } c \ \text{nil} \ \text{nil}) = [\text{nil} \ c]$
$(\text{numb } [s_0 \ c]) = (\text{numb } (\text{full-adder } c_0 \ x_0 \ y_0))$	
$= (\text{numb } [c_0]) + (\text{numb } [x_0]) + (\text{numb } [y_0])$	{full-adder-ok} *(fig. 7.5, p.143)*

Inductive Case

$R(n+1) \equiv ((\text{numb } [s_0 \ s_1 \ \ldots \ s_{n+1} \ c]) =$
$(\text{numb } [c_0]) + (\text{numb } [x_0 \ x_1 \ \ldots \ x_{n+1}]) + (\text{numb } [y_0 \ y_1 \ \ldots \ y_{n+1}]))$
where $[[s_0 \ s_1 \ \ldots \ s_{n+1}] \ c] = (\text{adder } c_0 \ [x_0 \ x_1 \ \ldots \ x_{n+1}] \ [y_0 \ y_1 \ \ldots \ y_{n+1}])$

$(\text{numb } [c_0]) + (\text{numb } [x_0 \ x_1 \ \ldots \ x_{n+1}]) + (\text{numb } [y_0 \ y_1 \ \ldots \ y_{n+1}])$	
$= (\text{numb } [c_0]) + (\text{numb } [x_0]) + 2 \times (\text{numb } [x_1 \ \ldots \ x_{n+1}])$	{*nmb1*}
	(p. 136)
$+ (\text{numb } [y_0]) + 2 \times (\text{numb } [y_1 \ \ldots \ y_{n+1}])$	{*nmb1*}
$= (\text{numb } [c_0]) + (\text{numb } [x_0]) + (\text{numb } [y_0]) +$	
$2 \times ((\text{numb } [x_1 \ \ldots \ x_{n+1}]) + (\text{numb } [y_1 \ \ldots \ y_{n+1}]))$	{*algebra*}
$= (\text{numb } (\text{full-adder } c_0 \ x_0 \ y_0)) +$	{full-adder-ok}
$2 \times ((\text{numb } [x_1 \ \ldots \ x_{n+1}]) + (\text{numb } [y_1 \ \ldots \ y_{n+1}]))$	
$= (\text{numb } [s_0 \ c_1]) + 2 \times ((\text{numb } [x_1 \ \ldots \ x_{n+1}]) + (\text{numb } [y_1 \ \ldots \ y_{n+1}]))$	{*add.bit0*}
	(p.143)
$= (\text{numb } [s_0]) + 2 \times (\text{numb } [c_1]) +$	{*nmb1*}
$2 \times ((\text{numb } [x_1 \ \ldots \ x_{n+1}]) + (\text{numb } [y_1 \ \ldots \ y_{n+1}]))$	
$= (\text{numb } [s_0]) +$	{*algebra*}
$2 \times ((\text{numb } [c_1]) + (\text{numb } [x_1 \ \ldots \ x_{n+1}]) + (\text{numb } [y_1 \ \ldots \ y_{n+1}]))$	
$= (\text{numb } [s_0]) + 2 \times (\text{numb } [s_1 \ \ldots \ s_{n+1} \ c])$	{$R(n)$}
$= (\text{numb } [s_0 \ s_1 \ \ldots \ s_{n+1} \ c])$	{*nmb1*}

Figure 7.9
Theorem {adder-ok}: proof by mathematical induction.

2. Define in ACL2 a theorem about the operator add-bin (exercise 1) that is analogous to theorem {adder-ok} (figure 7.8, page 146).

Box 7.2
Adder Circuit and Numerals of Different Lengths

A *word* in a computer is a collection of bits that the computer treats as a whole in certain operations, such as arithmetic operations. A circuit to perform arithmetic will carry out the operation on words denoting binary numerals. Since all words have the same number of bits, both of the numerals supplied as inputs to the circuit for an arithmetic operator will have the same number of bits.

We could change the design of the circuit for the adder to accommodate input numerals of differing lengths. However, since we are modeling a circuit in which the input numerals have the same length, the model does not need to account for that possibility.

7.5 Numerals for Negative Numbers

So far, all the numerals we've seen have denoted positive numbers. Arithmetic circuits also need to deal with negative numbers, and there is more than one way to do that. The most common scheme is known as the *two's-complement* system.

Two's-complement numerals are a special interpretation of ordinary binary numerals. For the numbers $0, 1, \ldots (2^{w-1} - 1)$, where w is the *word size* of the circuits for arithmetic operations, two's-complement numerals are ordinary binary numerals. All of the numerals for this set of numbers have $(w - 1)$ or fewer bits, not counting leading zeros (theorem {len-bits≤}, page 136). To make the numerals match the word size, the two's-complement system pads them with leading zeros to make them have exactly w bits. Leading zeros don't change the number that a numeral denotes (theorem {leading-0s}, page 136), but, as with the ripple-carry adder (figure 7.7, page 145), circuits to perform arithmetic on two's-complement numbers will require exactly w bits for each input numeral because there are w input lines for each addend, and each input line must carry a signal. The nonnegative numbers, $0, 1, \ldots (2^{w-1} - 1)$ consume half of the 2^w bit-patterns available with w-bit words.

For negative numbers, the two's-complement system uses the remaining bit-patterns. These are the numerals that would normally denote the numbers $2^{w-1}, (2^{w-1}+1), \ldots (2^w - 1)$. If $(-n)$ is a negative number in the range $-2^{w-1} \leq (-n) < 0$, then the two's-complement numeral for $(-n)$ is the ordinary binary numeral for $(2^w - n)$. Since $2^{w-1} = (2^w - 2^{w-1}) \leq (2^w - n) < 2^w$, this numeral has exactly w bits (theorem {len-bits}, page 136). We also know that its high-order bit is a one-bit (theorem {hi-1}, page 136), so there is an easy way to recognize numerals that denote negative numbers.

For example, a computer with 32-bit words that uses two-complement numerals has arithmetic circuits that deal with numbers n in the range $-2^{w-1} \leq n < 2^{w-1}$. In the posi-

7.5 Numerals for Negative Numbers

tive part of the range, it represents numbers as ordinary binary numerals but with enough leading zeros to fill the 32-bit word. For a number $(-n)$ in the negative part of the range, the two's-complement system uses the ordinary binary numeral for the positive number $(2^{32} - n)$ to represent the number $(-n)$. Since $(-n)$ is in the range $-2^{31} \le -n < 0$, we can assert that $2^{31} = 2^{32} - 2^{31} \le 2^{32} - n < 2^{32}$. Therefore, the two's-complement binary numeral for the negative number $(-n)$ has exactly 32 bits (theorem {*len-bits*}, page 136).

Modular arithmetic makes two's-complement numerals for negative numbers act like the negative numbers they stand for when they are added to other numerals. For negative numbers $(-n)$ in the range $-2^{31} \le -n < 0$, the value of $((-n) \bmod 2^{32})$ is $(2^{32} - n)$. Therefore, since addition and subtraction in modular arithmetic is consistent with ordinary addition and subtraction (box 3.3, page 74), it follows that $(m + (-n)) \bmod 2^{32} = ((m \bmod 2^{32}) + ((-n) \bmod 2^{32})) \bmod 2^{32} = ((m \bmod 2^{32}) + ((2^{32} - n) \bmod 2^{32})) \bmod 2^{32}$.

That is, adding the numbers represented by two's-complement numerals, including numbers in the negative range, is just like adding ordinary numbers in modular arithmetic. Subtraction is handled by negating a number (that is, computing the two's-complement representation of its negative), then performing addition modulo 2^{32}.

This method works for any word size. With word size w, the two's-complement system handles addition and subtraction for numbers n in the range $-2^{w-1} \le n < 2^{w-1}$ by performing ordinary addition of numerals, as with the ripple-carry adder, but interpreting the numerals according to the two's-complement scheme. Circuits for performing addition (and subtraction, which uses the same circuit in a two's-complement system) take advantage of the consistency between modular arithmetic and ordinary arithmetic.

In summary, the two's-complement representation for computers with word size w deals with numbers n in the range $-2^{w-1} \le n < 2^{w-1}$. We will refer to this set of integers as $I(w)$.

$$I(w) = \{-2^{w-1}, \cdots -1, 0, 1, 2, \ldots 2^{w-1} - 1\}$$

Two's-complement numerals for numbers in the negative part of the range have exactly w bits, with a one-bit in the high-order slot. Two's-complement numerals for numbers in the positive part of $I(w)$ take the form of ordinary binary numerals, except that they are padded with enough leading zeros to fill out a w-bit word, where w is the word size of the computer. The **twos** operator, defined as follows, delivers the two's-complement numeral for a number n in the set $I(w)$:

```
(defun twos (w n)              ; w = word size
  (if (< n 0)                  ; -2^(w-1) <= n < 2^(w-1)
      (bits (+ (expt 2 w) n))  ; {2s-}
      (pad w 0 (bits n))))     ; {2s+}
```

The **twos** operator uses the operator **bits** (page 135) to compute binary numerals. For negative numbers, it adds 2^w (that is, (expt 2 w), using the ACL2 intrinsic operator **expt**) before computing the numeral. For nonnegative numbers, it computes the numeral, then uses the **pad** operator (exercise 5, page 136) to insert leading zeros to match the word size.

$n \in I(3)$	$2^3 + n$	(twos 3 n)	binary numeral
−4	4	[0 0 1]	100
−3	5	[1 0 1]	101
−2	6	[0 1 1]	110
−1	7	[1 1 1]	111
0		[0 0 0]	000
1		[1 0 0]	001
2		[0 1 0]	010
3		[1 1 0]	011

$I(w) = I(3) = \{-4, -3, -2, -1, 0, 1, 2, 3\}$
word size $w = 3$, $-2^{w-1} = -2^{3-1} = -4$, $2^{w-1} - 1 = 2^{3-1} - 1 = 3$

Figure 7.10
Two's-complement numerals for three-bit words.

There will always be some padding for numerals representing positive numbers because $0 \leq n < 2^{w-1}$ implies that $0 \leq$(len (bits n)) $< w$ (theorem {*len-bits≤*}, page 136).

Figure 7.10 displays two's-complement numerals for the numbers in the set $I(3)$. This example is just to illustrate the idea. No computer would have three-bit words, but the example gets the point across with a table of manageable size.

If the input carry is zero and the input numerals are interpreted as two's-complement numerals for numbers in the set $I(w)$, then the sum-bits of the output numeral from the ripple-carry adder (figure 7.7, page 145) form the two's-complement numeral for the sum of the input numerals. The carry output from the circuit can be used to determine whether or not the sum is in $I(w)$, the set of numbers representable by w-bit two's-complement numerals.[39]

Now, here is an interesting trick for computing the two's-complement numeral of a negative number without computing 2^w or doing subtraction. Let $[x_0 \ x_1 \ldots x_{w-1}]$ be the w-bit binary numeral for a number n in the range $1, 2, \ldots 2^{w-1}$, padded with leading zeros to fill out the w-bit word. Then, the two's-complement numeral for $(-n)$ can be computed in a two-step procedure. First, invert the bits: change the zero-bits to one-bits and change the one-bits to zero-bits. Then, use the ripple-carry adder to add the numeral for the number 1 to the numeral with the inverted bits. The result will be the two's-complement numeral

---

[39] The output carry can be used to perform multiword arithmetic or to detect overflow conditions. Adding two numbers, $m + n$, that are both in the top half of the positive range ($2^{w-2} \leq m, n < 2^{w-1}$) produces a number that is outside the set $I(w)$, so the sum has no w-bit two's-complement numeral. This outcome is known as an overflow. Similarly, adding two numbers in the bottom half of the negative range ($-2^{w-1} \leq m, n < -2^{w-2}$) produces a number outside the two's-complement range, an overflow in the negative direction.

Exercises

for $(-n)$. The same trick works to negate the two's-complement numeral of a number $(-n)$ from the range $-2^{w-1} < -n \le 0$.

The trick does not work for the number -2^{w-1} because the negative of that number (namely, 2^{w-1}) is outside of the set $I(w)$, so it doesn't have a w-bit two's-complement numeral. The trick does work for negating the two's-complement numeral for zero. In that case, the procedure delivers an output numeral identical to the input (namely, a numeral consisting of w zero-bits). It produces a one-bit for the carry-out, but that bit is not part of the numeral. Figure 7.11 (page 154) explains how inverting the bits and adding one leads to the negation of the input numeral. It's an exercise in algebra and modular arithmetic.[40]

Exercises

1. Prove theorem $\{len\text{-}2s\}$: $\forall w.((n \in I(w)) \to ((\text{len (twos } w \; n)) = w))$.

2. Prove theorem $\{minus\text{-}sign\}$: $\forall w.(((n \in I(w)) \wedge (n < 0)) \to ((\text{fin(twos } w \; n)) = 1))$. The operator fin is defined on page 136.

3. Prove theorem $\{plus\text{-}sign\}$: $\forall w.(((n \in I(w)) \wedge (n \ge 0)) \to ((\text{fin(twos } w \; n)) = 0))$.

4. Diagram a negation circuit whose input signals represent a two's-complement numeral for a number n in the range $-2^{w-1} < n < 2^{w-1}$ and whose output signals represent the two's-complement numeral for $(-n)$. In your diagram, rely on the two's-complement negation trick (figure 7.11), and draw a box labeled "adder" to depict a ripple-carry adder circuit (figure 7.8, page 146, depicts an adder circuit using such a box).
   Note: Your circuit will also produce the two's-complement numeral for -2^{w-1}, given the binary numeral for 2^{w-1} as input.

5. Define an ACL2 model for the negation circuit of exercise 4. Your model may refer to the ACL2 model of the ripple-carry adder in figure 7.7 (page 145).

6. Define and run a DoubleCheck property that tests the ACL2 model of the negation circuit of exercise 5 when $w = 32$. You may refer to the operator expt: (expt 2 31) = 2^{31}.

7. Diagram a circuit that subtracts two's-complement numerals. In your diagram, use the gate-like symbol in figure 7.8 (page 146) to depict the ripple-carry adder circuit, and use a similar, gate-like symbol for the negation circuit of exercise 4.

8. Define an ACL2 model of the two's-complement subtraction circuit of exercise 7.

9. Define and run a DoubleCheck property that tests the ACL2 model of the subtraction circuit in exercise 8.

10. Diagram a comparison circuit that takes a pair of two's-complement numerals as inputs and delivers a one-bit if the first number is less than the second and a zero-bit otherwise. In your

---

[40] The proof of equation $\{ys \text{ increment}\}$ in figure 7.11 cites the geometric progression (exercise 13, page 155).

Some facts, notation, and equations

$1 \leq n \leq 2^{w-1}$	range of numbers to negate
(len (bits n)) $\leq w$	{*len-bits$\leq$*} *(page 136)*
$xs = [x_0\ x_1\ \ldots\ x_{w-1}]$	xs = (pad w 0 (bits n)), *padded numeral*
(numb xs) = n	{*leading-0s*} *(page 136)*
$ys = [y_0\ y_1\ \ldots\ y_{w-1}]$	*inverted bits* $y_i = 1 - x_i$ *(0 for 1, 1 for 0)*
$1 +$ (numb ys) $= 2^w - n$	{*ys increment*} *equation (see proof below)*
(bits (+ 1 (numb ys))) = (twos w ($-n$))	{*2s trick*} *equation (see proof below)*

Proof of {*ys increment*} *equation:* $1 +$ (numb ys) $= 2^w - n$

$\ 1\ +$ (numb ys)
$= 1\ +\ y_0 2^0 + y_1 2^1 + \cdots + y_{w-1} 2^{w-1}$ {*Horner 2*}
$= 1\ +\ (1 - x_0) 2^0 + (1 - x_1) 2^1 + \cdots + (1 - x_{w-1}) 2^{w-1}$ $\quad \forall i.(y_i = 1 - x_i)$
$= 1\ +\ (2^0 + 2^1 + \cdots + 2^{w-1})$ {*algebra*}
$\ -\ (x_0 2^0 + x_1 2^1 + \cdots + x_{w-1} 2^{w-1})$
$= 1\ +\ (2^w - 1) - (x_0 2^0 + x_1 2^1 + \cdots + x_{w-1} 2^{w-1})$ {*geometric progression*}
$= 2^w\ -\ (x_0 2^0 + x_1 2^1 + \ldots x_{w-1} 2^{w-1})$ {*algebra*}
$= 2^w\ -$ (numb xs) {*Horner 2*}
$= 2^w\ -\ n$ (numb xs) $= n$

Proof of {*2s trick*} *equation:* (bits (+ 1 (numb ys))) = (twos w ($-n$))

$\ $ (bits (+ 1 (numb ys)))
$=$ (bits ($2^w - n$)) {*ys increment*} *equation*
$=$ (bits ($2^w + (-n)$)) {*algebra*}
$=$ (bits (+ (expt 2 w) ($-n$))) *ACL2 notation for* ($2^w + (-n)$)
$=$ (twos w ($-n$)) {*2s–*} ({*twos*}, page 151)

Figure 7.11
Two's-complement negation trick.

diagram, use a gate-like symbol for the subtraction circuit of exercise 7.
Hint: Apply theorems {*minus-sign*} and {*plus-sign*} from exercises 2 and 3.

11. Define an ACL2 model of the comparison circuit of exercise 10. Your model may refer to the subtraction circuit model of exercise 8.

12. Define and run a DoubleCheck property that tests the comparison circuit model of exercise 11. You may refer to the operator **expt**: (expt 2 (− w 1)) = 2^{w-1}.

13. Use induction to prove the following equation. You may assume $r > 0$ and $r \neq 1$.
$$(r - 1)(r^0 + r^1 + r^2 + \ldots r^n) = r^{n+1} - 1 \quad \{geometric\ progression\}$$

8 Multipliers and Bignum Arithmetic

Chapter 7 discussed circuits for adding binary numerals of a fixed word length. The ACL2 model (figure 7.7, page 145) assumed that both numerals had exactly w bits, w being the word size. The circuit diagram reflected this assumption by showing $2w$ input wires (w lines for each numeral) and w output wires for the numeral denoting the sum.

The diagram has an additional input wire for the carry-bit coming into the adder (normally a zero-bit, unless the circuit is being used for some sort of multiword arithmetic) and an additional output wire for the output carry-bit. The output carry-bit is normally ignored in single-word arithmetic when one addend is positive and the other is negative, but it can be used to detect overflow[41] when they have the same sign.

The ACL2 model received the input carry as its first operand and the two input numerals as lists of length w as second and third operands. It delivered its output as a list of two elements, the first element being a list of w sum-bits and the second element being the carry-out. The model, like the circuit, did not allow for input numerals of differing lengths. That is the usual case for physical circuits. They have a fixed number of wires and gates.

Software has no such constraint. A software component for adding binary numerals can accept numerals of any length, and the two numerals need not have the same length. The numeral representing the sum would have as many bits as required to represent the sum of the numbers denoted by the input.

An adder expressed in software that is able to deal with numerals of any length is often called a bignum adder. It performs precise arithmetic on numbers of any size rather than on a fixed range of numbers based on word size. To simplify the discussion, we will talk about arithmetic for nonnegative integers only. Similar ideas but with serious complications carry over to the domain of negative integers.

---

[41] Overflow occurs when the sum of the two input numerals falls outside the range of numbers representable in the arithmetic system.

8.1 Bignum Adder

To begin, let's see what it takes to convert our ACL2 model for the ripple-carry adder to software that performs addition on binary numerals with an arbitrary number of bits, so that it can deal with numbers of unlimited size. The first step is to work out a way to increment a binary numeral by one. That is, we want to define an operator add-1 that, given a binary numeral x for the natural number n, delivers the binary numeral for $(n + 1)$. The operator will have the following property with respect to the operators bits and numb from section 6.3 (figure 6.5, page 135), which convert numbers to binary numerals and vice versa. The property is stated in terms of numbers, but add1 will work directly with numerals, bypassing entirely the intrinsic numbers of the computer system.

(add-1 x) = (bits (+ (numb x) 1)) {add-1 *property*}

Following our usual practice when we are trying to define an operator, we assume that someone has already defined it, and all we have to do is to write some equations that it would have to satisfy if it worked. If we manage to come up with equations that are consistent, comprehensive, and computational (figure 4.10, page 91), we will have defined an operator, and it will be the only operator that makes all of those equations true.

A particularly simple situation occurs when the numeral to be incremented has no bits in it. The interpretation we settled on in chapter 6 is that the empty numeral stands for the number zero (equation {numb0} in the definition of numb, page 135). So, incrementing the empty numeral should produce a numeral for the number 1, which is the list [1].

(add-1 nil) = (list 1) = (cons 1 nil) {add1nil}

Another simple situation occurs when the low-order bit in the numeral to be incremented is a zero. In that case, the output numeral is just like the input numeral, except that its low-order bit is a one rather than a zero.

(add-1 (cons 0 x)) = (cons 1 x) {add10}

At this point, we have equations to cover all numerals that have either no bits at all or a low-order bit of zero. If we can work out an equation for numerals with a low-order bit of one, our equations will be comprehensive. To do this, let's think about the low-order bit of the incremented numeral. Since adding a one-bit to a one-bit produces a sum-bit of zero and a carry-bit of one (figure 7.3, page 141), we conclude that the low-order bit of the incremented numeral is zero.

But what about the carry-bit? What do we do with that? It will need to be added to the higher-order bits of the input numeral. But that is just a matter of incrementing the higher-order bits by one. The higher-order bits themselves form a numeral, and because of our usual assumption with inductive definitions that someone has already defined the operator add-1 for shorter operands, we can just use it to increment that numeral. That is, the add-1 operator, were it defined, would satisfy the following inductive equation:

(add-1 (cons 1 x)) = (cons 0 (add-1 x)) {add11}

8.1 Bignum Adder

Now we have three equations. They are consistent (no overlapping cases) and comprehensive (all cases covered). The equations are also computational because the input numeral on the right-hand side of the inductive equation {add11} is shorter than the numeral on the left-hand side of the equation. Therefore, the equations define the add-1 operator. All we need to do now is to combine them into an ACL2 definition.

```
(defun add-1 (x)
  (if (and (consp x) (= (first x) 1))
      (cons 0 (add-1 (rest x)))        ; {add11}
      (cons 1 (rest x))))              ; {add10}
```

It turns out that the {add1nil} equation and the {add10} equation can be expressed as one equation because the ACL2 formula (cons 1 (rest x)), which is the proper translation for the right-hand side of the {add10} equation, also works for the right-hand side of the {add1nil} equation because (cons 1 (rest nil)) = (cons 1 nil) = (list 1) ({$rst0$}, figure 4.4, page 82). This observation reduces the definition from three equations to two and completes the formal definition of the add-1 operator.

The ripple-carry adder propagated the carry from each bit position to the next higher-order bit position. Each bit position involved three input bits (a carry-bit and one-bit from each addend). Our bignum adder will do that too. Each bit in the sum will depend on the corresponding bits in the addends and the carry from the previous, lower-order bit.

We already have the apparatus for this: the full-adder operator (figure 7.4, page 142). We can use that operator to add two corresponding bits, x_n and y_n, from the addend numerals, incorporating the carry c_n from the lower-order bit position. The full-adder operator delivers the sum-bit s_n for the current bit position and the carry-bit c_{n+1} for the next bit position.

$$[s_n\ c_{n+1}] = (\text{full-adder}\ c_n\ x_n\ y_n)$$

This analysis provides the basis for one of the equations for the bignum add operator. The equation is inductive and applies when neither addend is nil, so that both have low-order bits.

(add c_0 [x_0 x_1 x_2 ...] [y_0 y_1 y_2 ...]) = [s_0 s_1 s_2 ...]   {addxy}
where
[s_0 c_1] = (full-adder c_0 x_0 y_0)
[s_1 s_2 ...] = (add c_1 [x_1 x_2 ...] [y_1 y_2 ...])

This equation covers all addends whose numerals have at least one bit. So, all we need to do to make our equations comprehensive is to have equations for the cases when one or the other addend is nil.

If either addend is nil, then that addend denotes the number zero. The sum, then, would be the other addend with the carry added to it. We already have an operator, add-1 (page 159), that we can use to add the carry if it's a one-bit. When the carry is a zero-bit, we

```
(defun add (c0 x y)
  (if (not (consp x))
      (add-c c0 y)                                  ; {add0y}
      (if (not (consp y))
          (add-c c0 x)                              ; {addx0}
          (let* ((x0 (first x))
                 (y0 (first y))
                 (a  (full-adder c0 x0 y0))
                 (s0 (first a))
                 (c1 (second a)))
            (cons s0 (add c1 (rest x) (rest y))))))) ; {addxy}
```

Figure 8.1
Bignum addition operator.

don't need to add the carry-bit because adding zero doesn't change the number. We define an operator **add-c** that uses **add-1** to add the carry-bit when it is a one.

```
(defun add-c (c x)
  (if (= c 1)
      (add-1 x)    ; {addc1}
      x))          ; {addc0}
```

We use **add-c** to complete the addition when either of the addends is **nil**.

$$(\text{add } c\ x\ \text{nil}) = (\text{add-c } c\ x) \quad \{add10\}$$
$$(\text{add } c\ \text{nil}\ y) = (\text{add-c } c\ y) \quad \{add01\}$$

That covers all of the cases. Figure 8.1 translates the equations into ACL2. Let's look at some examples of adding numerals with bignum **add**.

$$(\text{add } 0\ [0\ 1]\quad [0\ 1])\ = [0\ 0\ 1] \quad\quad 2 + 2 = 4$$
$$(\text{add } 0\ [0\ 1\ 1\ 1]\quad [1\ 0\ 1]) = [1\ 1\ 0\ 0\ 1] \quad\quad 14 + 5 = 19$$
$$(\text{add } 0\ [1\ 0\ 0\ 1\ 1]\ [0\ 1\ 1]) = [1\ 1\ 1\ 1\ 1] \quad\quad 25 + 6 = 31$$

The bignum **add** operator delivers a binary numeral for the sum of the numbers represented by the input numerals and the input carry. The following theorem, **bignum-add-ok**, expresses that property formally, and we saw some examples showing that the property holds for three particular pairs of addends. We would like to know that it holds for all input numerals.

```
(defthm bignum-add-ok
  (= (numb(add c x y))
     (+ (numb (list c)) (numb x) (numb y))))
```

The ACL2 system succeeds in proving this theorem, and we could do a paper-and-pencil proof following a strategy similar to the one for the {adder-ok} theorem about the ripple-carry adder (figure 7.9, page 149). In any case, we know now to a mathematical certainty that the bignum **add** operator delivers the sum of its two input numerals and the input carry-bit.

Exercises 161

Exercises

1. Do a paper-and-pencil proof by induction of the following theorem, which says that if the high-order bit in both input numerals is a 1, then the high-order bit of the numeral delivered by the bignum add operator is also a 1. The theorem refers to the fin operator (page 136).
   $$(((\text{fin } x) = 1) \wedge ((\text{fin } y) = 1)) \rightarrow ((\text{fin }(\text{add } 0 \ x \ y)) = 1)$$

2. State the theorem of exercise 1 in ACL2. You may submit the theorem to ACL2, but this exercise does not require you to pursue a successful proof by the ACL2 mechanized logic.

8.2 Shift-and-Add Multiplier

The grade-school method of multiplying multidigit numbers proceeds one digit at a time, from right (low-order digit of the decimal numeral) to left (higher-order digits). The first step is to multiply the entire multiplicand by the low-order digit of the multiplier. Then comes the next-to-last digit of the multiplier (the tens digit) for the second step. In the second step, the product is written below the one from the first step, but shifted left one position. This continues across all of the digits of the multiplier, one by one, writing the products shifted one more position to the left at each step. After the products are completed for all the digits in the multiplier, they are totaled, taking care to keep the digits lined up according to the left shifts that occurred at each stage.

Grade-school students learn this procedure without knowing the algebra behind it. However, we want to specify a multiplication operator in the form of equations, so we need to work out the algebra. Figure 8.2 (page 162) presents the multiplication procedure, and provides an algebraic argument that justifies it. The algebra relies on the equation that grade-school students use to check the correctness of their long-division problems.[42]

$$x = (\lfloor x \div d \rfloor \cdot d) + (x \bmod d) \qquad \{\text{check} \div \}^{43}$$

For decimal numerals, $d = 10$. The remainder ($x \bmod 10$) is the last digit in the numeral for the number x, and the other digits are those of the quotient $\lfloor x \div 10 \rfloor$. That goes for any number, and figure 8.2 parlays this idea to elucidate digit-by-digit multiplication.

Of course, the bignum multiplier will use binary numerals rather than decimal numerals. The procedure is the same, but this one permits some economies because the multiplication table for bits is much simpler than that for decimal digits. We are looking for some equations that define a multiplication operator. That is, an operator that delivers the binary numeral for the product of the numbers represented by its operands, which are also binary numerals. Let's call it mul, and let's call its operands, which are binary numerals, x and y

---

[42] We're not sure they teach long division in school anymore. With computers and calculators constantly at hand, it probably makes more sense to learn something else. Nevertheless, we will need this division-check equation to justify the multiplication procedure.

[43] As usual, $\lfloor x \rfloor$ stands for the greatest integer that is x or less (that is, x rounded down to an integer) and ($x \bmod d$) is the remainder when x is divided by $d \neq 0$ (modular arithmetic, box 6.3, page 126).

Nomenclature

x	number represented by the numeral of the multiplier
$x_0 = x \bmod 10$	low-order digit of the numeral for x
$\lfloor x \div 10 \rfloor$	number represented by the other digits of the numeral for x
y	number represented by the numeral of the multiplicand
$m = x_0 \cdot y$	product of y and the low-order digit of x
$p = \lfloor x \div 10 \rfloor \cdot y$	product of y and the number represented by other digits of x
$m_0 = m \bmod 10$	low-order digit of the numeral for m
$\lfloor m \div 10 \rfloor$	number represented by the other digits of the numeral for m
$s = \lfloor m \div 10 \rfloor + p$	sum of p and the number represented by the other digits of m
$xy = s \cdot 10 + m_0$	shift the numeral for s and bring down digit m_0

Note: In this figure x, x_0, y, m, m_0, p, and s denote numbers. In practice, a person or computer doing multiplication would use numerals.

Procedure

1. *Low-Order Digit*: Multiply the multiplicand, y, by the low-order digit, x_0, of the numeral for the multiplier, x: $m = x_0 \cdot y$.

2. *Digit by Digit*: Multiply y by the number represented by the other digits of the numeral for x: $p = \lfloor x \div 10 \rfloor \cdot y$.

3. *Shift and Add*: Add p to the number represented by the other digits of the numeral for m: $s = p + \lfloor m \div 10 \rfloor$. (In grade-school terminology, this is the step where you shift and add.)

4. *Bring It Down*: Observe that the low-order digit, m_0, of m is the low-order digit of the numeral for the product xy. (In grade-school terminology, this is the step where you bring down the digit m_0.)

5. *Deliver Numeral for xy*: Form a numeral whose low-order digit is m_0 (the one you brought down) and whose other digits are those of the numeral for s.

Justification

$$
\begin{aligned}
xy &= (\lfloor x \div 10 \rfloor \cdot 10 + (x \bmod 10))y & &\{check \div\} \text{ (page 126)} \\
&= (\lfloor x \div 10 \rfloor \cdot 10 + x_0)y & &\{x_0 = x \bmod 10\} \\
&= (\lfloor x \div 10 \rfloor \cdot y) \cdot 10 + x_0 y & &\{algebra\} \\
&= (\lfloor x \div 10 \rfloor \cdot y) \cdot 10 + m & &\{m = x_0 \cdot y\} \\
&= p \cdot 10 + m & &\{p = \lfloor x \div 10 \rfloor \cdot y\} \\
&= p \cdot 10 + (\lfloor m \div 10 \rfloor \cdot 10 + (m \bmod 10)) & &\{check \div\} \\
&= (p + \lfloor m \div 10 \rfloor) \cdot 10 + (m \bmod 10) & &\{algebra\} \\
&= s \cdot 10 + (m \bmod 10) & &\{s = p + \lfloor m \div 10 \rfloor\} \\
&= s \cdot 10 + m_0 & &\{m_0 = m \bmod 10\}
\end{aligned}
$$

Figure 8.2
Grade-school multiplication: digit by digit.

8.2 Shift-and-Add Multiplier

(section 6.3, page 134). If y is nil, it stands for zero, which makes the product zero (that is, nil). That analysis yields an equation that completes the computation when y is nil.

$$(\text{mul } x \text{ nil}) = \text{nil} \qquad \{\text{mulx0}\}$$

We can now focus on the case when y is not nil. To complete the multiplication, we are going to invoke another operator, mxy, that assumes that y is not nil. That leads to another equation for mul.

$$(\text{mul } x\ y) = (\text{mxy } x\ y), \text{ if } (\text{consp } y) \qquad \{\text{mulxy}\}$$

Now, we turn our attention to defining mxy, which multiplies binary numerals x and y when y is not nil. It may happen that x is nil, which stands for zero, so that the product is also zero, and in that case (mxy $x\ y$) is nil, which gives us one of the equations for mxy.

$$(\text{mxy nil } y) = \text{nil} \qquad \{\text{mul0y}\}$$

That leaves the case when $x = [x_0\ x_1\ x_2\ ...]$ is not nil. Follow along, again, in figure 8.2 (page 162), but this time, because we're in binary mode, think two where you see ten and think bits where you see digits. (Or, interpret the numeral 10 in the figure as if it were a binary numeral.) We'll discuss step 1 shortly, but first look at step 2, which will be necessary, regardless of what happens in step 1.

In figure 8.2, the variables x, y, m, p, and s stand for numbers, but in this discussion outside the figure, they stand for binary numerals. We will sometimes want to refer, instead, to numbers, and we will use underlines to make the distinction. For example, x is a binary numeral, but $\underline{x} = (\text{numb } x)$ is the number that x stands for. Similarly, x_0 is a symbol for a bit, but $\underline{x_0}$ is the number it denotes (zero or one).

Step 2 in figure 8.2 requires the computation of $\underline{p} = \lfloor \underline{x} \div 2 \rfloor \cdot \underline{y}$. The numeral for $\lfloor \underline{x} \div 2 \rfloor$ is just x without its low-order bit. Since the low-order bit comes first in our representation of binary numerals, the numeral for x without its low-order bit is (rest x). Observing that the numeral (rest x) is shorter than the numeral x, we can invoke mxy and rely on induction. That gives us a way to compute p, which, don't forget, is a numeral at this point, not a number.

$$p = (\text{mxy (rest } x)\ y) \qquad \{\text{mul.p}\}$$

Binary numerals make step 1 in figure 8.2 ($\underline{m} = \underline{x_0} \cdot \underline{y}$) simpler than it would be with decimal numerals. The bit x_0 is either a zero or a one. If $\underline{x_0} = 0$, then $\underline{m} = \underline{m_0} = 0$, so all we have to do is multiply p by 2 (just shift, no add). Multiplying a binary numeral by 2 is a matter of inserting a zero at the beginning. (This is where we bring down a zero.) So, when $x_0 = 0$, we get the following equation for mxy:

```
(defun mxy (x y) ; assumption: y is not nil
  (if (consp x)
      (let* ((p  (mxy (rest x) y)))            ; {mul.p}
        (if (= (first x) 1)
            (cons (first y) (add 0 p (rest y))) ; {mul1xy}
            (cons 0 p)))                        ; {mul0xy}
      nil))                                     ; {mul0y}
(defun mul (x y)
  (if (consp y)
      (mxy x y) ; {mulxy}
      nil))     ; {mulx0}
```

Figure 8.3
Bignum multiplication operator.

$$(\mathsf{mxy}\ [0\ x_1\ x_2\ \ldots]\ y) = (\mathsf{cons}\ 0\ p) \qquad \{\mathrm{mul0xy}\}$$

If $x_0 = 1$, then $m = y$, so $\underline{s} = \lfloor \underline{m} \div 2 \rfloor + \underline{p} = \lfloor \underline{y} \div 2 \rfloor + \underline{p}$. Again, y is a numeral, so the numeral for $\lfloor y \div 2 \rfloor$ is just like y but without the first bit: (rest y). That gives us another equation for mxy. We use the bignum add operator to compute the sum shown in step 3 of figure 8.3. The first operand of the add operator is the input carry-bit, which is zero because we only want to add p and (rest y). We bring down the low-order bit of m, but $m = y$ ($x_0 = 1$, you remember), so the low-order bit of m is also the low-order bit of y, namely, (first y).

$$(\mathsf{mxy}\ [1\ x_1\ x_2\ \ldots]\ y) = (\mathsf{cons}\ (\mathsf{first}\ y)\ (\mathsf{add}\ 0\ (\mathsf{rest}\ y)\ p)) \qquad \{\mathrm{mul1xy}\}$$

These equations for mul are comprehensive because y is either nil, in which case equation {mulx0} applies, or y is not nil, in which case {mulxy} applies, and the computation is left up to the operator mxy. Now, we have to analyze the definition of mxy.

The equations in the definition of mxy are comprehensive because either x is nil, in which case equation {m0y} applies, or x is not nil. If x is not nil, its low-order bit is either zero, in which case equation {mul0xy} applies, or it is one, in which case equation {mul1xy} applies.

The equations are consistent. In the one case where they could overlap, namely, when both x and y are nil, they deliver the same result, namely, nil.

The equations are computational because in the inductive equations {m0xy} and {m1xy}, the first operand in the invocation of the operator mxy on the right-hand side is the numeral (rest x). Since x is not nil in the inductive case, (rest x) has fewer bits than x, which is the first operand on the left-hand side. So, the inductive invocation is closer to a noninductive case than the formula on the left-hand side of the equation.

Putting this all together leads to the ACL2 definition of the bignum multiplication operator mul in figure 8.3. Most of the work is done by the operator mxy. From these definitions,

Exercises

ACL2 successfully finds an inductive proof of the following theorem, which confirms that (mul *x y*) is the binary numeral for the product of the numbers that the numerals *x* and *y* represent:

```
(defthm bignum-mul-ok
  (= (numb (mul x y)) (* (numb x) (numb y))))
```

Exercises

1. Do a paper-and-pencil proof of the bignum multiplier theorem, bignum-mul-ok.

III Algorithms

9 Multiplexers and Demultiplexers

9.1 Multiplexer

Suppose you want to take two lists and shuffle them into one. You're looking for a perfect shuffle, an element from one list, then one from the other list, back to the first list, and so on. This is sometimes called multiplexing. The term comes from signal transmission. When there are more signals than channels to send them on, one way to share a channel between two signals is to send a small part of one signal, then part of the other, then part of the first one again, and so on. There could be any number of signals sharing the channel, and the same kind of round-robin approach would work. We call the shuffle operator mux. It follows the pattern of the following equation:

$$(\text{mux } [x_1\ x_2\ x_3\ \ldots]\ [y_1\ y_2\ y_3\ \ldots]) = [x_1\ y_1\ x_2\ y_2\ x_3\ y_3\ \ldots] \quad \{mux\}$$

As usual, we want to define the mux operator in terms of a collection of comprehensive, consistent, and computational equations that it would have to satisfy if it worked properly (figure 4.10, page 91). If both lists are nonempty, then the first element of the multiplexed list is the first element of the first list, and the second element of the multiplexed list is the first element of the other list. So, the following formula would get the first two elements into the right places in the multiplexed list:

$$(\text{mux } (\text{cons } x\ xs)\ (\text{cons } y\ ys)) = (\text{cons } x\ (\text{cons } y\ \ldots \textit{rest of formula} \ldots))$$

Fortunately, there is no great mystery concerning the missing part of the formula. Multiplexing what's left of the two input lists will get all the elements in the right place for a perfect shuffle. That observation leads to an inductive equation that the mux operator would satisfy if it worked properly.

$$(\text{mux } (\text{cons } x\ xs)\ (\text{cons } y\ ys)) = (\text{cons } x\ (\text{cons } y\ (\text{mux } xs\ ys))) \quad \{mux11\}$$

The $\{mux11\}$ equation covers the case when both lists are nonempty. Since it's an inductive equation, we need to be careful to make sure that operands of mux on the right-hand side of the equation are closer to those in a noninductive equation than they are to the operands on the left-hand side. If not, the equation will fail to be computational and will not define the mux operator. We observe that the operands on the right are one element shorter than the operands on the left. Therefore, the equation {*mux11*} can be used as a defining axiom. It applies whenever both lists are nonempty.

If both lists are empty, there is nothing to multiplex, so mux would deliver the empty list in that case, but what should it deliver if one list is empty but the other isn't? There is more than one reasonable choice, and each leads to a different operator. One choice is to incorporate the elements in the nonempty list, just as they are, into the multiplexed list that mux delivers. That would make mux satisfy the following equations:

Axioms mux

(mux nil *ys*) = *ys*	{*mux0x*}
(mux *xs* nil) = *xs*	{*mux0y*}
(mux (cons *x xs*) (cons *y ys*)) = (cons *x* (cons *y* (mux *xs ys*)))	{*mux11*}

The three equations, {*mux0x*}, {*mux0y*}, and {*mux11*}, are comprehensive (either both operands are nonempty or at least one of them is empty) and computational (as discussed). They are consistent because the only overlapping case occurs when both lists are empty, and in that case, the overlapping equations ({*mux0x*} and {*mux0y*}) specify the same result (namely, the empty list). We can, therefore, take the equations as axioms defining the mux operator. Converting the axioms to ACL2 notation leads to the following definition:

```
(defun mux (xs ys)
  (if (not (consp xs))
      ys                                              ; {mux0x}
      (if (not (consp ys))
          xs                                          ; {mux0y}
          (cons (first xs)
                (cons (first ys)
                      (mux (rest xs) (rest ys))))))) ; {mux11}
```

As always, the axioms that define an operator determine not only the properties they specify directly but all other properties of the operator, too. What properties would we expect the mux operator to have? Surely the number of elements in the multiplexed list would be the sum of the lengths of its operands. The following theorem states this property formally, and ACL2 succeeds in finding a proof:

```
(defthm mux-length-thm
  (= (len (mux xs ys))
     (+ (len xs) (len ys))))
```

For practice, let's construct a paper-and-pencil proof of the theorem. Our strategy will be an induction on the length of the first operand. We are trying to prove that the following equation holds for all natural numbers, n:

Theorem {mux-length}: $\forall n. L(n)$
where $L(n) \equiv ((\text{len}(\text{mux } [x_1\ x_2\ \ldots\ x_n]\ ys)) = n + (\text{len } ys))$
proof by induction

9.1 Multiplexer

Base case (first operand empty): $L(0) \equiv ((\text{len}(\text{mux nil } ys)) = 0 + (\text{len } ys))$

$$\begin{aligned}(\text{len}(\text{mux nil } ys)) & \\ = \quad (\text{len } ys) & \quad \{mux0x\} \\ = \quad 0 + (\text{len } ys) & \quad \{algebra\}\end{aligned}$$

Inductive case (first operand has $n + 1$ elements):

$L(n + 1) \equiv ((\text{len}(\text{mux } [x_1 \ x_2 \ \ldots \ x_{n+1}] \ ys)) = (n + 1) + (\text{len } ys))$

We split the inductive case, $L(n + 1)$, into two parts. The second operand of mux is either nil or it's not. We derive the conclusion from both possibilities, and that completes the proof because we can infer that the conclusion holds in all circumstances.[44]

Box 9.1
Formal Version of Mux-Val Theorem

In exercise 2 (page 173) the mux-val theorem and the axioms of the occurs-in predicate have been stated in the form we use for paper-and-pencil proofs. These proofs are rigorous but not formal in the sense of the mechanized proofs of ACL2. Below is an ACL2 formalization of these ideas, which the ACL2 system succeeds in admitting. The iff operator is Boolean equivalence (box 2.5, page 32).

```
(defun occurs-in (x xs)
  (if (consp xs)
      (or (equal x (first xs))
          (occurs-in x (rest xs)))
      nil))
(defthm mux-val-thm
  (iff (occurs-in v (mux xs ys))
       (or (occurs-in v xs)
           (occurs-in v ys))))
```

The proof of the inductive case when ys is nil is like the proof when xs is nil, except that it cites $\{mux0y\}$ instead of $\{mux0x\}$. Figure 9.1 (page 172) presents a proof of the inductive case when both operands are nonempty. That is, when the second operand has the form (cons y ys) and the first operand has $n + 1$ elements. This completes the proof by mathematical induction of theorem {mux-length}: $\forall n.L(n)$.

The next section discusses an operator that goes in the other direction. It "demultiplexes" a list into two lists, reversing the perfect shuffle. We will prove that dmx undoes the effect of mux and vice versa. That is, the two operators invert each other.

---

[44] Case-by-case proofs of this kind would cite the $\{\vee\ \text{elimination}\}$ inference rule (page 39) if they took the form of natural deduction. The proof here is rigorous but not formal. ACL2 carries out a formal proof.

$L(n + 1) \equiv (\text{len}(\text{mux } [x_1 \; x_2 \; \ldots \; x_{n+1}] \; (\text{cons } y \; ys))) = (n + 1) + (\text{len } (\text{cons } y \; ys))$

$$
\begin{aligned}
&\quad (\text{len}(\text{mux } [x_1 \; x_2 \; \ldots \; x_{n+1}] \; (\text{cons } y \; ys)))) \\
&= (\text{len}(\text{mux } (\text{cons } x_1 \; [x_2 \; \ldots \; x_{n+1}] \; (\text{cons } y \; ys)))) && \{cons\} \; (page \; 82) \\
&= (\text{len}(\text{cons } x_1 \; (\text{cons } y \; (\text{mux } [x_2 \; \ldots \; x_{n+1}] \; ys)))) && \{mux11\} \; (page \; 170) \\
&= 1 + (1 + (\text{len}(\text{mux } [x_2 \; \ldots \; x_{n+1}] \; ys))) && \{len1\} \; twice \; (page \; 84) \\
&= 1 + (1 + (n + (\text{len } ys))) && \{L(n)\} \; induction \; hypothesis \\
&= (n + 1) + (1 + (\text{len } ys)) && \{algebra\} \\
&= (n + 1) + (\text{len}(\text{cons } y \; ys)) && \{len1\}
\end{aligned}
$$

Figure 9.1
Theorem {mux-length}: inductive case when both operands are nonempty.

Box 9.2
Multiplexer: A Two-Equation Definition

The multiplexer operator can be defined with two equations instead of three by swapping the operands in the inductive equation. When the first operand is nonempty, mux satisfies the following equation.

$\quad$ (mux (cons x xs) ys) = (cons x (mux ys xs))   {mux1y}

The inductive invocation, (mux ys xs), on the right-hand side of {mux1y} delivers a list that starts with the first element of ys, then alternates between xs and ys. Perfect shuffle. It's a two-equation definition.

```
(defun mux2 (xs ys) ; declare induction scheme
   (declare (xargs :measure (+ (len xs) (len ys))))
   (if (consp xs)
       (cons (first xs) (mux2 ys (rest xs))) ; {mux2-1x}
       ys))                                  ; {mux2-0x}
```

The equations define an operator mux2 that produces the same results as mux. However, the new definition makes reasoning more complicated because the operands switch roles in the inductive invocation. A declare directive in the definition helps ACL2 cope with this wrinkle in its proof that mux2 terminates, which the mechanized logic must complete before admitting the operator to the logic.

Exercises

1. Our proof of the inductive case, $L(n + 1)$, of the mux-length theorem (page 171) glossed over the part when the second operand is empty. Complete that part of the proof. That is, prove the following equation:
   $\quad$ (len(mux $[x_1 \; x_2 \; \ldots \; x_{n+1}]$ nil)) = $(n + 1)$ + (len nil)

9.2 Demultiplexer

2. Prove that the mux operator neither adds nor loses values from its operands. That is, a value that occurs in either *xs* or *ys* also occurs in (mux *xs ys*) and, vice versa, a value that occurs in (mux *xs ys*) also occurs in either *xs* or *ys*.
   Theorem {mux-val}: $\forall v.(((\text{occurs-in } v \ xs) \lor (\text{occurs-in } v \ ys)) \leftrightarrow (\text{occurs-in } v \ (\text{mux } xs \ ys)))$
   Note: The "↔" operator is Boolean equivalence (box 2.5, page 32).
   Note: The occurs-in predicate is defined as follows:
   $$(\text{occurs-in } v \ xs) = (\text{consp } xs) \land ((v = (\text{first } xs)) \lor (\text{occurs-in } v \ (\text{rest } xs))) \quad \{occurs\text{-}in\}$$
   Hint: For the inductive case of your proof (that is, the case when *xs* is nonempty), split the proof into two parts, as in the proof of the mux-length theorem (page 170). In one part, the value *v* will be equal to the first element of *xs*: $v = (\text{first } xs)$. In the other part, *v* will occur in (rest *xs*). That is, (occurs-in *v* (rest *xs*)) will be true. Prove each part separately. Since the two parts cover all the possibilities, you can infer that the inductive case is true.

3. Give an example of lists [*x*], [*y*], and [*u w*] for which [*u w*] ≠ (mux [*x*] [*y*]) but the following formula is true:
   $$\forall v.(((\text{occurs-in } v \ [x]) \lor (\text{occurs-in } v \ [y])) \leftrightarrow (\text{occurs-in } v \ [u \ w]))$$

4. Do a paper-and-pencil proof that (mux2 *xs ys*) is the same as (mux *xs ys*).
   Note: mux2 is defined in box 9.2, page 172, and mux is defined on page 170.

5. Formalize the theorem of exercise 4 in ACL2.

9.2 Demultiplexer

A demultiplexer transforms a list of signals that alternate between *x*-values and *y*-values into two lists, with the *x*-values in one list and *y*-values in the other.
$$(\text{dmx } [x_1 \ y_1 \ x_2 \ y_2 \ x_3 \ y_3 \ \ldots]) = [[x_1 \ x_2 \ x_3 \ \ldots] \ [y_1 \ y_2 \ y_3 \ \ldots]] \quad \{dmx\}$$
The following equations form an inductive definition of dmx. The inductive equation covers the case when the operand has at least two elements (that is, it starts with an *x* and then a *y*), and the noninductive equations cover the cases when the operand has just one element or none.

Axioms dmx
(dmx [x_1 y_1 x_2 y_2 ... x_{n+1} ...]) = [(cons x_1 xs) (cons y_1 ys)] {*dmx2*} where [xs ys] = (dmx [x_2 y_2 ... x_{n+1} ...])
(dmx [x_1]) = [[x_1] nil] {*dmx1*}
(dmx nil) = [nil nil] {*dmx0*}

```
(defun dmx (xys)
  (if (consp (rest xys)) ; 2 or more elements?
      (let* ((x (first xys))
             (y (second xys))
             (xsys (dmx (rest (rest xys))))
             (xs (first xsys))
             (ys (second xsys)))
        (list (cons x xs) (cons y ys)))    ; {dmx2}
      (list xys nil)))  ; 1 element or none ; {dmx1}
```

The informal axioms for dmx provide a basis for a formal definition. The formal version takes advantage of the fact that if the operand has less than two elements, then it is the first component of the result and the second component is the empty list. Like the multiplexer, the demultiplexer preserves the total length and preserves the values in its operand. ACL2 succeeds in verifying these facts without assistance, and the paper-and-pencil proofs are similar to the corresponding theorems for the multiplexer.

The two operators also satisfy some round-trip properties that bolster our confidence that they do what we expect them to do. Demultiplexing a list of x-y values into the list of x-values and the list of y-values and then multiplexing those two lists reproduces the original list of x-y values. It works the other way around too if the operands of the mux operator are lists of the same length.[45]

```
(defthm mux-inverts-dmx-thm
   (implies (true-listp xys)
            (equal (mux (first  (dmx xys))
                        (second (dmx xys)))
                   xys)))
(defthm dmx-inverts-mux-thm
   (implies (and (true-listp xs) (true-listp ys)
                 (= (len xs) (len ys)))
            (equal (dmx (mux xs ys))
                   (list xs ys))))
```

The dmx operator delivers every other element of the operand in one component of a list and the remaining elements in the other component. That means that each component of the result is half as long as the operand. If the operand has an odd number of elements, the extra one goes into the first component. These length properties can be specified in terms of the floor and ceiling operators (box 3.5, page 77). The length of the first component is the length of the operand divided by two and rounded up to the next integer if the operand has an odd number of elements. The second component is also half the length of the operand but rounded down if necessary. The mechanized logic of ACL2 succeeds in proving these theorems, but it needs the help of some theorems about arithmetic.

```
(include-book "arithmetic-3/top" :dir :system)
(defthm dmx-len-first
   (= (len (first (dmx xs)))
      (ceiling (len xs) 2)))
(defthm dmx-len-second
   (= (len (second (dmx xs)))
      (floor (len xs) 2)))
```

---

[45] Both of the round-trip properties require the operands to be true lists because the multiplexer can lose information if its operands aren't true lists. (The term "true list" is defined on page 105.)

> **Box 9.3**
> Cleverness Sometimes Complicates Reasoning
>
> An alternate definition of a demultiplexer observes that if the operand alternates between x and y values, starting with an x, then the same list without its first element also alternates but starting with a y. The definition is shorter, but it complicates reasoning.
>
> $$\text{Axioms } \mathsf{dmx2} \text{ (maybe too clever by half)}$$
>
> (dmx2 (cons x yxs)) = [(cons x xs) ys]   {$dmx2\text{-}1x$}
> where [ys xs] = (dmx2 yxs)
> (dmx2 nil) = [nil nil]   {$dmx2\text{-}0x$}

Exercises

1. Prove that the dmx operator preserves total length. That is, prove the theorem stated formally in ACL2 as follows:

   ```
   (defthm dmx-length-thm
     (= (len xys)
        (+ (len (first (dmx xys)))
           (len (second (dmx xys))))))
   ```

2. Do paper-and-pencil proofs of the dmx-len-first and dmx-len-second theorems (page 174). You may find the proofs easier if you split them into two cases, one when the operand has an even number of elements (that is, $2n$ for some natural number n), the other when it has an odd number of elements ($2n + 1$).

3. Give an example of lists [x y], [u], and [w] for which [[u] [w]] ≠ (dmx [x y]) but the following formula is true:
   $\forall v.(((\mathsf{occurs\text{-}in}\ v\ [u]) \lor (\mathsf{occurs\text{-}in}\ v\ [w])) \leftrightarrow (\mathsf{occurs\text{-}in}\ v\ [x\ y]))$
   Note: Such an example demonstrates that length and value preservation are not enough to guarantee that dmx delivers the right value. They aren't enough for the mux operator either (exercise 3, page 173).

4. State formally, in ACL2, the dmx-val theorem analogous to the mux-val theorem (page 171).

5. The dmx-val theorem (exercise 4) says that dmx neither adds nor drops values from its operand. Do a paper-and-pencil proof of the dmx-val theorem.

6. Do a paper-and-pencil proof of the mux-inverts-dmx theorem (page 174).

7. Do a paper-and-pencil proof of the dmx-inverts-mux theorem (page 174).

8. Do a paper-and-pencil proof that (dmx2 xs) = (dmx xs).
   Note: dmx2 is defined in box 9.3, page 175, and dmx is defined on page 173.

10 Sorting

The task of sorting records into a desired order (alphabetical order, for example, or chronological order, or numeric order by an identifying key) is one of the most studied problems in computing. Solutions abound, and a good one can save an enormous amount of time. A sorting operator that is twice as fast as a slower operator when rearranging a few hundred records will typically be many times faster than the slow operator for thousands of records and thousands of times faster when there are millions of records to be arranged. Data archives with thousands or millions of records are common, and that makes the sorting process important.[46]

This chapter will discuss two sorting operators that deliver the same results but differ greatly in the amount of time they take to do the job. Since they deliver the same results, they are equivalent operators in a mathematical sense, but they are vastly different computationally. We will discuss both the computational differences and the mathematical equivalence.

Deriving the resource requirements of an operator from the equations that define it is similar to deriving other properties. Previously, we have been mostly concerned with meeting expectations with regard to the form of the results of an operation, not the time it takes to deliver those results. Now we will discuss engineering choices that affect the usefulness of software as the amount of data increases. Engineering requires not only producing the expected results but also dealing with scale in effective ways.

---

[46] The difference between using a fast sorting operator and a slow one can be dramatic. Some years ago, one of the authors helped the US Forest Service figure out why their central computing system was bogged down. The culprit turned out to be about two dozen lines of code in their road design system. Those lines defined a slow sorting method known as bubble sort. Replacing it with a fast sorting method known as quicksort cut the amount of computation attributable to the road design system from over a hundred hours a week on each of eight mainframe computers to a few hours a week on one.

10.1 Insertion-Sort

To focus our attention on the essentials of arranging records in order by a key, we will assume that the entire content of a record resides in its key. In practice, there is usually a lot of information in a record, not just an identifying key, but the process of arranging the records in order by key is the same, regardless of what information is associated with each key. To simplify the discussion, we will use numbers for keys and discuss operators that rearrange lists of numbers into increasing order. For example, if the operand of the sorting operator were the list [5 9 4 6 5 2], the operator would deliver the list [2 4 5 5 6 9], which contains the same numbers, but arranged so that the smallest one comes first, increasing up the line to the largest at the end.

In practice, keys need not be numbers, but they do need to be comparable to determine an ordering (alphabetical, chronological, and so on). If the keys aren't numbers, then the numeric comparisons (<, >) in our discussion would be replaced by other operators designed to compare keys to see which one precedes the other in the desired ordering. The sorting method is the same, regardless of how keys are compared.

Suppose someone has defined an operator that, given a list of numbers that has already been arranged into increasing order, along with a new number to put in the list, delivers a list with the new number inserted in a place that preserves the ordering. If we call the operator insert, then the formula (insert 8 [2 4 5 5 6 9]) would deliver [2 4 5 5 6 8 9].

What are some equations that we would expect the insert operator to satisfy? If the list were empty, then the operator would deliver a list whose only element would be the number to be inserted in the list.

(insert x nil) = (cons x nil)   {$ins0$}

If the number to be inserted is less than or equal to the first number in the list, the operator could simply insert the number at the beginning of the list.

(insert x (cons x_1 xs)) = (cons x (cons x_1 xs)) if $x \leq x_1$ {$ins1$}

If the number to be inserted is greater than the first number in the list, we don't know where it will go in the list, but we do know it won't come first. The first number in the list will still be the first number after the new one is inserted somewhere down the line. If we trust the operator to put it in the right place, we can make a new list starting with the same first number and then let the insertion operator put the new number where it belongs among the numbers after the first one. That leads to an inductive equation for the operator insert.

(insert x (cons x_1 xs)) = (cons x_1 (insert x xs)) if $x > x_1$ {$ins2$}

The equations {$ins0$}, {$ins1$}, and {$ins2$} are comprehensive, consistent, and computational, so they define the operator insert (three C's, figure 4.10, page 91). The following

10.1 Insertion-Sort 179

definition in ACL2 transliterates the three equations but consolidates {*ins0*} and {*ins1*} into one equation by observing that the right-hand sides of both equations are the same formula: (cons *x s*), where *s* is the second operand on the left-hand side (which is nil in equation {*ins0*} and (cons x_1 *xs*) in equation {*ins1*}).

```
(defun insert (x xs)  ; assume x1 <= x2 <= x3 ...
  (if (and (consp xs) (> x (first xs)))
      (cons (first xs) (insert x (rest xs)))   ; {ins2}
      (cons x xs)))                            ; {ins1}
```

Now suppose someone has defined a sorting operator called isort (insertion-sort). Empty lists and one-element lists already have their elements in order, by default. Therefore, the formula (isort nil) would deliver nil and the formula (isort (cons *x* nil)) would deliver (cons *x* nil). That is, (isort *xs*) = *xs* when *xs* has one element or none.

$$\text{(isort nil)} = \text{nil} \qquad \{isrt0\}$$
$$\text{(isort (cons } x \text{ nil))} = \text{(cons } x \text{ nil)} \qquad \{isrt1\}$$

If the list to be sorted has two or more elements, it has the form (cons x_1 (cons x_2 *xs*)) ({consp} axiom, page 81). If the isort operator works properly, the formula (isort (cons x_2 *xs*)) would be a list made up of the number x_2 and all the numbers in the list *xs* taken together and arranged in increasing order. Given that list, the insert operator can put the number x_1 in the right place, producing a list made up of all the numbers in the original list rearranged into increasing order.

$$\text{(isort(cons } x_1 \text{ (cons } x_2 \text{ } xs\text{)))} = \text{(insert } x_1 \text{ (isort(cons } x_2 \text{ } xs\text{)))} \qquad \{isrt2\}$$

The equations {*isrt0*}, {*isrt1*}, and {*isrt2*} are comprehensive (the operand is either empty, has one element, or has more than one element) and consistent (no overlapping cases). They are computational because the operand of isort on the right-hand side of the inductive equation {*isrt2*}, namely (cons x_2 *xs*), has fewer elements than the operand on the left-hand side, which is (cons x_1 (cons x_2 *xs*)). Therefore, the operand on the right-hand side is closer than the operand on the left-hand side to a list of the form (cons *x* nil), which is the operand on the left-hand side of the noninductive equation {*isrt1*}. Therefore, the equations satisfy the three C's requirements (figure 4.10, page 91), which means that they define the operator isort. The three equations can be consolidated into two because {*isrt0*} and {*isrt1*} are both the same equation: (isort *xs*) = *xs*, where *xs* is nil in equation {*isrt0*} and (cons *x* nil) in equation {*isrt1*}.

```
(defun isort (xs)
  (if (consp (rest xs))  ; xs has 2 or more elements?
      (insert (first xs) (isort (rest xs)))  ; {isrt2}
      xs))                  ; (len xs) <= 1   ; {isrt1}
```

We expect the insertion-sort operator to preserve the number of elements in its operand and to neither add nor drop values from the list. Theorems stating these properties would be similar to the corresponding theorems for the multiplex and demultiplex operators discussed in chapter 9. The theorem on preservation of values is stated as a Boolean equivalence and uses the **occurs-in** predicate (box 9.1, page 171) for determining whether or not a value occurs in a list (box 9.1, page 171, and exercise 2, page 173).[47]

```
(defthm isort-len-thm
  (= (len (isort xs)) (len xs)))

(defthm isort-val-thm
  (iff (occurs-in e xs)
       (occurs-in e (isort xs))))
```

We also expect the numbers in the list that the **isort** operator delivers to be in increasing order. To state that property, we need a predicate to distinguish between lists containing numbers in increasing order and lists that have some numbers out of order. A list with only one element or none is automatically in order. A list with two or more elements is in order if its first element doesn't exceed its second and if all the elements after the first element are in order. These observations lead to the following ACL2 definition of a predicate **up**, which is true when its operand is a list of numbers that is in increasing order and false otherwise:

```
(defun up (xs)        ; (up[x1 x2 x3 ...]): x1 <= x2 <= x3 ...
  (or (not (consp (rest xs)))    ; (len xs) <= 1
      (and (<= (first xs) (second xs))  ; x1 <= x2
           (up (rest xs)))))     ; x2 <= x3 <= x4 ...
```

Our expectations about ordering in the list that the **isort** operator delivers can be expressed formally in ACL2 in terms of the **up** predicate. ACL2 succeeds without assistance in proving all three properties: length preservation, value preservation, and ordering. The proof can induct on the length of the list supplied as the operand of **isort**.

```
(defthm isort-ord-thm
  (up (isort xs)))
```

Later, we will analyze the computational behavior of the **isort** operator and will find that it is extremely slow for long lists. The next section begins a discussion of a sorting operator that is fast, even on long lists.

Exercises

1. Do a paper-and-pencil proof that the **isort** operator preserves the values in its operand (**isort-val-thm**, above).

---

[47] Preservation of length and values does not guarantee that the operator delivers the correct result. For example, the lists [1 1 2] and [1 2 2] have the same length and the same values but (isort [1 1 2]) ≠ [1 2 2]. The sorted list must be a permutation of the original list. The permutation property is not much harder to prove than length and value preservation, but it does require a definition of permutation (exercise 6, page 181).

Exercises

2. Do a paper-and-pencil proof that the isort operator preserves the length of its operand (isort-len-thm, page 180). You may assume theorem {*insert-len*}: (len(insert x xs)) = 1 + (len xs).

3. Do a paper-and-pencil proof that the isort operator delivers a list arranged in increasing order (isort-ord-thm, page 180).

4. Suppose (ct x xs) delivers a count equal to the number of times the value x occurs in the list xs.

   a) What value should the operator ct deliver if xs has no elements?

   b) State a theorem in ACL2 that expresses the number of occurrences of a value x in the list (cons x xs) in terms of the number of occurrences of x in xs.

   c) State a theorem in ACL2 that expresses the number of occurrences of a value x in the list (cons y xs) when y is not equal to x.

   d) Use the above observations to define the operator ct.

```
(defun ct (x xs)  ; number of occurrences of x in xs
   ...)
```

5. The del operator deletes an occurrence of x in xs if x occurs in xs.

```
(defun del (x xs)
   (if (not(consp xs))
       nil
       (if (equal x (first xs))
           (rest xs)
           (cons (first xs) (del x (rest xs))))))
```

Define a theorem in ACL2 that expresses the number of occurrences of x in (del x xs) in terms of the number of occurrences of x in xs. Refer to the operator ct (exercise 4).
Hint: Be careful to take into account the possibility that x does not occur in xs.

6. The predicate permp, defined as follows, is true if its second operand is a permutation of its first operand and false otherwise:[48]

```
(defun permp (xs ys)
   (if (not(consp xs))
       (not(consp ys))
       (and (occurs-in (first xs) ys)
            (permp (rest xs) (del (first xs) ys)))))
```

Define a theorem in ACL2 stating that (isort xs) is a permutation of xs, and get ACL2 to prove the theorem. Since the theorem will refer to the predicate permp and permp refers to the operators occurs-in and del, ACL2 will need to admit definitions of those operators to its logic before it can attempt to prove the theorem.

---

[48] The predicate occurs-in is defined in box 9.1 (page 171).

10.2 Order-Preserving Merge

The multiplex operator (mux, section 9.1) combines two lists into one in a perfect shuffle. The merge operator is another way to combine lists. It combines ordered lists in a way that preserves order. If both lists contain numbers arranged in increasing order, the merge operator mrg will combine the two lists into one in which all of the elements from both lists are arranged in increasing order. Two of the equations for mrg specify the results when one of the lists is empty. The equations for mrg in those cases are the same as the corresponding equations for the multiplex operator ($\{mux0x\}$ and $\{mux0y\}$, page 170).

$$(\text{mrg nil } ys) = ys \quad \{mg0\}$$
$$(\text{mrg } xs \text{ nil}) = xs \quad \{mg1\}$$

When both lists are nonempty, the merged list will start with either the first element of the first operand or the first element of the second operand, depending on which is smaller. The remaining elements in the merged list come from merging what's left of the list whose first element is smaller with all of the elements in the other list. That divides the nonempty case into two subcases, one when the first operand starts with a smaller number than the second operand and the other when the second operand begins with the smaller number.

$$(\text{mrg } (\text{cons } x \text{ } xs) \text{ } (\text{cons } y \text{ } ys)) = (\text{cons } x \text{ } (\text{mrg } xs \text{ } (\text{cons } y \text{ } ys))) \text{ if } x \le y \quad \{mgx\}$$
$$(\text{mrg } (\text{cons } x \text{ } xs) \text{ } (\text{cons } y \text{ } ys)) = (\text{cons } y \text{ } (\text{mrg } (\text{cons } x \text{ } xs) \text{ } ys)) \text{ if } x > y \quad \{mgy\}$$

The four equations, taken as a whole, are comprehensive because either one list is empty or the other one is empty or both lists are nonempty, in which case the first element of one of them is less than or equal to the first element of the other. They are consistent because, as with the mux operator, the only overlapping situation is when both lists are empty, in which case equation $\{mg0\}$ delivers the same result as equation $\{mg1\}$.

Two of the equations ($\{mgx\}$ and $\{mgy\}$) are inductive, so we need to make sure they are computational. In both equations, there are fewer elements in the operands on the right-hand side than on the left-hand side. That is, the total number of elements to be merged on the right-hand side of the inductive equation $\{mgx\}$ is less than the total on the left-hand side. That makes the operands on the right-hand side closer to a noninductive case than the operands on the left-hand side. Therefore, the equations are computational. That covers the three C's (figure 4.10, page 91), so we can take the equations as axioms that define the mrg operator.

A formal definition in ACL2 can be constructed from the equations $\{mg0\}$, $\{mg1\}$, $\{mgx\}$, and $\{mgy\}$. However, ACL2 needs some help in finding an induction scheme to prove that the equations lead to a terminating computation. We reasoned that the merge equations are computational because the total number of elements in the two operands is smaller on the right-hand side of the inductive equations than on the left-hand side. The declare directive in the following ACL2 definition suggests basing the proof by induction on this total, and that suggestion is enough to get the mechanized logic on the right track.

```
(defun mrg (xs ys)
  (declare (xargs :measure (+ (len xs) (len ys)))) ; induction scheme
  (if (and (consp xs) (consp ys))
      (let* ((x (first xs)) (y (first ys)))
        (if (<= x y)
            (cons x (mrg (rest xs) ys))    ; {mgx}
            (cons y (mrg xs (rest ys)))))) ; {mgy}
      (if (not (consp ys))
          xs         ; ys is empty          ; {mg0}
          ys)))      ; xs is empty          ; {mg1}
```

The mrg operator preserves the total length of its operands, and it neither adds nor drops any of the values in those operands. The equations specifying these properties are like those of the corresponding properties of the mux operator, namely, the mux-length theorem (page 170) and the mux-val theorem (page 173).[49]

The mrg operator also preserves order. If the numbers in both operands are in increasing order, the numbers in the list it delivers are in increasing order. A formal statement of this property can employ the same order predicate (up, page 180) that was used to specify a similar property of the isort operator. However, in the case of the mrg operator, the property is guaranteed only under the condition that both operands are already in order, so the property is stated as an implication.

```
(defthm mrg-ord-thm
  (implies (and (up xs) (up ys))
           (up (mrg xs ys))))
```

ACL2 can verify this property without assistance. The induction scheme for proving that the mrg operator terminates, namely, induction on the total number of elements in the operands, also works in the proof of the merge order theorem. A paper-and-pencil proof could follow the same strategy.

Exercises

1. Using the mux-length theorem (page 170) as a model, make a formal ACL2 statement of the mrg-length theorem.

2. Do a paper-and-pencil proof of the mrg-length theorem of exercise 1.

3. Do a paper-and-pencil proof of mrg-ord-thm (see defthm, above).

4. Using the mux-val theorem (page 173) as a model, make a formal ACL2 statement of the mrg-val theorem.

5. Do a paper-and-pencil proof of the mrg-val theorem of exercise 4.

---

[49] As with the theorems about mux, dmx, and isort, length and value preservation do not guarantee that mrg delivers a correct result. The result must be a permutation of the elements of the lists to be merged, which is a more restrictive property than preservation of length and values (see exercise 6, page 181).

10.3 Merge-Sort

We can use the mrg operator (page 183), together with the demultiplexer (dmx, page 173), to define a sorting operator msort (merge-sort) that is fast for long lists. The msort operator uses dmx to split the list into two parts, sorts each part inductively into increasing order, and finally uses the mrg operator to combine the sorted parts into one list.

If the operand of msort has only one element or none, it is already in increasing order, so the equations in that case, like those for isort (page 179), are not inductive. If the operand of msort has two or more elements, the defining equation is inductive and involves two sorting operations, one for each of the two lists delivered by applying dmx to the operand.

$$(\text{msort nil}) = \text{nil} \hspace{3cm} \{msrt0\}$$
$$(\text{msort (cons } x \text{ nil})) = (\text{cons } x \text{ nil}) \hspace{2cm} \{msrt1\}$$
$$(\text{msort (cons } x_1 \text{ (cons } x_2 \text{ xs)})) = (\text{mrg (msort odds) (msort evns)}) \hspace{0.5cm} \{msrt2\}$$
where
$$[\text{odds, evns}] = (\text{dmx (cons } x_1 \text{ (cons } x_2 \text{ xs)}))$$

The inductive equation will be computational only if both of the lists that dmx delivers are strictly shorter than the operand of msort. We expect this to be true because half of the elements go into each list (dmx length theorems, page 174). The following formal definition expresses the msort equations ($\{msrt0\}$, $\{msrt1\}$, $\{msrt2\}$) in ACL2 but consolidates the equations for lists with one element or none in the manner of the ACL2 definition of isort (page 179):

```
(defun msort (xs)
  (declare (xargs
            :measure (len xs)
            :hints (("Goal"
                     :use ((:instance dmx-shortens-list-thm))))))
  (if (consp (rest xs))         ; 2 or more elements?
      (let* ((splt (dmx xs))
             (odds (first splt))
             (evns (second splt)))
        (mrg (msort odds) (msort evns))) ; {msrt2}
      xs))                      ; (len xs) <= 1   ; {msrt1}
```

The definition of msort includes a declare directive to help ACL2 verify that msort terminates. The directive suggests basing the induction on the length of the operand. To apply this inductive measure successfully, ACL2 needs a hint suggesting the use of a lemma[50] stating that the dmx operator splits its operand into two lists, both of which are strictly

---

[50] Since the theorem about the lengths of the lists delivered by dmx is cited in the proof of another theorem (namely, the theorem stating that msort terminates), we refer to it as a lemma. The lemma could be derived from length theorems about dmx proven in section 9.2 (page 174), but a weaker form of those theorems turns out to be just what ACL2 needs for its proof that msort terminates.

Exercises

```
(defthm msort-order-thm
  (up (msort xs)))
(defthm msort-len-lemma-base-case
  (implies (not (consp (rest xs)))
           (= (len (msort xs)) (len xs))))
(defthm msort-len-lemma-inductive-case
  (= (len (msort (cons x xs)))
     (1+ (len (msort xs)))))
(defthm msort-len-thm
  (= (len (msort xs))
     (len xs)))
(defthm msort-val-thm
  (iff (occurs-in e xs)
       (occurs-in e (msort xs))))
```

Figure 10.1
Theorems and lemmas about merge-sort.

shorter than its operand. ACL2 proves the lemma without assistance, and it then admits (with the help of the **declare** directive) the definition of **msort** to its mechanized logic.

```
(defthm dmx-shortens-list-thm  ; lemma helps ACL2 admit def of msort
  (implies (consp (rest xs))   ; can't shorten 0- or 1-element lists
           (let* ((odds (first  (dmx xs)))
                  (evns (second (dmx xs))))
             (and (< (len odds) (len xs))
                  (< (len evns) (len xs))))))
```

Like **isort**, the **msort** operator puts the elements of its operand in increasing order and preserves length and values. The following statements of these properties, like those for **isort**, use the predicates **up** (page 180) and **occurs-in** (page 173).[51] Like **isort**, (**msort** *xs*) delivers a permutation of *xs* (exercise 6, page 181), but the proof is trickier for **msort**. It might make a good project for ambitious readers. ACL2 verifies the **msort** ordering property without help, but it needs some lemmas stating the base case and the inductive case for the proof of the length property. ACL2 fails on the value property, so we will settle for a paper-and-pencil proof of that one (exercise 3, page 186). Figure 10.1 (page 185) states the **msort** theorems and lemmas in ACL2.

Exercises

1. Do a paper-and-pencil proof that, under certain conditions, the **dmx** operator delivers lists that are shorter than its operand (**dmx-shortens-list-thm**, above).

2. Do a paper-and-pencil proof that the **msort** operator delivers a list that is in increasing order (**msort-order-thm**, above). You may cite **mrg-ord-thm** (exercise 3, page 183).

---

[51] The theorem statements employ the ACL2 operator iff, which is Boolean equivalence (box 2.5, page 32).

3. Do a paper-and-pencil proof that the msort operator preserves the values in its operand (msort-val-thm, page 185). You may cite the mrg-val theorem (exercise 4, page 183).

4. Do a paper-and-pencil proof that the msort operator preserves the length of its operand (msort-len-thm, page 185). You may cite the merge-length theorem (exercise 1, page 183).

5. Do a paper-and-pencil proof that the msort operator delivers a permutation of its operand (see exercise 6, page 181). *Caveat*: This is a project, not an exercise.

10.4 Analysis of Sorting Algorithms

In this section, we discuss a computation model for ACL2 that gives us a way to count the number of computation steps required to compute the value denoted by a formula. We use the equations defining the msort operator to derive inductive equations for the number of computation steps that msort requires to rearrange lists into increasing order. Then, we assert a formula for the number of computation steps required by msort and prove, by mathematical induction, that the formula is correct. It turns out that the number of steps in the computation (msort $[x_1, x_2, \ldots x_n]$) is proportional to $n\ log(n)$.

We do the same for isort. The number of steps in the isort computation varies widely, depending on the order of the values in the operand, so we estimate the average over randomized lists. That average turns out to be proportional to the square of the number of elements in the operand. Finally, we compare the computation steps required by the two operators, msort and isort, and find that msort is much faster for long lists.

10.4.1 Counting Computation Steps

A definition of an operator in ACL2 is a collection of equations that reduces invocations of the operator to computations of the result. Predicates determine which equations to select from the definition. Both the formula that computes the result and the predicate formulas that control its selection invoke other operators. In the case of an inductive equation, the selected formula may invoke the operator defined by the definition. The computation eventually boils down to a sequence of basic, one-step operations. Analyzing the number of steps required to compute a result amounts to counting the number of steps in that sequence.

A detailed analysis would allow a different computation time for each basic operator. That is, a model for detailed analysis could associate several computation steps with one basic operator and associate only a few steps with another operator. There would also be a scale establishing a relationship between computation steps and computation time.

Our analysis will provide a less refined picture than such a model because we will assume that each basic operator delivers its result in just one computation step. In the worst case, this would throw comparisons between the number of steps in different computations off by the ratio between the time required by the slowest basic operator and the fastest. That is,

10.4 Analysis of Sorting Algorithms

Insertion: (cons *x xs*)
Extraction: (first *xs*), (rest *xs*)   *operators that add one step*
Arithmetic: (+ *x y*), (− *x y*), (∗ *x y*), ...   *to a computation*
Boolean: (and *x y*), (or *x y*), (not *x*), ...   *(after computing needed operands)*
Comparison: (< *x y*), (<= *x y*), (= *x y*), ...
Cons predicate: (consp *xs*)
Selection: (if *p x y*)   *computes p, then x or y, but not both*

Figure 10.2
Basic one-step operators.

formula	steps	step 1, step 2, ...
(cons 1 (cons 2 (cons 3 (cons 4 nil))))	4	(cons 4 nil), (cons 3 ...), cons, cons
(second [1 2 3])	2	(rest [1 2 3]), (first [2 3])
(if (> 7 3) (+ 3 (∗ 5 4)) (+ 2 2))	4	(> 7 3), (if T □ □), (∗ 5 4), (+ 3 20)
(if (< 7 3) (+ 3 (∗ 5 4)) (+ 2 2))	3	(< 7 3), (if nil □ □), (+ 2 2)

Figure 10.3
Computation steps in formulas with basic operators.

comparisons based on our crude model will be off by a small factor, but they will provide a rough estimate of the ratio between the computation speed of one operator and another.

Figure 10.2 specifies the basic, one-step operators of the computation model. Each basic operator in a formula contributes one step to the computation, so counting the number of steps in a computation is straightforward if the definitions of the operators it invokes refer only to the operators listed in the figure. For example, analyzing the formula for constructing the list [1 2 3 4] reveals a four-step computation.

[1 2 3 4] denotes (cons 1 (cons 2 (cons 3 (cons 4 nil))))
Four Steps: (cons 4 nil), (cons 3 [4]), (cons 2 [3 4]), (cons 1 [2 3 4])

Figure 10.3 displays a similar analysis on some formulas composed of basic operations. The same kind of analysis applies when the formulas invoke defined operators rather than intrinsic ones like cons, first, and rest. For example, the operator F-from-C, defined as follows, converts a temperature from degrees Celsius to degrees Fahrenheit. It multiplies the temperature by the ratio 180/100 (to adjust from Celsius degrees to the more refined scale of Fahrenheit degrees), then adds 32 (to adjust the freezing point from zero to 32).[52] That makes two basic operations in all, so the formula (F-from-C 100) represents a two-step computation.

```
(defun F-from-C (C)
  (+ (* 180/100 C) 32))
```

The formula (list (F-from-C 0) (F-from-C 100)) makes a list in Fahrenheit degrees of two important points on the temperature scale: the freezing point of water (0 °C) and the boiling point (100 °C). To count the number of steps in this computation, we need to write the formula in terms of basic operations. The operator list is shorthand for a sequence of nested cons operations to build a list, so in terms of basic operations, the formula is (cons (F-from-C 0) (cons (F-from-C 100) nil)). The total step-count comes to six: two for each F-from-C invocation and one for each cons.

Another example: the operator swap2, defined as follows, interchanges the first two elements of a list if the list has at least two elements. If not, it leaves the list as is.

```
(defun swap2 (xs)
  (if (consp (rest xs))
      (cons (second xs) (cons (first xs)) (rest (rest xs)))
      xs))
```

It refers to the operator that extracts the second element from a list, which is shorthand for using the basic operator rest to drop the first element, then the operator first to extract the first element of what remains. So, the formula (second *xs*) would add two steps to the computation. The number of steps in the computation (swap2 *xs*) depends on how many elements *xs* has. If *xs* has two or more elements, then (swap2 *xs*) takes ten steps: if, consp, rest, cons, two steps for second, cons again, first, and rest again, twice. If *xs* has fewer than two elements, then (swap2 *xs*) takes three steps: if, consp, and rest.

Exercises

1. Count the number of steps in (− (F-from-C 100) (F-from-C 0)).

2. Count the number of steps in (swap2 (list 1 2 3)).
   Note: (list 1 2 3) is shorthand for nested cons operations (figure 4.1, page 80).

3. Count the number of steps in (swap2 (list 1)).

4. Count the number of steps in (list (third *xs*) (second *xs*) (first *xs*)), where the formula (third *xs*) is shorthand for (first (rest (rest *xs*))).

5. Define an operator C-from-F that converts degrees Fahrenheit to degrees Celsius, and count the number of operations required to compute (C-from-F (F-from-C 20)).

---

[52] Ratios in ACL2 are designated by two integers separated by a slash. The notation represents the number itself, not a computation: 1/2 represents one-half, just as 2 represents two. No computation is involved.

Exercises

6. What is (C-from-F (F-from-C x))? What is (F-from-C (C-from-F x))?

7. Define a theorem about (C-from-F (F-from-C x)) and get ACL2 to prove it.
Note: The complicated formula must be on the left-hand side of the equation.[53]
Note: The predicate ACL2-numberp is true if its operand is a number and false otherwise. The theorem must constrain the domain to numbers (use ACL2-numberp and implies).

10.4.2 Computation Steps in Demultiplex

The demultiplex operator, dmx (page 173), parcels out the elements of a list into two separate lists, with every other element going into one list and the remaining elements going into the other list. We repeat its definition here for convenience in counting steps.

```
(defun dmx (xys)
  (if (consp (rest xys))      ; 2 or more elements?
      (let* ((x (first xys))
             (y (second xys))
             (xsys (dmx (rest (rest xys))))
             (xs (first xsys))
             (ys (second xsys)))
        (list (cons x xs) (cons y ys)))   ; {dmx2}
      (list xys nil)))  ; 1 element or none  ; {dmx1}
```

From the inductive equations for dmx, we will derive corresponding equations for counting computation steps. Let D_n stand for the number of steps required to compute (dmx xs) when xs has n elements. If n is zero or one, (consp (rest xs)) is false, so dmx selects the third operand of the if operator as the result. The computation takes five steps: one step each for selection (if), consp, and rest, plus two steps for (list xys nil) because it is shorthand for (cons xys (cons nil nil)). Therefore, $D_0 = D_1 = 5$.

If xs has two or more elements (that is, $n + 2$ elements for some natural number n), the computation will require D_{n+2} steps. From the definition of dmx, we see that the computation has several parts: selection (if), one step; consp, one step; extraction (first), one step; a two-step extraction (second), two steps; extraction (rest) twice, two steps; computation of (dmx (rest (rest xs))), D_n steps because (rest (rest xs)) has n elements; another extraction (first), one step; another two-step extraction (second), two steps; a double cons (the operator list with two operands), two steps; and two insertions (cons), two steps. Altogether, that comes to $D_n + 14$ steps. Putting the two cases together, we come up with the following recurrence equations:[54]

$$D_0 = D_1 = 5 \quad \{d1\}$$
$$D_{n+2} = D_n + 14 \quad \{d2\}$$

---

[53] Constraints on the order of operands in equations have to do with strategic considerations in the ACL2 theorem-proving engine that are beyond the scope of the treatment here. If you want to track down those ideas, read about rewrite rules in the ACL2 documentation.

[54] Inductive equations in the numeric domain are called *recurrence equations*.

Theorem {dmx computation steps}:
$D_n \equiv$ *number of computation steps in* $(\text{dmx } [x_1 \; x_2 \; \ldots \; x_n]) = 14\lfloor n/2 \rfloor + 5$
proof
Base case $(n = 0)$
$\quad D_0 \; = 5 \qquad\qquad \{d1\}$
$\qquad\; = 14\lfloor 0/2 \rfloor + 5 \quad \{\lfloor 0/2 \rfloor = 0\}$
Inductive case for $n = 1$
$\quad D_1 \; = 5 \qquad\qquad \{d1\}$
$\qquad\; = 14\lfloor 1/2 \rfloor + 5 \quad \{\lfloor 1/2 \rfloor = 0\}$
Inductive case for $n + 2 \geq 2$
$\quad D_{n+2} \; = D_n + 14 \qquad\qquad\qquad \{d2\}$
$\qquad\quad\; = 14\lfloor n/2 \rfloor + 5 + 14 \qquad \{\textit{induction hypothesis}\}$
$\qquad\quad\; = 14(\lfloor n/2 \rfloor + 1) + 5 \qquad \{\textit{algebra}\}$
$\qquad\quad\; = 14\lfloor n/2 + 1 \rfloor + 5 \qquad\;\; \{\lfloor x \rfloor + 1 = \lfloor x + 1 \rfloor\}$
$\qquad\quad\; = 14\lfloor (n+2)/2 \rfloor + 5 \qquad \{\textit{algebra}\}$

Figure 10.4
Computation steps in demultiplex.

Sometimes it's possible to guess a closed-form formula for the numbers in a sequence defined by recurrence equations and then prove by induction that the formula is correct.[55] For equations {d1} and {d2}, $D_n = 14(\lfloor n/2 \rfloor + 1) + 5$ is the right guess.[56] Figure 10.4 proves this conjecture using strong induction (figure 6.4, page 133).

Exercises

1. Derive recurrence equations for the number of steps in the computation of (len xs) from the axioms for the len operator (figure 4.5, page 84). Assume that selecting between the two axioms is a two-step computation (one step to determine whether or not xs has any elements and one step to use that determination to select the appropriate axiom).

2. Use the recurrence equations from exercise 1 to guess a formula for the number of steps in the computation of (len xs). Prove that the formula is correct.

3. Derive recurrence equations for the number of steps in the computation of (append xs ys) from the definition of append in figure 4.11 (page 93).

4. Use the recurrence equations from exercise 3 to guess a formula for the number of steps in the computation of (append xs ys). Prove that the formula is correct.

---

[55] A closed-form formula for D_n is a formula that doesn't refer to D_m for any m.
[56] We will use floor brackets $\lfloor x \rfloor$ and ceiling brackets $\lceil x \rceil$ (box 3.5, page 77) extensively in this chapter.

Exercises   191

10.4.3 Computation Steps in Merge

Our next goal is to estimate the number of steps in the computation of (mrg xs ys) (page 183). We will not try to count the exact number of steps in the computation, but we will look for an upper bound. Our analysis will ensure that the number of steps does not exceed an amount that we can compute from the number of elements in the operands.

We begin by defining $M_{j,k}$ to be the maximum number of steps required to merge a list of j elements with a list of k elements.[57] We also define A_n to be the maximum number of steps required to merge two lists with a combined total of n elements.

$M_{j,k} \equiv$ *maximum steps in computation of* (mrg $[x_1\ x_2\ \ldots x_j]\ [y_1\ y_2\ \ldots y_k]$)

$A_n \equiv \text{maximum}\{M_{j,k} \mid j + k = n\}$

We will prove $\forall n.(A_n \leq 10(n + 1))$ by induction. For the base case, $A_0 = M_{0,0}$ because the only pair of natural numbers j and k with $j + k = 0$ is $j = k = 0$. So, A_0 is the number of steps in the computation of (mrg xs ys) when (consp xs) and (consp ys) are false. Let's look at the definition of mrg and count those steps.[58] In this case, the computation consists of seven or fewer one-step operations (if, and, consp twice, if again, not, and consp again).[59] Therefore, $A_0 \leq 7 < 10 \cdot (0 + 1)$, which proves the base case.

Now, consider the inductive case: $\forall n.((A_n \leq 10(n + 1)) \rightarrow (A_{n+1} \leq 10((n + 1) + 1)))$. The induction hypothesis is $A_n \leq 10(n + 1)$. A_{n+1} is a maximum of a set of numbers $M_{j,k}$, where $j + k = n + 1$ and $M_{j,k}$ represents the maximum number of steps in the computation of (mrg $[x_1\ x_2\ \ldots x_j]\ [y_1\ y_2\ \ldots y_k]$).

```
(defun mrg (xs ys)
  (if (and (consp xs) (consp ys))
      (let* ((x (first xs)) (y (first ys)))
        (if (<= x y)
            (cons x (mrg (rest xs) ys))     ; mgx
            (cons y (mrg xs (rest ys)))))   ; mgy
      (if (not (consp ys))
          xs        ; ys is empty           ; mg0
          ys)))     ; xs is empty           ; mg1
```

---

[57] There are an infinite number of such lists, and the maximum of an infinite set of numbers is problematic. However, the merge computation depends only on the ordering of the numbers in the operands, not on their specific values. There are a finite number of permutations of the elements of a list, so the set of combinations to be considered in computing the maximum is finite. A similar caveat applies to most of our proofs. We have defined properties, including step-counting formulas, in terms of the number of list elements without considering the values of those elements. Properties of the form P(n) ≡ (... $[x_1\ x_2\ \ldots x_n]$ ...) more properly would take the form P(n) ≡ $\forall x_1.\forall x_2 \ldots \forall x_n.(\ldots [x_1\ x_2\ \ldots x_n] \ldots)$. However, even though property definitions have glossed over this issue, the proofs themselves have been independent of the values, x_k. That is, the proofs are correct and rigorous but with certain details omitted. Fortunately, proofs carried out by an engine of mechanized logic such as ACL2 take the details into account.

[58] We repeat the definition of mrg here to make counting operations more convenient. The previous definition (page 183) declared an induction scheme to help ACL2 admit the definition into its mechanized logic. We have omitted the declaration here because we are analyzing the computation without the assistance of ACL2.

[59] Actually, it's six steps. The operator and does not compute its second operand if its first operand is false.

If one of the operands is empty, there are, as in the base case, at most seven steps in the computation. If neither operand is empty, there are eight one-step operations (if, and, consp twice, first twice, if again, and the comparison $x \leq y$) that culminate in the selection of one of two inductive formulas: (cons x (mrg (rest xs) ys)) or (cons y (mrg xs (rest ys))). Both of these formulas have two one-step operations (cons and rest) for a total, counting the eight previous steps, of ten one-step operations. There is also an inductive invocation to account for: (mrg (rest xs) ys) or (mrg xs (rest ys)).

The total number of elements in the operands of the inductive invocation of mrg is n because there are $n + 1$ elements in the two lists xs and ys, taken together, and in the inductive invocation the operator rest has dropped an element from one of the lists. That is, there are a total of n elements in the operands in the invocation (mrg (rest xs) ys), and the same is true of (mrg xs (rest ys)). Therefore, by the definition of A_n, the number of steps in the computation of the inductive invocation, no matter which one of them is selected, cannot exceed A_n. We conclude that $A_{n+1} \leq A_n + 10$ (A_n steps for the inductive invocation plus ten one-step operations). By the induction hypothesis, $A_n \leq 10(n + 1)$. Therefore, $A_{n+1} \leq 10(n + 1) + 10$. Factoring out the 10 algebraically, we find that $A_{n+1} \leq 10((n+1)+1)$. That completes the proof of the inductive case, and we conclude by induction that $\forall n.(A_n \leq 10(n + 1))$.

Theorem {mrg computation steps}:

$$((\text{len } xs)+(\text{len } ys) = n) \to (\textit{steps to compute } (\text{mrg } xs\ ys) \equiv A_n \leq 10(n + 1))$$

Exercises

1. The double index of $M_{j,k}$ made the proof a little tricky. Another approach would have been to use the principle of double induction (figure 10.5, page 193), which reframes mathematical induction for double-index predicates. Suppose that P is a predicate whose universe of discourse is pairs of natural numbers. That is, for each pair of natural numbers m and n, $P(m, n)$ selects a proposition in the predicate. Use the predicate P to define another predicate Q that has the natural numbers as its universe of discourse and that has the following properties:
   a. $Q(0)$ is the base case for double induction on P.
   b. $(\forall n.(Q(n) \to Q(n + 1)))$ is the inductive case for double induction on P.
   Note: This exercise is difficult and not particularly instructive. The point is that double induction can be reduced to ordinary mathematical induction.

10.4.4 Computation Steps in Merge-Sort

We have been working towards an upper bound on the number of steps needed to arrange the elements of a list into increasing order using the msort operator (pages 184 and 193).[60]

---

[60] As with the definition of mrg, we repeat the definition of msort to facilitate counting steps, but we omit the hints provided in the previous definition to help ACL2 prove termination.

Exercises   193

> Prove $(\forall m.P(m,0)) \land (\forall n.P(0,n))$ *base case*
> --
> Prove $(\forall m.(\forall n.((P(m+1,n) \land P(n,m+1)) \rightarrow P(m+1,n+1))))$ *inductive case*
> ─── {dbl ind}
> Infer $(\forall m.(\forall n.P(m,n)))$

Figure 10.5
Double induction: a rule of inference.

```
(defun msort (xs)
  (if (consp (rest xs))      ; 2 or more elements?
      (let* ((splt (dmx xs))
             (odds (first splt))
             (evns (second splt)))
        (mrg (msort odds) (msort evns))) ; {msrt2}
      xs))                   ; (len xs) <= 1 ; {msrt1}
```

Let S_n stand for the number of steps in the computation (msort $[x_1\ x_2\ \ldots x_n]$). If n is zero or one, msort selects a noninductive formula, and this requires three one-step operations: if, consp, and rest. If n is two or bigger, it's more complicated. Figure 10.6 (page 194) derives the recurrences from the definition of msort.

The crucial steps are the two inductive invocations of msort. The operand in the first invocation is a list delivered as the first component in a demultiplexed list of n elements. (It is called odds in the let* formula.) According to theorem {dmx-len-first} (page 174), odds is a list containing $\lceil n/2 \rceil$ elements. So, by the definition of S, there are $S_{\lceil n/2 \rceil}$ steps in the computation of (msort odds). Similarly, theorem {dmx-len-second} (page 174) says that the second component of the demultiplexed list (evns) has $\lfloor n/2 \rfloor$ elements, which means that there are $S_{\lfloor n/2 \rfloor}$ steps in the computation of (msort evns).

The (dmx xs) computation takes $14\lfloor n/2 \rfloor + 5$ steps (theorem {dmx computation steps}, figure 10.4, page 190). The mrg computation takes at most $A_n = 10(n+1)$ steps (theorem {mrg computation steps}, page 192). There are six additional steps in (msort xs) when xs has two or more elements, all of which are basic operations: one step each for if, consp, rest, and first, plus two steps for second. Adding all this up (two msorts, dmx, mrg, and the six basic steps), we find that $S_n \leq S_{\lceil n/2 \rceil} + S_{\lfloor n/2 \rfloor} + (14\lfloor n/2 \rfloor + 5) + 10(n+1) + 6$.

Figure 10.6 (page 194) summarizes this recurrence analysis. The recurrences are upper bounds rather than equations because our analysis of the mrg operator put an upper bound on the number of steps in the computation, not an exact count. The simplified formula $7n+5$ is a bit larger than $14\lfloor n/2 \rfloor + 5$ when n is odd, but we're deriving an upper bound anyway, so the inequality still holds. A final algebraic simplification, $(14\lfloor n/2 \rfloor + 5) + 10(n+1) + 6 \leq (7n+5) + 10(n+1) + 6 = 17n + 21$, leads to inequality {s2}.

Now, we are going to guess a closed-form formula (see footnote, page 190) for an upper bound on S_n and then use strong induction (figure 6.4, page 133) to prove that the formula

operator	steps≤	step-counts for $n \geq 2$
if	1	figure 10.2, page 187
consp	1	figure 10.2
rest	1	figure 10.2
dmx	$14\lfloor n/2 \rfloor + 5$	{dmx computation steps}, n elements, page 190
first	1	figure 10.2
second	2	figure 10.3, page 187
mrg	$10(n + 1)$	{mrg computation steps}, n elements, page 192
msort	$S_{\lceil n/2 \rceil}$	{dmx-len-first}, page 174
msort	$S_{\lfloor n/2 \rfloor}$	{dmx-len-second}, page 174
total = $1 + 1 + 1 + (14\lfloor n/2 \rfloor + 5) + 1 + 2 + 10(n+1) + S_{\lceil n/2 \rceil} + S_{\lfloor n/2 \rfloor}$		

$S_n \equiv$ steps to compute (msort $[x_1\ x_2\ \ldots x_n]$)
$S_0 = S_1 = 3$ {s1}
$S_n \leq S_{\lceil n/2 \rceil} + S_{\lfloor n/2 \rfloor} + 17n + 21$, if $n \geq 2$ {s2}

Figure 10.6
Recurrence inequalities for merge-sort computation steps.

is correct. The right-hand side of the recurrence inequality {s2} expresses an upper bound on S_n in terms of $S_{n/2}$. People experienced in solving recurrences take this as an indication of $n\ log(n)$ growth for S_n. Accordingly, we can expect to be able to find a multiplier α such that $\forall n.(S_{n+2} \leq \alpha \cdot (n + 2)\ log_2(n + 2))$.[61] Finding a multiplier that works is mostly a matter of fiddling around with the recurrences to get some intuition about the numbers they produce. Here's the multiplier we came up with: $\alpha = 42$. Figure 10.7 (page 195) proves that $\forall n.(S_{n+2} \leq 42(n + 2)\ log_2(n + 2))$ by strong induction.[62]

Exercises

1. Use the recurrences in figure 10.6 to define an operator S in ACL2 with (S n) = S_n.

2. For any nonzero positive number x, $\lfloor log_2(x) \rfloor$ is the biggest integer n such that $2^n \leq x$. Define an operator log2 in ACL2 that computes $\lfloor log_2(x) \rfloor$ for $x \geq 1$.

---

[61] The formula $\alpha \cdot n\ log_2(n)$ cannot be an upper bound for S_n when n is zero or one because $n\ log_2(n) = 0$ when n is zero or one, and we know (figure 10.6) that $S_0 = S_1 = 3$, which is more than zero.

[62] A smaller multiplier can be found, but we're not too concerned with the size of the multiplier, especially since we don't have a scale for the amount of time a computation step takes in our model. It's the order of growth, the $n\ log_2(n)$ part, that interests us. The coincidence that 42 works may amuse Douglas Adams fans.

Theorem {msort $n \log(n)$}: $\forall n.(S_{n+2} \leq 42(n+2) \log_2(n+2))$
Proof by strong induction
Base Case ($n = 0$)
$$\begin{aligned}
S_{0+2} &\leq S_{\lceil(0+2)/2\rceil} + S_{\lfloor(0+2)/2\rfloor} + 17(0+2) + 21 &&\{s2\} \\
&= S_1 + S_1 + 55 &&\{algebra\} \\
&= 3 + 3 + 55 &&\{s1\} \\
&< 42(0+2) \log_2(0+2) &&\{arithmetic, \log_2(0+2) = 1\}
\end{aligned}$$

Inductive Case for $1 \leq n \leq 18$

Use recurrences (figure 10.6, page 194) to calculate S_{n+2}, $n = 1, 2, \ldots 18$.
Observe that $S_{n+2} \leq 42(n+2) \log_2(n+2)$, $n = 1, 2, \ldots 18$.

Inductive Case for $n \geq 19$ *(using* $m \equiv n + 2$ *to save space)*

$$\begin{aligned}
S_{n+2} &= S_m &&\{definition\ m \equiv n+2\} \\
&\leq S_{\lceil m/2 \rceil} + S_{\lfloor m/2 \rfloor} + 17m + 21 &&\{s2\} \\
&\leq 42\lceil m/2 \rceil \log_2\lceil m/2 \rceil + &&\{induction\ hypothesis,\ twice \\
&\quad 42\lfloor m/2 \rfloor \log_2\lfloor m/2 \rfloor + 17m + 21 &&\ (\lceil m/2 \rceil < m, \lfloor m/2 \rfloor < m)\} \\
&\leq 42\lceil m/2 \rceil \log_2\lceil m/2 \rceil + \\
&\quad 42\lfloor m/2 \rfloor \log_2\lceil m/2 \rceil + 17m + 21 &&\{\lfloor x \rfloor \leq \lceil x \rceil \to \log_2\lfloor x \rfloor \leq \log_2\lceil x \rceil\} \\
&= 42(\lceil m/2 \rceil + \lfloor m/2 \rfloor)\log_2\lceil m/2 \rceil + \\
&\quad 17m + 21 &&\{algebra\ (factor\ out\ 42\ \log_2\lceil m/2 \rceil)\} \\
&= 42m \log_2\lceil m/2 \rceil + 17m + 21 &&\{\lceil m/2 \rceil + \lfloor m/2 \rfloor = m\} \\
&\leq 42m \log_2((m+1)/2) + 17m + 21 &&\{\log_2\lceil m/2 \rceil \leq \log_2((m+1)/2)\} \\
&\leq 42m \log_2((m+1)/2) + 17m + m &&\{m = n+2 \geq 19+2 = 21\} \\
&\leq 42m \log_2((m+1)/2) + 18m &&\{17m + m = 18m\} \\
&= 42m(\log_2((m+1)/2) + (18/42)) &&\{algebra\ (factor\ out\ 42m)\} \\
&= 42m(\log_2(m+1) \\
&\quad -\log_2(2) + (18/42)) &&\{\log_2(x/y) = \log_2(x) - \log_2(y)\} \\
&= 42m(\log_2(m+1) - 1 + (18/42)) &&\{\log_2(2) = 1\} \\
&< 42m(\log_2(m+1) + \log_2(m/(m+1))) &&\{m \geq 3 \to \log_2\tfrac{m}{m+1} > -1 + \tfrac{18}{42}\} \\
&= 42m \log_2((m+1) \cdot m/(m+1)) &&\{\log_2(x) + \log_2(y) = \log_2(xy)\} \\
&= 42m \log_2(m) &&\{algebra\} \\
&= 42(n+2) \log_2(n+2) &&\{definition\ m \equiv n+2\}
\end{aligned}$$

Figure 10.7
Bound on steps in merge-sort computation.

```
(defun insert (x xs) ; assume x1 <= x2 <= x3 ...
  (if (and (consp xs) (> x (first xs)))
      (cons (first xs) (insert x (rest xs))) ; {ins2}
      (cons x xs)))                           ; {ins1}
(defun isort (xs)
  (if (consp (rest xs)) ; xs has 2 or more elements?
      (insert (first xs) (isort (rest xs))) ; {isrt2}
      xs))                ; (len xs) <= 1     ; {isrt1}
```

Figure 10.8
Formal definition of isort.

3. If $S_{n+2} \leq 42(n+2)\lfloor log_2(n+2) \rfloor$, then $S_{n+2} \leq 42(n+2)log_2(n+2)$. Use the operators from exercises 1 and 2 to compare S_{n+2} with $42(n+2)\lfloor log_2(n+2) \rfloor$ for each natural number n between 1 and 18. Explain any anomalies. *Hint*: $log_2(3) > 3/2$.

10.4.5 Computation Steps in Insertion-Sort

We want to compare the performance of the msort operator (merge-sort) with that of the isort operator (insertion-sort). The difference can be dramatic (see footnote, page 177). The isort operator almost always takes much more time than msort to arrange a list of numbers, and the difference grows rapidly with the number of elements in the list.

However, the number of steps in the computation (isort *xs*) varies widely depending on the arrangement of the numbers in the operand *xs*. For a few arrangements, (isort *xs*) is faster than (msort *xs*), and when *xs* is a short list, isort can be faster for any list. In fact, high-speed, general purpose software for sorting usually combines a method similar to insertion-sort with a method comparable to merge-sort. This hybrid strategy treats the list to be sorted as a collection of short lists (usually up to about eight elements), applies an isort-like operator to the short lists, and then combines them using an msort-like operator.

The number of computation steps required by the isort operator varies widely, but an estimate of the average number of computation steps that isort requires for randomly arranged lists is useful for comparison purposes. We repeat the definition of isort (figure 10.8) for convenience in the analysis.

The computation of (insert *x* nil) takes four steps (if, and, consp, cons).[63] In the computation (insert *x* [x_1 x_2 ... x_{n+1}]), the insert operator assumes that $x_1 \leq x_2 \leq ... x_{n+1}$. In the worst case, $x > x_{n+1}$ and the insert computation will deliver the list [x_1 x_2 ... x_{n+1} x]. A careful analysis shows that this worst case insertion takes $8(n + 1) + 4$ steps (exercise 2, page 199).

It seems reasonable to expect that, on average, for random data, *x* will get inserted about half way down the list *xs*. That means it would take, on average, about half as many steps to compute (insert *x xs*) as it takes in the worst case, which is when *x* exceeds all the numbers

---

[63] The operator and does not compute its second operand if its first operand is false.

Exercises

in xs, putting x at the end of the list that (insert x xs) delivers. Proving the assertion about the average requires understanding the nature of random data and probabilistic effects, so we are not going to pursue a proof, but we will assume that it's true. With that assumption and $8(n + 1) + 4$ computation steps for worst case insertion, the average number of steps for (insert x $[x_1\ x_2\ \ldots x_{n+1}]$) would be $G_{n+1} = (8(n + 1) + 4)/2 = 4(n + 1) + 2 = 4n + 6$.

What does this mean for (isort xs)? From the definition of isort, we see that when (len xs) is zero or one, isort performs three one-step operations (if, consp, rest). When xs has $n + 2$ elements (that is, two or more elements), there are five one-step operations (if, consp, rest, first, rest again) plus an insert operation with a second operand that has $n + 1$ elements, which takes $G_{n+1} = (4n + 6)$ computation steps, on average. Finally, there remains the inductive invocation of isort with an operand that has $n + 1$ elements. This analysis yields the recurrence equations in figure 10.9 (page 198) for the average number of steps in the computation (isort $[x_1\ x_2\ \ldots x_n]$), which we denote by I_n.

We could guess a closed-form formula for I_n and prove it by induction, but instead let's carry out an analysis that applies the recurrence equations for I_n in stages. Figure 10.9 (page 198) displays the recurrence equations and presents the analysis, using equation $\{i2\}$ to build a formula for I_{n+2} step by step. First, replace I_{n+2} by the right-hand side of equation $\{i2\}$. Then, if $(n + 1) \geq 2$, replace I_{n+1} by the right-hand side of $\{i2\}$ again and continue using equation $\{i2\}$ in this way until the formula comes down to $(I_1 + \ldots)$. At that point, replace I_1 by 3, citing equation $\{i1\}$.

Then, some algebraic reorganization reveals a factor of $(1 + 2 + 3 + \ldots n)$ in one of the terms in the sum. This is the well-known triangular number:[64] $(1+2+3+\ldots n) = n(n+1)/2$. In the end, with a little more algebraic reorganization, we find that (isort $[x_1\ x_2\ \ldots x_{n+2}]$) requires, on average, $I_{n+2} = (2n + 11)(n + 1) + 3$ computation steps. Another way to say this is $I_n = (2n + 7)(n - 1) + 3$ for $n \geq 2$.

Theorem $\{$msort $n\ log(n)\}$ (figure 10.7, page 195) provides an upper bound[65] on the number of steps S_n in the computation (msort $[x_1\ x_2\ \ldots x_{n+2}]$): $S_n \leq 42n\ log_2(n)$ when $n = $ (len xs) ≥ 2. To compare the time it takes to sort a list of numbers using isort versus msort, we can compute a lower bound on the ratio between the number of computation steps that insertion-sort requires versus merge-sort: $I_n/S_n \geq (2n + 7)(n - 1)/(42n\ log_2(n))$ when $n \geq 2$. (We dropped +3 from the numerator, but the ratio is a lower bound anyway, so it's still a lower bound.) This ratio gets big fast with increases in n, the number of elements in the list to be sorted.

---

[64] The formula for the triangular number might win a contest for most popular example of proof by mathematical induction in a poll of discrete math textbooks. It's either that or the geometric progression (exercise 13, page 155). Proving the triangular number formula is exercise 4 (page 199).

[65] It turns out that this upper bound is close to the average for merge-sort on randomized lists.

Recurrence Equations for I_n	
$I_0 = I_1 = 3$	{i1}
$I_{n+2} = I_{n+1} + (4n + 6) + 5 = I_{n+1} + 4n + 11$	{i2}

Closed-Form Formula for I_n	
I_{n+2}	
$= I_{n+1} + (4n + 11)$	{i2} I_{n+2}
$= I_n + (4(n-1) + 11) + (4n + 11)$	{i2} I_{n+1}
$\vdots$	
$= I_3 + (4 \cdot 2 + 11) + (4 \cdot 3 + 11) + \ldots (4n + 11)$	{i2} I_4
$= I_2 + (4 \cdot 1 + 11) + (4 \cdot 2 + 11) + (4 \cdot 3 + 11) + \ldots (4n + 11)$	{i2} I_3
$= I_1 + (4 \cdot 0 + 11) +$	{i2} I_2
$\quad (4 \cdot 1 + 11) + (4 \cdot 2 + 11) + (4 \cdot 3 + 11) + \ldots (4n + 11)$	
$= 3 + (4 \cdot 0 + 11) +$	{i1}
$\quad (4 \cdot 1 + 11) + (4 \cdot 2 + 11) + (4 \cdot 3 + 11) + \ldots (4n + 11)$	
$= 3 + 11 + 4 \cdot (1 + 2 + 3 + \ldots n) + 11n$	{algebra}
$= 3 + 11 + 4 \cdot (n(n+1)/2) + 11n$	{triangular number}
$= (2n + 11)(n + 1) + 3$	{algebra}
$I_n = (2(n-2) + 11)(n - 2 + 1) + 3 = (2n + 7)(n - 1) + 3$	if $n \geq 2$

Figure 10.9

I_n = average number of steps to compute (isort x [x_1 x_2 ... x_n]).

The break-even point occurs at about a hundred elements. For 1,000 elements, the ratio is about five. That is, msort, as we have defined it (page 193), is about five times faster than isort (page 196) for a list with 1,000 elements. For 10,000 elements, msort is about 40 times faster. For 100,000 elements, 300 times faster, and for 1,000,000 elements, about 2,000 times faster. These estimates are on the conservative side because we made no attempt to get a tight upper bound on the number of computation steps for msort, and we have not paid close attention to computational details in the definitions of isort and msort. Serious software for sorting is loaded with performance tweaks, and that can make the ratios more extreme in practice.

Beyond a comparison of the performance of merge-sort versus insertion-sort, the main thing to take away from this discussion is that inductive definitions provide a straightforward way to derive recurrence equations for the number of steps that the defined operator takes to carry out a computation. If a solution to the recurrences can be found, whether by guessing or by using one of a host of solution methods that are beyond the scope of the treatment here, then there is a good chance that an inductive proof can verify that the solution is correct. In this way, defining operators with inductive equations facilitates analyzing the number of computational steps the operators require to perform an operation.

Exercises

1. Derive, from the definition of insert (page 196), recurrence equations for the number of steps in the computation (insert x $[x_1\ x_2\ \ldots x_{n+1}]$) if $x_1 \leq x_2 \leq \ldots x_{n+1} < x$.

2. Using the recurrences from exercise 1, prove that the number of steps in the computation of (insert x $[x_1\ x_2\ \ldots x_{n+1}]$) is $8(n + 1) + 6$ or less if $x_1 \leq x_2 \leq \ldots x_{n+1} < x$.

3. The multiplier 42 in our bound on S_n gets sloppier as n increases. Find $\beta < 42$ that works for lists with over 100 elements: $S_n \leq \beta \cdot n\ log_2(n)$ if $n > 100$.

4. Use induction to prove the formula for the triangular number.
   $(1 + 2 + 3 + \ldots n) = n(n + 1)/2$
   You may be the billionth person to prove it. Knock yourself out.

11 Search Trees

11.1 Finding Things

Consider the problem of keeping things organized so you can find them. It's not hard if you don't have many things. If you have a dozen pairs of socks, you can just throw them in a drawer and find the pair you want quickly, without really having to search. However, if you run an office supply store that carries thousands of types of paper, envelopes, pencils, and erasers, you'll need to arrange them carefully so people can find them.

The importance of being organized increases with the number of items. If you run a warehouse, you might have to keep track of many thousands of things. One solution is to associate a stock number with each kind of item, set up a bin for each, and arrange the bins by stock number. That works reasonably well, even if there are many thousands of kinds of items. When the warehouse decides to stock a new kind of item, you just set up a bin at the end of the line and give it a stock number higher than the one before it.

However, what if an item gets discontinued? You can empty out its bin, but what do you do with the space? It might be okay to leave the space empty for the first few discontinuations, but with many thousands of items, there will be, over time, thousands of discontinuations, and you'll end up wasting a lot of space.

Furthermore, stock numbers solve the problem of finding the right bin, but they don't solve the problem of remembering the stock number that goes with, say, 20-pound white paper versus color ink cartridges. To be able to find the stock number associated with an item, you'll need some kind of index that arranges the items in categories, subcategories, and the like, and perhaps in alphabetical order by name within each category. When the warehouse discontinues an item, you need to cross it out in the index, and when a new kind of item comes in, you need to write it into the index at the appropriate place. Eventually, the index is a mess and needs reprinting.

You might be able to make this work, even for many thousands of items, but you can see that the idea doesn't scale up well, and when you're trying to keep track of many millions of things, as you might need to do, for example, if you were the data manager for a large company, keeping lists in order by name or stock number becomes untenable. New kinds of data need to be added every day, or even every few seconds, and obsolete items need to be deleted. So, keeping the items in a list starting with the first item, then the second, third, and so on cannot work in practice. It will require too much data movement to expand spaces between items to insert new ones or collapse space between items to delete old ones. The problem needs a better solution, and that's what this chapter is about. Let's recap the problem.

Pile-of-socks method. If you have only a few items to keep track of, the pile-of-socks method works. When you get a new pair of socks, you just throw it in the drawer. When an old pair wears out, you just throw it away. If you want to locate a particular pair of socks, you just look through the drawer, pair by pair, until you find the one you're looking for. If you have ten pairs, the first pair you look at might be the one you want, or the second, or third. At worst, you'll have to look through all ten pairs. On average you'll probably find the right pair after looking through about half of them.

Binary search method. If you had a thousand books on your bookshelves, you could arrange them in alphabetical order by title. When you get a new book, you just insert it in the right place, maybe sliding a few books around to accommodate it. If you lose one, you can just leave the slot empty, or slide the books around to fill in the space.

To find a particular book, you would probably just guess about where it would be, look there, and adjust your next guess accordingly. However, there is an efficient way to manage your search that guarantees finding it in just a few steps. It's called *binary search*. First, you look at the middle book. If it's the one you want, great. You're done. If not, the title you're looking for will either come after the middle one or before it in alphabetical order. In either case, the number of possibilities has been cut in half, and you can repeat the procedure to find the book you're looking for among the remaining possibilities. Since you cut the number of books in half each time, it will not take more than ten steps to locate a particular book among a thousand books because cutting a thousand in half, then cutting five hundred in half, and so on, you get down to one book in ten steps.

Looking at it from the other direction, starting from one and doubling at each stage, it goes 1, 2, 4, 8, 16, 32, 64, 128, 256, 512, 1024. These are the powers of two: $2^0, 2^1, 2^2, 2^3$ ... 2^{10}. So, the number of items you need to examine in the binary search procedure when there are n items altogether is the first integer k that makes $2^k \geq n$, or what amounts to the same thing, $k \geq log_2(n)$. That is, the number of steps in a binary search through n items is $log_2(n)$ rounded up to the next integer.

When you round a number up to the next integer, you get the *ceiling* of the number: $\lceil x \rceil$ = ceiling of x = x rounded up to the next integer. With this notation, the formula for the maximum number of steps required to locate a particular item using binary search with a collection of n items is $\lceil log_2(n) \rceil$.

This gives binary search an amazing advantage in speed over the pile-of-socks method. It reduces the number of steps required to find a particular item in a collection of n items

11.2 The AVL Solution

from an average of $n/2$ in the pile-of-socks approach to a maximum of $\lceil log_2(n) \rceil$ using binary search. That's $(n/2)/\lceil log_2(n) \rceil$ times faster, and $(n/2)/\lceil log_2(n) \rceil$ turns out to be a very big number when there are a lot of items.

For 1,000 items, binary search is, on average, $(1,000/2)/\lceil log_2(1,000) \rceil$ times faster than the pile-of-socks method. That's about 50 times faster. For 10,000 items, it's about 350 times faster. The speed-up factor rises to over 3,000 for 100,000 items, and for 1,000,000 items, binary search is, on average, over 25,000 times faster than the pile-of-socks method. It gets better and better the more items you have, and in the computer search business, it is common to search for one item among millions, billions, or even trillions of items.[66]

So, binary search, or something as good or better, is indispensable. The good news is that we can rely on it. The bad news is that, while it works well for a few thousand items, the problem of keeping the items arranged in order becomes unwieldy, to say the least, when there are millions of items.

That's what happens when the items are entries in an array of n elements, numbered, say, 0 to $n - 1$. Whenever you add a new item or delete an old one, you have to move things around, one way or the other, to make space or fill in gaps to keep the numbers that select array elements in a contiguous sequence, 0, 1, 2, 3, ... So, keeping items in order with indexes in sequence to facilitate using binary search works fine when the items don't change but becomes infeasible when they do. In practice, items tend to come and go, so we need to organize the items in some way other than just numbering them sequentially.

11.2 The AVL Solution

Not long after computers began to have enough memory to keep track of many thousands of items, the mathematicians Adelson-Velskii and Landis devised a structure that eliminated the problem of sliding things around to accommodate new items (the *insertion problem*) or get rid of old ones (the *deletion problem*). This structure, known as the AVL tree ("AV" for Adelson-Velskii and "L" for Landis), makes it possible to find an item or insert a new one in only about $log(n)$ steps, where n is the number of items stored in the tree. It takes about the same number of steps to delete an item, so the AVL solution provides a practical way to do all three operations: search, insertion, and deletion. Adelson-Velskii and Landis solved a very difficult problem when they figured out how to do this, but their solution isn't hard to explain and that is the goal of this chapter.

We will need some terminology to discuss the idea. The term *tree* will mean either an *empty tree* or a *node* called the *root*. A node consists of a sequence of one or more

---

[66] For the record, binary search is 17 million times faster for a billion items and 12 billion times faster for a trillion items. The pile-of-socks method would not be a feasible way to find one item among billions if you had to make such a search millions of times every day.

trees known as *subtrees*.[67] We will focus on *binary trees* in which each node has four components:

1. a *key*,
2. some data associated with the key,
3. a search tree known as the *left subtree*, and
4. a search tree known as the *right subtree*.

Search trees are special trees that facilitate finding keys. The empty tree is a search tree, by default. The two subtrees in a node are called *siblings*. A key is said to *occur in* a search tree if it is the key at the root or if the key occurs in either subtree of the root. No keys occur in an empty tree.

Keys in search trees must come from a domain that has a *total ordering*. That is, for any two different keys, there must be some way to determine which one precedes the other in the ordering. For example, when the keys are words made up of lowercase letters, alphabetic ordering is a total order. If the keys are numbers, then ordinary numeric ordering is a total order.

As a consequence of having a total order on the keys, it is possible, given any three keys, no two of which are the same, to determine which one comes first, which comes second, and which comes third. More generally, any collection of distinct keys can be arranged in a sequence in increasing order: each key in the sequence precedes, in the ordering, the next key in the sequence.

To support binary search, each node in a search tree has the property that all of the keys in the left subtree precede (in the ordering on the domain of keys) the key of the node, and the key of the node precedes all of the keys in its right subtree. A tree with this property is called an *ordered tree*, so search trees are ordered trees.

To find a key in a search tree, look first at the key at the root (unless the tree is empty, in which case you can conclude that the key does not occur in the tree). If the key you are looking for is the same as the key at the root, then you have found the key you're looking for. If it isn't the same as the key at the root, it must either precede or follow the root key in the ordering. If the key you are looking for precedes the root key, look for it in the left subtree. If it follows the root key, look for it in the right subtree.

To build a search tree that has a single key, just construct a node consisting of the key, its associated data, an empty left subtree, and an empty right subtree. That amounts to inserting the key into an empty search tree. To insert a key in a nonempty search tree, put it in the left subtree if the key precedes the root key in the ordering and in the right subtree if it follows the root key. To build a search tree containing all the keys from a list, simply

---

[67] Yet another usefully circular (inductive) definition.

11.2 The AVL Solution

insert the first one into an empty search tree, then insert each successive one into the tree produced by inserting the previous key from the list.

That's the simple part. The hard part is in making sure the tree doesn't get too tall. Nodes in a search tree have subtrees, and the subtrees have subtrees, and so on. Eventually, in a binary search, the desired key is encountered if it occurs in the tree. If the key doesn't occur in the tree, the search will come to an empty subtree. At that point, the search stops with the conclusion that the key does not occur in the tree.

The *height* of a search tree is the maximum number of keys that binary search might need to examine to find a key in the tree. Later, we will define the height of a tree formally, but for now, just think of it as the maximum number of steps in a binary search for a key in the tree.

AVL trees maintain, in every node, a balance between the heights of the subtrees in the node. The AVL method of inserting new nodes or deleting old ones maintains balance, which keeps the tree from getting too tall on one side or the other. This preserves a high ratio between the number of keys in the tree and the height of the tree. That is, the insertion process makes sure to keep the height of the tree small compared to the number of keys. The AVL deletion method does this too.

When a key is inserted or deleted, the AVL method keeps the height of the tree from exceeding the base-2 logarithm of the number of keys by more than 50 percent.[68] Therefore, the number of steps in a binary search of an AVL tree containing n keys will not exceed $\frac{3}{2} \cdot log_2(n)$ steps. To have an inkling of how ingenious the AVL solution is, try to figure out how to insert keys into a search tree in a way that has the following properties:

1. The key at each node follows every key that occurs in its left subtree.
2. The key at each node precedes every key in the right subtree.
3. Sibling subtrees differ in height by zero or one.

The first two requirements make the search tree ordered, so that binary search will work. The last requirement maintains *balance*, which keeps the tree from getting too tall. To match the effectiveness of AVL trees, you would need to define an insertion method such that neither the height of the tree nor the number of steps in the insertion process exceeds the logarithm of the number of nodes in the tree by more than a fixed percentage (that is, a percentage that doesn't depend on the size of the tree). The number of steps in your method for insertion might be, for example, up to twice the height of the tree but not more. After struggling with the problem of finding a reliable way to build search trees that don't get too tall, you may begin to appreciate the contribution of Adelson-Velskii and Landis to the problem of storing large amounts of data so that individual items can be found quickly and so that new items can be quickly inserted or old ones quickly deleted.

---

[68] The proof of this fact would take us too far afield, so we are leaving the proof out of this discussion.

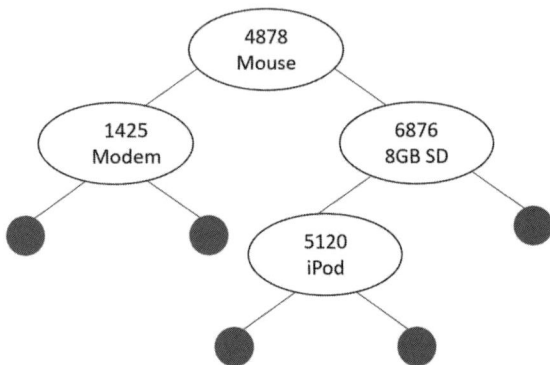

[4878 "Mouse" [1425 "Modem" nil nil] [6876 "8GB SD" [5120 "iPod" nil nil] nil]]

Figure 11.1
Search tree diagram and corresponding formula.

11.3 Representing Search Trees

Any detailed discussion of the AVL solution requires a way to represent search trees. In our representation, keys will be natural numbers ordered in the usual way. That is, one key, k_1, precedes another key, k_2, if $k_1 < k_2$ in the usual numeric ordering. That means that the key at a node of a search tree is numerically greater than all the keys that occur in the left subtree of the node and numerically less than all the keys that occur in the right subtree of the node. The empty list (nil) will represent the empty tree. A node in a search tree will be a list of four elements: key (a number), data (of any kind), left subtree, and right subtree. Those are the essentials of the representation.

The *root node* of a search tree represents the entire tree. The formula for the root node, like the formula for any node, has a key, a left subtree, and a right subtree. However, unlike other nodes, the root node is not a subtree of any other node, and all keys in the tree occur either at the root or in one of its subtrees. The key at the root is the first element of the list that represents the tree, followed by the data associated with the root, which is the second element of the list. The third element is the left subtree, and the fourth element of the list is the right subtree.

Figure 11.1 shows a formula for a search tree in which the data are strings. The figure also displays a diagram of the tree that the formula represents. This pictorial way of looking at search trees will clarify discussions of the insertion process. In a tree diagram, the root node is the one at the top, and the subtrees dangle from lines going down to the left and right. To follow the discussion, you will need to be able to diagram a tree given its formula and vice versa.

```
(defun mktr (k d lf rt)              ; make tree from
  (list k d lf rt))                  ;   key, data, subtrees
(defun key (s) (first s))            ; key at root
(defun dat (s) (second s))           ; data at root
(defun lft (s) (third s))            ; left subtree
(defun rgt (s) (fourth s))           ; right subtree
(defun emptyp (s) (not (consp s)))   ; empty tree?
(defun height (s)                    ; tree height
  (if (emptyp s)
      0                                                    ; {ht0}
      (+ 1 (max (height (lft s)) (height (rgt s))))))      ; {ht1}
(defun size (s)                      ; number of keys
  (if (emptyp s)
      0                                                    ; {sz0}
      (+ 1 (size (lft s)) (size (rgt s)))))                ; {sz1}
(defun iskeyp (k)
  (natp k))
(defun treep (s)                     ; search tree?
  (or (emptyp s)
      (and (= (len s) 4) (natp (key s))
           (treep (lft s)) (treep (rgt s)))))
(defun keyp (k s)                    ; key k occurs in s?
  (and (iskeyp k) (treep s) (not (emptyp s))
       (or (= k (key s)) (keyp k (lft s)) (keyp k (rgt s)))))
```

Figure 11.2
Search tree operators and predicates.

The *height* of an empty tree is zero. The height of a nonempty tree is one more than that of the taller of its left and right subtrees. The size of a search tree is the number of keys that occur in the tree. No keys occur in an empty tree, so its size is zero. The size of a nonempty tree is one more than the sum of the sizes of its left and right subtrees.

We use ACL2 to formalize these definitions (figure 11.2), starting with an operator mktr to build a tree from its four components. Then we define operators to extract keys, data, and subtrees from nodes (key, dat, lft, and rgt). Finally, we define predicates to recognize keys and search trees (iskeyp, treep, emptyp) and to find out whether a key occurs in a tree (keyp).

The only tree of height zero is the empty tree because the height of a nonempty tree is one more than the maximum of two other numbers, which makes it at least one. Theorem {*ht-emp*} states this fact more rigorously.

Theorem {*ht-emp*}: (treep *s*) → (((height *s*) = 0) = (emptyp *s*))

11.4 Ordered Search Trees

Since keys are natural numbers, search trees are ordered (page 204) if, for each node, its key is greater than all of the keys that occur in the left subtree and less than all the keys that occur in the right subtree. An empty tree is ordered by default. The following equations define the predicate ordp so that (ordp s) is true if s is ordered and false otherwise:

(ordp s) = (emptyp s) ∨ {*ord*}
((treep s) ∧
($\forall x$.((keyp x (lft s))) → x < (key s))) ∧ (ordp (lft s)) ∧
($\forall y$.((keyp y (rgt s))) → y > (key s))) ∧ (ordp (rgt s)))

Duplicate keys do not occur in ordered search trees. A more rigorous statement of this fact can be based on the following observations:

1. A key at a node does not occur in either of its subtrees.
2. Any key in one subtree of a node is not equal to the key of the node and does not occur in the other subtree.

Theorem {*keys unique*}:
((iskeyp k) ∧ (ordp s)) →
(((k = (key s)) → ((¬(keyp k (lft s))) ∧ (¬(keyp k (rgt s))))) ∧
(((keyp k (lft s))) → ((k ≠ (key s)) ∧ (¬(keyp k (rgt s))))) ∧
(((keyp k (rgt s))) → ((k ≠ (key s)) ∧ (¬(keyp k (lft s))))))

Stating this theorem is more complicated than proving it. By equation {*ord*}, if (keyp x (lft s)), then x < (key s). Since k = (key s), we conclude that x ≠ k. That is, (k = (key s)) → (¬(keyp k (lft s))). That proves one of the implications in the theorem. The others are as easily dispatched by citing parts of the definition of the predicate ordp.

Exercises

1. Prove: (ordp s) → ((keyp k (lft s)) → ((k ≠(key s)) ∧(¬ (keyp k (rgt s))))).
2. Prove: (ordp s) → ((keyp k (rgt s)) → ((k ≠(key s)) ∧(¬ (keyp k (lft s))))).

11.5 Balanced Search Trees

Search trees must be ordered to make it convenient to find things. However, order is not enough. Trees must also be short relative to the number of items in the tree. Otherwise, order doesn't help. It can take as long, on average, to find an item in an ordered but unbalanced tree as it would if the data were completely unorganized. Figure 11.3 (page 209) compares some extremes.

The tree of height seven in figure 11.3 is unbalanced at every level. A binary search on this tree would have no advantage over looking through a pile of socks one by one. The unbalanced tree of height four in the figure is not much better. It has the same number of nodes in the left subtree as in the right subtree, but all of the subtrees are maximally unbalanced, like a pile of socks.

11.5 Balanced Search Trees

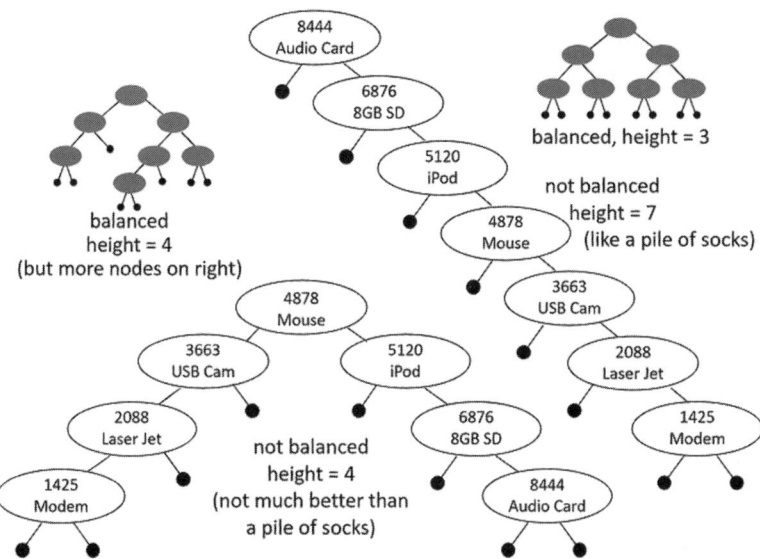

Figure 11.3
Balance shortens trees.

Balance is what prevents time-consuming searches, and the unbalanced examples in figure 11.3 show how bad it can get. A search tree that has two subtrees of the same size in every node is balanced in terms of both size and height. The tree of height three in figure 11.3 has this maximally balanced shape. However, the number of steps required to find a key in a search tree is determined by the heights of subtrees, not by the number of nodes they contain, and a tree can be balanced with respect to height even though some nodes contain more keys in one subtree than the other. The balanced tree of height four in the figure shows how this can happen. The shape of this tree isn't symmetric at any level, but no node has subtrees whose heights differ by more than one, and that's good enough. We are trying to keep the height of a tree with n nodes within a fixed percentage of $log_2(n)$, and height balancing is sufficient to accomplish this goal. Full symmetry isn't necessary.

As we mentioned before (page 205), the height of a balanced search tree with n keys is less than $\frac{3}{2}log_2(n)$. That makes finding things take more steps than the optimal case in which both subtrees of each node have the same number of keys, but searches are still fast. Finding a particular key among a billion nodes might require 45 steps instead of 30, but that's still plenty fast compared with half a billion steps, on average, for unorganized data.

What we get for a few extra steps in finding things is an astonishing improvement in the number of steps required to insert a new key or delete an old key. Instead of $n/2$ steps, on average, when keys beyond the point of insertion all need to be moved to make space

```
(defun balp (s) ; tree s is balanced?
  (or (emptyp s)
      (and (<= (abs (- (height (lft s)) (height (rgt s)))) 1)
           (balp (lft s)) (balp (rgt s)))))
```

Figure 11.4
Balance predicate.

for the new one, insertion can be done in logarithmic time. That is, the number of steps required to insert a node in a search tree will be proportional to $log_2(n)$, giving us the same advantage in insertion speed that binary search provides in look-up speed. Deletion can be handled in a similar way, and with the same effectiveness. That is, search, insertion, and deletion can all be done in logarithmic time.

Exercises

1. Find a way to put $2^n - 1$ keys in a binary search tree of height n.

11.6 Inserting a New Item in a Search Tree

To make search, insertion, and deletion efficient, search trees must be both ordered and balanced. With regard to balance, we must make sure that in every node the heights of the left and right subtrees differ by one or less. The predicate **balp** expresses this notion formally (figure 11.4).

It isn't difficult to maintain order. You can do this by moving left or right down the tree according to whether the new key is less than or greater than the key at the node under consideration. When you arrive at an empty tree, insert a node that has the new key with its associated data and has empty trees as its left and right subtrees. The new tree will be properly ordered because of the way the procedure located a place to hook the new key on the tree, but the new tree will not be balanced if the location of the new key increases the height of a subtree that was already taller than its sibling. Figure 11.5 (page 211) provides an inductive definition of this insertion method. The definition uses a noninductive formula to put the new key into an empty tree and an inductive formula for the nonempty case.[69]

Inserting a new key in this way can throw the tree out of balance. That happens when placement of the new key increases the height of a subtree that was already taller than its sibling. Then the subtree is two units taller than its sibling, making the tree unbalanced. In this case, a rearrangement brings it back into balance without getting keys out of order.

---

[69] In case the **hook** operator encounters a key that is the same as the one it is inserting, it delivers a tree with that key, as it should. However, the data associated with the new key will be the data supplied as the second operand in the invocation of **hook**. The old data is lost. This provides a way to associate new data with a key. The definition might have chosen a different alternative, but this is a viable one for our purposes.

11.6 Inserting a New Item in a Search Tree

```
(defun hook (x a s) ; put a new key x with data a into tree s
   (if (empty s)     ; preserve order, but not necessarily balance
       (mktr x a nil nil)                     ; {hook0}
       (let* ((k  (key s)) (d (dat s))
              (lf (lft s)) (rt (rgt s)))
         (if (< x k)
             (mktr k d (hook x a lf) rt)      ; {hook<}
             (if (> x k)
                 (mktr k d lf (hook x a rt))  ; {hook>}
                 (mktr x a lf rt))))))        ; {hook=}
```

Figure 11.5
Insert new key, preserving order but not balance.

Box 11.1
Inserting New Nodes in Small Trees

The following example starts with a tree containing one item, then inserts three new items, one at a time. We use the formula (ins *x a s*) to denote the tree produced by inserting the key *x* and associated data *a* into the search tree *s*.

The end result is an ordered, balanced tree containing four items. It will aid your understanding of the insertion process if you draw diagrams similar to figure 11.3 (page 209) for the trees denoted by the formulas in the example. Verify, as you go, that each tree is both ordered and balanced.

(ins 1125 "Modem"
  [8444 "Audio Card" nil nil])
  ⇓
[1125 "Modem" nil [8444 "Audio Card" nil nil]]

(ins 4878 "Mouse"
  [1125 "Modem" nil [8444 "Audio Card" nil nil]])
  ⇓
[4878 "Mouse" [1125 "Modem" nil nil]
              [8444 "Audio Card" nil nil]]

(ins 2088 "Laser Jet"
  [4878 "Mouse" [1125 "Modem" nil nil]
                [8444 "Audio Card" nil nil]])
  ⇓
[2088 "Laser Jet" [1125 "Modem" nil nil]
                  [4878 "Mouse" nil [8444 "Audio Card" nil nil]]]

In small trees, it's easy to find an ad hoc rearrangement that works, as illustrated in box 11.1 (page 211), but we need a procedure that works for all search trees, not just the small ones where it's easy to see what to do. Figuring out a rearrangement procedure is what the rest of this chapter is mostly about.

Putting the new node at the bottom may make the tree taller but not necessarily. For example, the insertion point might be on the empty side of a node that has a tree of height one on the other side, in which case the insertion would leave the height of the tree unchanged. But, if the height of the tree changes, how much could it change? Not by more than one (theorem $\{i\text{-}ht\}$, exercise 4, below).

If the tree with the new key is taller than the old tree, the new tree could be unbalanced. However, because a height of the left subtree of a balanced tree does not differ from the height of the right subtree by more than one and because the insertion of a new node cannot increase the height of either subtree by more than one, the heights of the left and right subtrees in the new tree cannot differ by more than two. So, if we can figure out how to rebalance trees where one subtree is two units taller than its sibling, we will have found a way to preserve balance while inserting a new node.

Exercises

1. Given any three distinct keys, there is only one search tree that is ordered, balanced, and contains those three keys but no others. Explain why.

2. Box 11.1 (page 211) displays insertions leading to an ordered, balanced search tree containing four items. The trees resulting from the insertions were, in each case, chosen from some equally suitable alternatives. Write formulas for ordered, balanced trees different from the ones in the example but still containing the same keys.

3. Use induction on height to prove the following theorem (ordp is defined on page 208):

   Theorem $\{i\text{-}ord\}$: (ordp s) $\rightarrow$ (ordp (hook k a s))

4. Prove by induction on tree height that insertion of a new node does not increase height by more than one. That is, assuming that x is a key, s is a search tree, and hook is the operator defined in figure 11.5 (page 211), prove the following theorem:

   Theorem $\{i\text{-}ht\}$: (height (hook x a s))$\leq$(height s) + 1

11.7 Insertion, Case by Case

Balancing small trees is easy because there are only a few possibilities to consider. Search trees of height two or less are always balanced.

Theorem $\{bal\text{-}ht2\}$: ((height s) $\leq$ 2) $\rightarrow$ (balp s)

11.7 Insertion, Case by Case

Working through all the possibilities, one by one, leads to a proof of this theorem. A tree of height zero is empty (theorem {*ht-emp*}, page 207), and (balp nil) = (emptyp nil) is true, by definition (figure 11.4, page 210). Any tree of height one will consist of a single node, [*k d* nil nil], which is balanced because both subtrees have the same height (namely, zero). The formula for a tree of height two must match one of the following templates: [*k d* [*j c* nil nil] nil], [*k d* nil [*i b* nil nil]], or [*k d* [*j c* nil nil] [*i b* nil nil]]. Applying the predicate balp confirms that all of these trees are balanced, and that completes the proof.

With big trees, there are more possibilities, but we can reduce part of the problem to a shorter tree, rely on induction to deal with that tree, and use the solution produced on the shorter tree to put together a full solution. We want to define an insertion operator, ins, to put a new key in an ordered, balanced search tree, producing a new search tree that is ordered, balanced, and contains the new key as well as all of the old ones. The operator hook (figure 11.5, page 211) does the job for trees of height zero or one. The new tree is ordered (theorem {*i-ord*}, page 212) and being of height two or less, it is also balanced (theorem {*i-ht*}, page 212, together with theorem {*bal-ht2*}, page 212). So, for trees of height zero or one, the hook operator, by itself, is adequate for inserting new keys.

That leaves us with trees of height two or more. We want to define an insertion operator, ins, so that if *s* is an ordered, balanced search tree of height $n + 2$ (where n is a natural number), then the tree (ins *x a s*) is ordered, balanced, contains all the keys in *s*, contains the key *x*, contains no keys other than *x* and those in *s*, and has height $n + 2$ or $n + 3$. To do this, we will start with the hook procedure (which has already the order and height properties that we need for ins), then find ways to rebalance when it produces a tree with subtrees whose heights differ by more than one.[70] Our inductive definition of ins will assume that it has the desired properties when operating on trees of height less than $n + 2$ and prove that, with that assumption, it also has those properties on trees of height $n + 2$.

Diagrams of search trees will help us work through the analysis case by case. Figure 11.6 (page 214) displays the three configurations that a search tree of height $n+2$ can have. Each diagram shows the root node as a circle labeled with a name for its key and shows the left and right subtrees as triangles dangling from the root. Each triangle represents an ordered,

---

[70] Our primary concerns will be the issues of height and balance. The other issues (presence of the new key, preservation of all the old keys, and so on) are easy to work through from the definitions. An issue that we will gloss over throughout the discussion is the treatment of data associated with a key. We include the data in the operator definitions because, as a practical matter, search trees need some way to associate data with keys. Usually, keys just provide a way to find the data. The operator definitions keep each data item with its associated key. Whenever we use mktr to build a tree, we put the key in the first operand and the associated data in the second operand. This keeps the key with its data. However, that is pretty much the extent of our analysis of key/data associations. Doing more is tricky because there are no constraints on the domain of the data. The data could even come from a domain that doesn't support reasoning about equality. This would be the case, for example, if the data items were themselves operators and the search tree were being used to provide organized access to those operators. There is no algorithm for determining, in general, whether two operators denote the same operation, so it would be difficult to reason about whether or not key/data associations stay the same throughout the process.

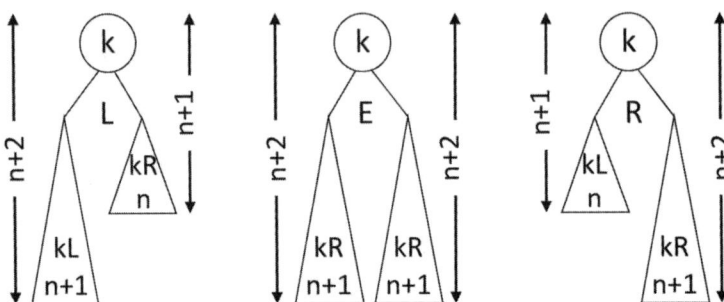

Figure 11.6
Balanced trees of height $n + 2$.

balanced subtree. A label inside the triangle simplifies references to the subtree, and a formula near the bottom of the triangle specifies the height of the subtree. Each tree as a whole is labeled with a name near the top of its diagram (L, E, and R). The tree L is one unit taller on the left than on the right, R is taller on the right, and both subtrees in E have the same height.

An insertion into tree E cannot cause the tree to go out of balance because, at worst, insertion will make one side of the tree one unit taller than the other side, which leaves the tree balanced. (The subtrees have height less than $n + 2$, so the induction hypothesis guarantees that they remain balanced after insertion.) Therefore, we do not concern ourselves with trees like E. However, if we insert a new key that is less than k into tree L, the key would end up in subtree kL, the left subtree of L, and the tree could go out of balance because the height of its left subtree could increase to $n + 2$, whereas the right subtree would remain at height n. Similarly, inserting a key greater than k into tree R could make its right subtree kR too tall. Tree L and tree R represent the cases we need to look into.

Figure 11.7 (page 215) will guide deeper analysis. In the figure, certain subtrees of L and R from figure 11.6 are expanded to show additional details. The new diagrams expand subtree kL in tree L and subtree kR in tree R. Both of these subtrees have height $n + 1$. That makes them nonempty, so each of them must have at least one key. The expanded diagrams of these subtrees reveal details about their keys and subtrees.

In tree L (figure 11.6), the subtree kL has height $n + 1$, so at least one of its subtrees has height n. Its sibling might also have height n, but unless n is zero, it could alternatively have height $n - 1$. The diagrams in figure 11.7 and other diagrams in this chapter use the notation n^- to denote a value that could be either n or, if n isn't zero, $n - 1$. The value denoted by n^- is not necessarily the same in all of the tree diagrams in the figure, but the value will be, in every instance, either n or $n - 1$.

11.7 Insertion, Case by Case

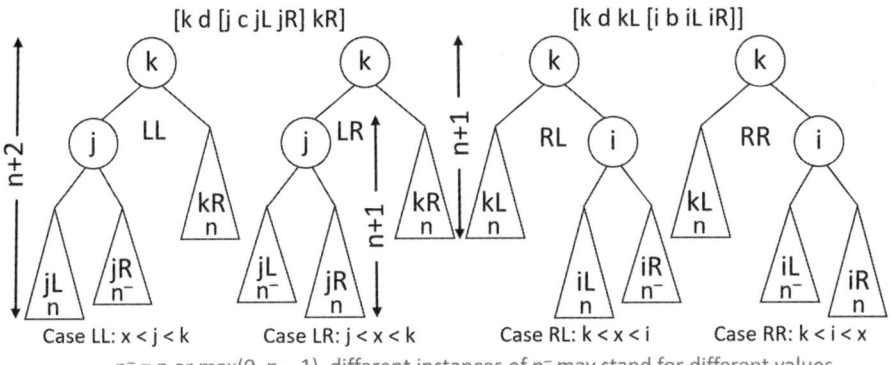

Figure 11.7
Balanced trees of height $n + 2$, subtrees expanded.

Either subtree may be the taller one, so we draw two diagrams representing the two possibilities (trees *LL* and *LR*). Similarly, there are two diagrams for tree *R*, which makes a total of four tree diagrams in figure 11.7: two diagrams (*LL* and *LR*) expanding tree *L* from figure 11.6 (page 214) and two diagrams (*RL* and *RR*) expanding tree *R*.

Consider the problem of inserting a new key, x, into an ordered, balanced search tree of height $n + 2$ with the key k at its root, the key j at the root of its left subtree, and the key k at the root of its right subtree. The value of x will fall into one of four intervals:[71]

1. case *LL*: $x < j < k$
2. case *LR*: $j < x < k$ *refer to*
3. case *RL*: $k < x < i$ *figure 11.7*
4. case *RR*: $k < i < x$

The names of the cases correspond to the tree diagram in figure 11.7. For example, if the new key x is less than k, insertion cannot cause the tree to go out of balance unless the left subtree is taller than the right subtree, which focuses the analysis on trees *LL* and *LR* in the figure. We use *LL* for the case when the new key x is smaller than the key j ($x < j < k$) because that pushes the key into subtree *jL*, and tree *LL* will definitely go out of balance if insertion increases the height of *jL*. The analysis of case *LL* covers subtrees *jR* both of height $(n - 1)$ and of height n.

If the new key x were between j and k ($j < x < k$), it would go into subtree *jR*. That might not cause tree *LL* to go out of balance, even if it increases the height of subtree *jR*,

---

[71] We can ignore the possibility that x is the same as k, j, or i because inserting a key that is the same as one already in the tree does not change the tree, except to replace the data associated with the key. Therefore, the tree will remain balanced and require no further attention.

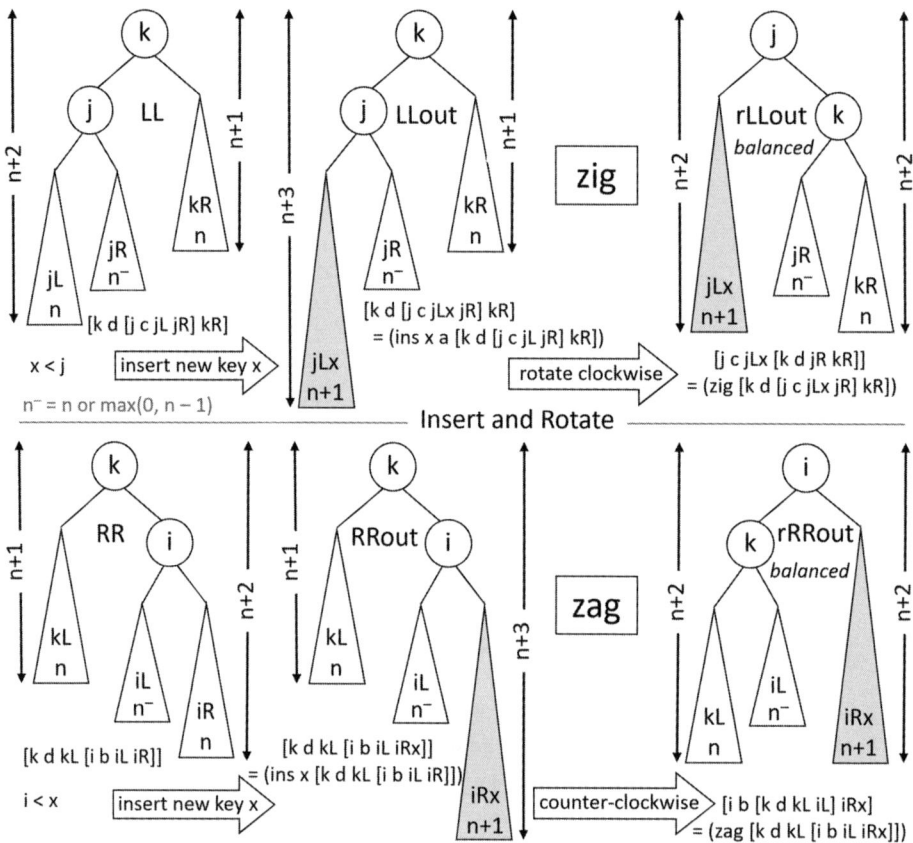

Figure 11.8
Insertion, unbalanced outcome, and rebalancing.

because in tree *LL*, subtree *jR* may be shorter than its sibling. So, we use the diagram of tree *LR* (where an insertion into subtree *jR* that increases its height will definitely produce a tree that is out of balance) to analyze the insertion of a key x that lies between j and k. The analysis of case *LR* covers subtrees *jL* both of height $(n - 1)$ and of height n.

Case *RR* mirrors case *LL* and deals with the situation when the new key x is greater than i ($k < i < x$). Case *RL* mirrors case *LR* and deals with the situation when the new key x is between k and i ($k < x < i$). To summarize, we start with $j < k < i$ and a new key x. The new key has to be in one of four places: $x < j < k$ (case *LL*), $j < x < k$ (case *LR*), $k < x < i$ (case *RL*), or $k < i < x$ (case *RR*). The four cases cover all of the possibilities, so an analysis of all four cases is a complete analysis of the insertion procedure.

Case *LL* is the problem of inserting a key x that is less than j into tree *LL*. The new key will go into subtree *jL*. Since the height, n, of *jL* is less than $n + 2$, the induction hypothesis says that the new subtree, after insertion, will be balanced and have height n or $n + 1$. If it has height n, the tree as a whole remains balanced and nothing further needs to be done. If it has height $n + 1$, the tree, as a whole, will be out of balance (too tall on the left). It will need rebalancing.

Figure 11.8 (page 216) displays tree *LL* before insertion (as it appeared in figure 11.7, page 215) and a new, unbalanced tree, *LLout*, which represents the outcome of hooking a new key x ($x < j$) into *LL*. In the diagram of *LLout*, subtree *jLx* is the one with the new key. Subtree *jLx* has height $n + 1$, which makes *LLout* too tall on the left. Figure 11.8 also presents the mirror insertion into *RR* of a key x ($i < x$), producing a tree labeled *RRout* that is too tall on the right.

Because these insertions lead to unbalanced trees, we need to rearrange them in some way, and figure 11.8 shows how to do that. In the case of *LLout*, the key j can go at the root, and the former root key, k, can hang from the right side of the new root. If we then plug the subtrees into the only places they can go to preserve order, the result, tree *rLLout* (figure 11.8), is balanced and has height $n + 2$. Voila! Like magic.

This clockwise rotation of tree *LLout*, as shown in the figure, is traditionally called zig. The mirror operation, zag, rotates counterclockwise to fix tree *RRout*, which is too tall on the right. Formal definitions of these rotation operators reside in figure 11.9 (page 218). The definitions emerge in a straightforward way from the diagrams in figure 11.8.

The rotation trick fixes unbalanced trees that come from cases *LL* and *RR* (page 215). That leaves us with the other two cases: case *LR* ($j < x < k$) and case *RL* ($k < x < i$). We turn our attention to these problems in the next section.

Exercises

1. Prove theorem {*unbal-ht3*}: If s is an unbalanced tree of height three, then one subtree of s is empty and the other has height two. That is, prove the following implication:

```
(defun zig (s) ; rotate clockwise
  (let* ((k   (key s))  (d  (dat s))
         (j   (key (lft s)))  (c  (dat (lft s)))
         (jL  (lft (lft s)))  (jR (rgt (lft s)))
         (kR  (rgt s)))
    (mktr j c jL (mktr k d jR kR))))
(defun zag (s) ; rotate counterclockwise
  (let* ((k   (key s))  (d  (dat s))
         (kL  (lft s))
         (i   (key (rgt s)))  (b  (dat (rgt s)))
         (iL  (lft (rgt s)))  (iR (rgt (rgt s)))))
    (mktr i b (mktr k d kL iL) iR)))
```

Figure 11.9
Rotation operators zig and zag.

$((($height $s) = 3) \wedge (\neg ($balp $s))) \rightarrow$
$(((($height (lft $s)) = 2) \wedge ($emptyp (rgt $s))) \vee (($emptyp (lft $s)) \wedge (($height (rgt $s)) = 2)))$

11.8 Double Rotations

Cases *LR* and *RL* (page 215) are the only two remaining cases that we need to cover to arrive at a complete solution of the insertion problem. The cases we already solved (cases *LL* and *RR*) are known as the "outside cases" because the subtrees that get the new keys and become too tall are on the outside borders of the diagram. Cases *LR* and *RL* are "inside cases" because the problematic subtrees are on the inside portion of the tree diagram, away from the borders. Another way to look at it is that x is outside of the interval between j and i in the outside cases and inside the interval in the inside cases.

Inside Left Case *LR*:   (ins x a *LR*), when $j < x < k$ *refer to*
Inside Right Case *RL*: (ins x a *RL*), when $k < x < i$ *figure 11.7, page 215*

The inside cases are trickier. Figure 11.10 (page 219) shows how insertions into trees *LR* and *RL* can make an inside subtree too tall.

Consider tree *LRin* in figure 11.10. A first guess might be that tree *LRin* could benefit from a clockwise rotation. However, that produces a tree with a left subtree of height n or $n + 1$ and a right subtree of height $n + 3$ (see figure 11.11, page 220). That is, applying the rotation operator zig to *LRin* leads to a tree that is at least as out of balance as *LRin*. Clockwise rotation doesn't help. We need a new trick to rebalance *LRin*.

The height of subtree *jRx* (figure 11.10) is $n + 1$, so at least one of its subtrees has height n. The diagram on the left side of figure 11.12 (page 221) shows *jRx* expanded to reveal its key y, left subtree *yL*, shown with height n^-, and right subtree *yR*, shown with height n. The heights of subtrees *yL* and *yR* are either the same or *yL* is one unit shorter. It could have been the other way around, with *yL* having height n and *yR*, height n^-, but the analysis is the same either way. We will pursue the alternative displayed in figure 11.12 and leave

11.8 Double Rotations

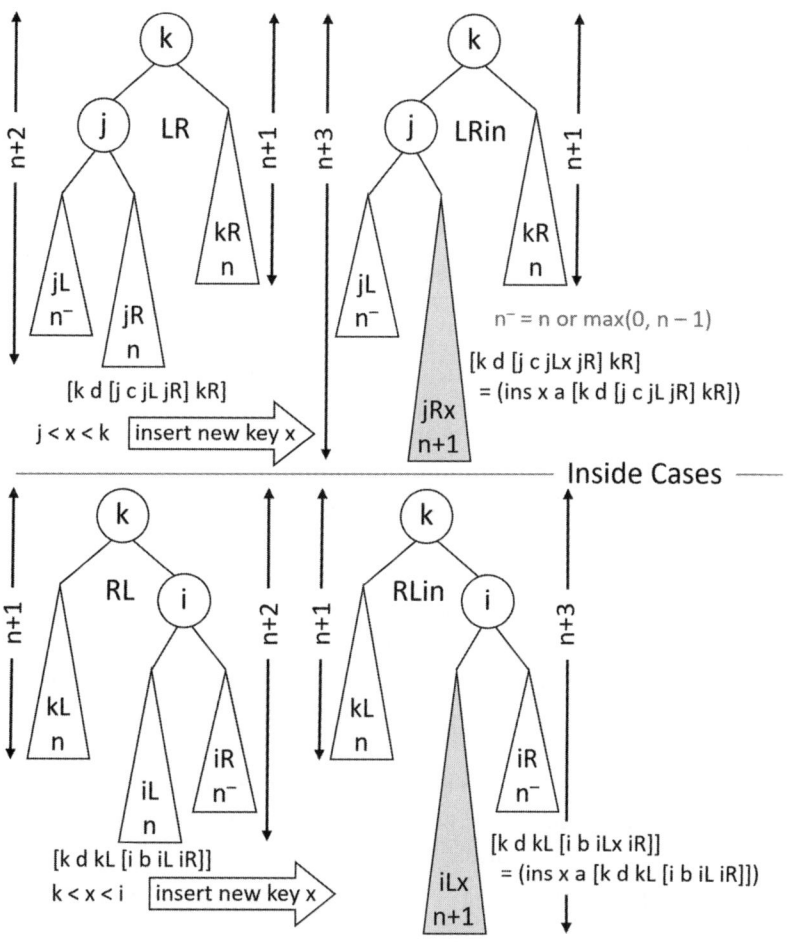

Figure 11.10
Inside cases.

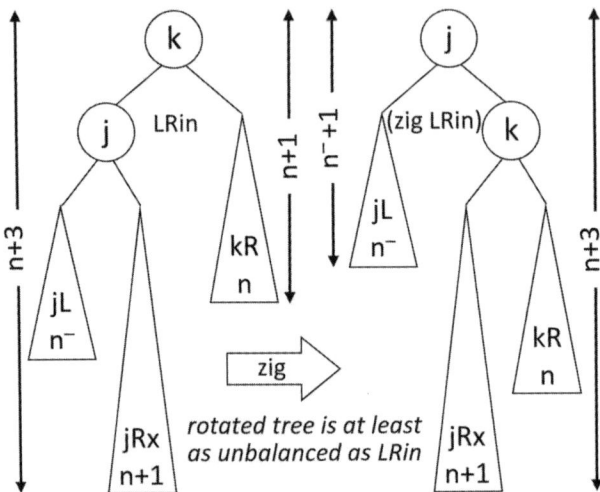

Figure 11.11
Using the wrong rotation can make it worse.

it to you to draw the diagram the other way and make sure it works. The exercise will be good practice.

The diagram in figure 11.12 (page 221) makes it clear that we can apply zag (counterclockwise rotation) to the left subtree of *LRin*. It's a counterintuitive move, might not help, might even make things worse, but it's possible, nonetheless.[72] The algebraic formula [*y p yL yR*] in figure 11.12 is the expansion of subtree *jRx* (figure 11.10, page 219) to display its key (*y*) and its subtrees (*yL* and *yR*). The formula uses *p* to denote the data associated with key *y*.

After applying the zag operator to the left subtree of tree *LRin* (figure 11.12), the new tree *rLRin* is displayed in the middle of the figure. The left subtree of the new tree *rLRin* has the following algebraic formula:

$$(\text{zag }(\text{lft } LRin)) = (\text{zag } [j \ c \ jL \ [y \ p \ yL \ yR]]) = [y \ p \ [j \ c \ jL \ yL] \ yR]$$

To repeat, starting from the unbalanced search tree *LRin* on the left side of figure 11.12, the middle portion of the figure displays tree *rLRin*, which is *LRin* with zag applied to its left subtree. Tree *rLRin* is not balanced, but the subtree that is too tall (namely, [*y p* [*j c jL yL*] *yR*]) is on the outside left part of *rLRin*. Therefore, we can apply zig to *rLRin* as a whole to get it back in balance.

---

[72] It would not be possible to do a zag rotation if *jRx* were empty.

11.8 Double Rotations

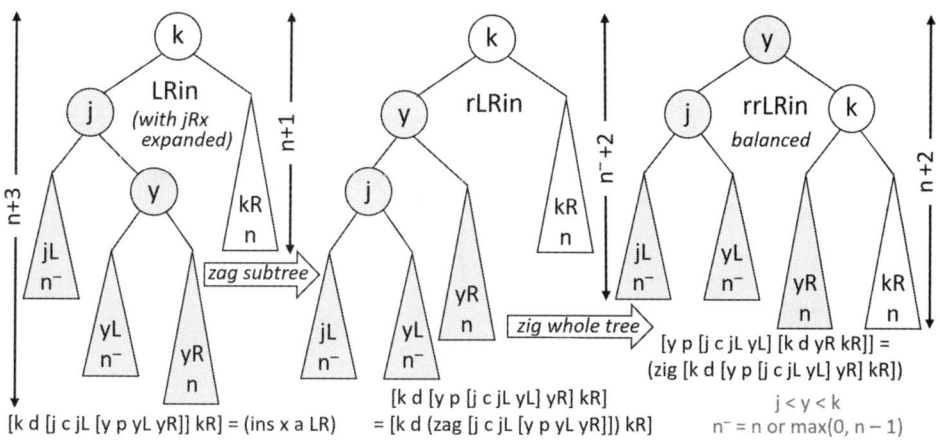

Figure 11.12
Double rotation rebalances inside cases.

```
(defun rot+ (s)                      ; rotate clockwise if too tall on left
  (let* ((k  (key s)) (d  (dat s))   ; rot+ assumes s is not empty
         (kL (lft s)) (kR (rgt s)))
    (if (> (height kL) (+ (height kR) 1))                ; unbalanced?
        (if (< (height (lft kL)) (height (rgt kL)))      ; inside lft?
            (zig(mktr k d (zag kL) kR))                  ; dbl rotate
            (zig s))                                     ; sngl rotate
        s)))                                             ; no rotate
```

Figure 11.13
Formal definition of the clockwise rotation operator.

$$(\text{zig } rLRin) = (\text{zig } [k\ d\ [y\ p\ [j\ c\ jL\ yL]\ yR]\ kR]) = [y\ p\ [j\ c\ jL\ yL]\ [k\ d\ yR\ kR]] = rrLRin$$

This second rotation leads to tree *rrLRin*, diagrammed on the right-hand side of figure 11.12, which is balanced and has height $n + 2$. Hallelujah! Wonders never cease.

Figure 11.13 formalizes the rebalancing of trees that have become too tall on the left due to the insertion of a key that is less than the key at the root. The operator **rot+**, which is defined in the figure, chooses whether to apply a single rotation, double rotation, or no rotation (depending on subtree heights) and uses the **zig** and **zag** operators (figure 11.9, page 218) to carry out the necessary rotations if there are any.

The operator **rot+** performs a clockwise rotation on a tree that is too tall on the left (that is, a tree with a left subtree whose height exceeds that of its right subtree by two). Assuming that, before rotation, the tree is ordered and that all of its subtrees are balanced, the tree that **rot+** delivers will be ordered and balanced, as our case-by-case analysis showed. If **rot+** is applied to a tree whose left subtree is the same height as its right subtree, or possibly one

```
(defun ins (x a s)
  (if (emptyp s)
      (mktr x a nil nil)                              ; one-node tree
      (let* ((k  (key s)) (d  (dat s))
             (kL (lft s)) (kR (rgt s)))
        (if (< k x)
            (rot+ (mktr k d (ins x a kL) kR))         ; insert left
            (if (> k x)
                (rot- (mktr k d kL (ins x a kR)))     ; insert right
                (mktr x a kL kR))))))                 ; new root data
```

Figure 11.14
Formal definition of the insertion operator.

unit taller, rot+ delivers its operand as-is because it is already a balanced tree. Likewise, rot+ delivers its operand as-is if the left subtree is shorter than the right subtree. Counterclockwise rotation is the mirror image of rot+. We will use the name rot- for the counterclockwise rotation operator and leave its definition as an exercise.

There are no other cases in which insertion can cause the tree to go out of balance, so the proof by induction of the order, balance, and height properties of the insertion operator, ins, is complete, and figure 11.14 displays the formal definition of ins.

The definition is inductive. The new key goes into the left subtree if it is smaller than the key at the root and into the right subtree if it is greater. If the new key is the same as the one at the root, the insertion operator puts the new data at the root with that key.[73] If the insertion produces, at first, an unbalanced tree, it is rebalanced by rot+ or rot-.

To make search trees really useful, we need to be able to delete items as well as insert them. Deletion is a little more complicated than insertion, but it uses the same basic rotation operators. You know enough now to explore the AVL deletion operator on your own. You can check it out online or in a textbook on algorithms.

Exercises

1. A footnote on page 220 points out that it is not possible to apply the zag operator to a tree whose right subtree is empty. Similarly, zig does not apply to a tree whose left subtree is empty. Explain why.

2. Prove that (ins *k d* nil) delivers a tree that is ordered and balanced.

3. Prove that (ins *k d s*) delivers an ordered, balanced tree if (height *s*) = 1.

4. Define rot- (see the definition of rot+, page 221).

---

[73] This treatment of the key/data association is as with the hook operator and for the same reasons (page 210).

11.9 Fast Insertion

```
(defun ht (s) ; extract height of tree s
  (if (empty s)
      0
      (fifth s)))
(defun mktr (k d lf rt) ; make tree from key, data, and subtrees
    (list k d lf rt h (+ 1 (max (ht lf) (ht rt)))))
(defun treep (s) ; search tree?
  (or (emptyp s)
      (and (n-element-list 5 s) (iskeyp (key s))
           (treep (lft s)) (treep (rgt s))
           (= (ht s) (+ 1 (max (ht(lft s)) (ht(rgt s)))))))) 
(defun rot+ (s) ; rotate clockwise if too tall on left
  (let* ((k    (key s)) (d    (dat s))  ; rot+ assumes s is not empty
         (kL (lft s)) (kR (rgt s)))
    (if (> (ht kL) (+ (ht kR) 1))                 ; unbalanced?
        (if (< (ht (lft kL)) (ht (rgt kL)))       ; inside lft?
            (zig(mktr k d (zag kL) kR))           ; dbl rotate
            (zig s))                              ; sngl rotate
        s)))                                      ; no rotate
```

Figure 11.15
Revised operators to avoid height computation.

11.9 Fast Insertion

The rotation operators rot+ and rot-, as discussed so far, compute the heights of the subtrees to choose the proper rotation. This takes a lot of time because it requires looking through all the ways to get from the root to the leaves.[74] To do this, all of the nodes must be examined, so the number of steps in the computation is proportional to the number of nodes in the tree. Furthermore, since ins has to check for possible rotations at every level in the tree, there are a great many height computations to perform, and this makes the insertion process take way too many computation steps.

Fortunately, there is a way to avoid the height computation by recording tree height in each node along with keys, data, and subtrees. When a new tree is formed from a key, data, and two subtrees, its height can be recorded quickly by extracting the heights of the subtrees, adding one to the larger of those heights, and recording the result with the key, data, and subtrees. We use the term "extracting" instead of "computing" because the height of a subtree is right there with the key. Extracting the height is a noninductive operation, just like extracting the key.[75]

---

[74] A leaf is a search tree whose left and right subtrees are both empty.

[75] The situation is even better than it seems because it's not the height that the rotation operators need to know. It's the difference between the heights of the subtrees, which is known as the *balance factor*. Because AVL trees are balanced, the balance factor will always be -1, 0, or $+1$. Keeping track of balance factors is no more difficult than keeping track of heights, and balance factors are a little more efficient in terms of time and space. However,

To use the approach that avoids the time-consuming computation of height, a few of the operators implementing the AVL solution (figure 11.2, page 207; figure 11.13, page 221) must change.

1. The mktr operator must put the height in the tree it builds.
2. The new height operator, ht, will simply extract the height that mktr records in the tree (rather than computing the height).
3. The treep predicate that is used to determine whether or not a given entity is a tree must account for the height record in the new representation.
4. The rot+ and rot- operators must refer to ht instead of to the old, slow height operator.

Figure 11.15 (page 223) formalizes the new versions of mktr, ht, treep, and rot+. The other operators remain unchanged, and the only changes needed in rot+ and rot- are that the invocations of the height function must refer to ht, the new height extraction operator, instead of to the operator that computed height in a time-consuming way.

The new definitions are no more complicated than the originals, but they make a huge difference in the speed of insertion. The number of computation steps required to insert a new element with the original definitions grows faster than the number of items in the tree. Recording heights directly in the tree rather than computing the height of the tree makes the number of computation steps for insertion proportional to the logarithm of the number of items in the tree.

To get a feeling for the difference, invoke the operator time-chk (defined as follows) for larger and larger trees, and measure the time it takes. Start with, say, 100 elements, then 200, 400, 800, and so on, doubling the size of the tree each time. Chart the number of elements against the time it takes to build the tree.

```
(defun build (n)
  (if (zp n)
      nil
      (ins n nil (build (- n 1)))))
(defun time-chk (n)
  (ht (build n)))
```

After you complete the timings with the slow version of ins, switch to the fast version and do the timings again. You will see that the slow version takes a long time when there are many keys. Inserting items one by one, starting from an empty tree, to build a tree with n keys takes time proportional to $n\ log(n)$ with the fast version of insertion. With the slow version, building a tree with n keys takes an amount of time that grows faster than n^2. These are the kinds of improvements in speed that make software useful in practice compared to software that would be infeasible to use in large applications.

going this route would require several more changes in the formal definitions, so we'll just stick with heights to avoid complicating the presentation.

Exercises

AVL trees are one of several kinds of self-balancing trees that support fast retrieval of data associated with keys. The various solutions to the problem have different advantages, but all of them make fast insertion, deletion, and retrieval possible.

Box 11.2
Checking Heights of AVL Trees

The reason the operator time-chk delivers only the height of the tree it builds rather than the tree itself is because it would take a lot of time and space to print the tree. We're just interested in the amount of time it takes, not the tree itself. We don't have a way to deliver the computation time directly, so we just write a formula that will cause the computation to take place. Then, we measure the amount of time the computer takes to do it with a stopwatch. This is a crude way to measure the performance of a piece of software, but we are only interested in ballpark estimates, so it will do for our purposes.

The height of the tree that (time-chk n) delivers should be less than $\frac{3}{2} \cdot log_2(n + 1)$. If it's not, something is wrong. An approximate way to use ACL2 to check tree heights for plausibility is to invoke the predicate height-rightp, defined as follows, whose value should be true:

```
(defun log2-ceiling (n)
  (if (posp (- n 1))
      (+ 1 (log2-ceiling(floor (+ n 1) 2)))
      0))
(defun height-rightp (n)
  (let* ((h   (ht (build n))))
    (<   (/ (* 3 (log2-ceiling n)) 2))))
```

Exercises

1. Carry out the timing experiment described in this section and report the results.

2. Observe, experimentally, using the operators defined in box 11.2, that the height of an AVL tree containing n keys does not exceed $\frac{3}{2} \cdot log_2(n + 1)$. Your experiment should include many observations.

12 Hash Tables

In previous chapters, we have seen how lists can be used to store multiple values, such as all the students in a class or all the books in a library. Operators provide convenient access to the first element in a list and to the elements after it, in sequence one by one. This is fine when you want to process all the elements in the list, but what if you're only interested in the student with ID #93574 or Mary Shelley's *Frankenstein; or, The Modern Prometheus*? Computer scientists have designed many ways to get fast and convenient access to individual records out of large collections. In this chapter we'll study a solution known as *hashing*.

12.1 Lists and Arrays

Suppose we have a list of states and their capitals.

[["Alabama"   "Montgomery"]
  ["Alaska"   "Juneau"]
  ["Arizona"   "Phoenix"]
       ⋮
  ["Wyoming"   "Cheyenne"]]

Figure 12.1 (page 228) defines an operator to find the capital of any given state given such a list. The operator works, but it's slow. It takes only a handful of steps to find the capital of Alabama, but it takes fifty steps to find the capital of Wyoming. Some states are near the front of the list, others far down the list. If a query is as likely to ask about one state as another, then on average it will take twenty-five steps to find the state capital requested by a query. The situation would be worse for a bigger problem, such as finding the population of a city in a list of city/population pairs. In other words, the solution works but doesn't scale well.

Chapter 11 discusses the binary search method (page 202), which is a way to speed up the process of finding an item (such as a capital city) associated with a search key (such as a state name) in a collection of items. The chapter goes on to discuss a data structure

Axioms for Finding State Capitals

(capital *s* nil) = nil		{*cap0*}
(capital *s* (cons [*state city*] *caps*)) = *city*	if *s* = *state*	{*cap1*}
(capital *s* (cons [*state city*] *caps*)) = (capital *s caps*)	if *s* ≠ *state*	{*cap2*}

```
(defun capital (s caps)
  (if (consp caps)
      (if (equal (first (car caps)) s)
          (car (rest (first caps)))   ; {cap1}
          (capital s (rest caps)))    ; {cap2}
      nil))                           ; {cap0}
```

Figure 12.1
Operator to find state capitals.

known as a binary tree that provides an effective solution to the search problem.[76] Binary trees would make it possible to find state capitals faster than the operator of figure 12.1. The states and capitals would need to be stored in a special way, not in an ordinary list, but the number of steps required to find a capital would drop from an average of twenty-five to a maximum of six. That's four times faster, but there is a method known as hashing that cuts the search to something close to one step, or at least to a small number of steps, regardless of how many search keys are in the data. Hashing works well for data sets that don't change often, such as state capitals. It works less well for large data sets that change frequently, such as tweets. You have to choose a solution that fits the problem. Hashing provides one alternative.

Part of the problem with using ACL2 lists for data is that it is easy and fast to retrieve the first element of a list but slower to retrieve the nth element of a list. However, that's not the whole problem. It also takes time to compare search keys (state names, in the problem at hand). Hashing addresses that problem by converting the search key into a numeric index and putting the data in an array providing one-step access by index.

There is an ACL2 operator called nth that, given an index and a list, delivers the nth element of the list. Indexes start at zero, so the index zero selects the first element, the index one selects the second element, and so on.[77] If we let **states** stand for the list of

---

[76] Chapter 11 discusses a particular kind of binary tree known as an AVL tree, but that is just one of many kinds of trees that can be used to solve the problem of searching for data associated with keys.

[77] Most people count things starting from the number one (1, 2, 3, ...), but computer scientists usually start from zero (0, 1, 2, 3, ...) for reasons that have to do with the simplicity of formulas for indexes in some contexts, such as when an index selects a polynomial coefficient or when elements are grouped in blocks. It seems odd to count from zero instead of from one, but that's the way the operator nth does its indexing. We're stuck with it.

12.2 Hash Operators

states and capitals (page 227), then (nth 1 states) is ["Alaska" "Juneau"], whereas (nth 0 states) is ["Alabama" "Montgomery"].

$$\text{Axioms for nth operator} \quad \begin{array}{ll} (\text{nth } 0 \ xs) = (\text{first } xs) & \{nth0\} \\ (\text{nth } n+1 \ xs) = (\text{nth } n \ (\text{rest } xs)) & \{nth1\} \end{array}$$

However, the nth operator is a slow way to reach a list-element that has a high index. An array is a data structure that is similar to a list, but it provides fast access to its elements, including those with a high index. If the list of states and capitals were recorded in an array instead of a list, it would be just as fast to find the entry for the fiftieth state (Wyoming) as it is to find the entry for the first state (Alabama).

Suppose the operator nth could be implemented in such a way that it would not take any longer to deliver the element with index 1,000,000 than it would take to deliver the element with index 0. Let's not worry about how that could be done, but just assume for the moment that ACL2 could do it. If nth were such an operator, how much faster would the operator capital be?

A fast implementation of nth is part of the answer but not a complete solution to finding state capitals with a one-step operation. We still need to find the right state. That is, it still takes many steps to find Wyoming. Hashing solves the problem of finding any search key fast.

Box 12.1
Arrays and ACL2

Lists are one way to maintain collections of data in ACL2, but not the only way. A data structure known as an array is another. It comes with an operator called aref1 that retrieves values from array elements. It is similar to the hypothetical "fast nth" operator discussed in this section. However, using ACL2 arrays in a way that ensures that aref1 retrieves elements fast is a complicated technical procedure. Acquiring the required ACL2 expertise to use arrays would take us too far afield, so we will rely on a descriptive approach that omits the details.

12.2 Hash Operators

We are looking for a definition of the operator capital that, given a state, finds the right entry in the list of state/capital pairs. We will assume that the formula (nth n xs) completes its retrieval of element n from the list xs in one step. That is, we are assuming that a hypothetical, fast version of nth exists. It doesn't, but the array facilities in ACL2 make an operator like that possible, so when we write a formula like (nth n xs), we think of xs as an array and assume that nth is a one-step operation.

Our first solution to the search problem is alphabetical. It's not a complete solution, but will help us move in that direction. We will use an array of 26 elements and put all states

Letterbox Array: fcaps

Index	State Capitals
0	[["Alabama" "Montgomery"] ["Alaska" "Juneau"] ... ...["Arkansas" "Little Rock"]]
1	[ ]
2	[["California" "Sacramento"] ["Colorado" "Denver"] ... ...["Connecticut" "Hartford"]]
3	[["Delaware" "Dover"]]
⋮	⋮

Axioms for Capital Search Operator: fcapital

(fcapital *s* fcaps) = (lookup *s* (nth (state-idx *s*) fcaps))		{*fcap*}
(lookup *s* nil) = nil		{*look0*}
(lookup *s* (cons (cons *state city*) *caps*)) = *city*	if *s* = *state*	{*look1*}
(lookup *s* (cons (cons *state city*) *caps*)) = (lookup *s caps*)	if *s* ≠ *state*	{*look2*}

Figure 12.2
Letterbox array and capital search operator.

that start with the letter A in the element with index zero, the ones that start with B in the element with index one, and so on. Since there are 50 states and only 26 elements in the array, some elements will have more than one state, so searching as before will still be necessary but shorter. Finding Kansas, for example, will be a matter of retrieving the array element associated with the letter K, then sorting through the states in that entry. There are two states that start with K, so it will take at most two more steps after retrieving the array element for K. Let's assume that's about average.[78] That is, after retrieving the array element corresponding to the first letter in the name of the state, there will be typically about two states in the box.

Let's call this an array of letterboxes (one array element for each letter in the alphabet) and give it the name fcaps. It represents the same information as the earlier list, states, but in a form that makes finding a state capital faster. We call the new operator for finding the capital of a state fcapital ("f" for fast). It delivers the same results as the previous operator, capital, but it does so in fewer steps because the information in its array operand,

---

[78] If the first letters in state names were evenly distributed across the alphabet, there would be two entries in most elements of the array, so the number of steps for Kansas would be more or less typical. Of course, some letters are associated with more states than others, so it's not always going to be a matter of finding the letter and picking between two states. There are eight states that start with M, so it can get bad, but the average for array elements with states in them is about 2.6 states per array element, so Kansas isn't far from average.

12.2 Hash Operators

fcaps, is arranged to shorten the search. Figure 12.2 (page 230) suggests the contents of a few of the letterboxes in fcaps and presents some equations defining fcapital.

The fcapital equations refer to an operator called lookup that, like the operator capital, returns the matching element from a list of state/capital pairs. We could have stuck with the previous operator, but lookup searches short lists, typically one to three elements, whereas capital had to deal with all fifty states. In a working system, a better definition of lookup would take advantage of the special nature of the data set.

We want to focus on the definition of fcapital, which does most of the work. The operator fcapital invokes lookup on the letterbox corresponding to the first letter in the name of the state. The operator state-idx computes the index of the letterbox given the name of a state: 0 for Alabama or Arkansas, 2 for California, 3 for Delaware, ... 22 for Wyoming. We will leave the concrete definition of state-idx as an exercise for people who want to delve into the arcane details of operations like retrieving the first letter from a string.[79] Those details are necessary to build a working system, but they don't add much to the discussion of ways to retrieve data quickly from a data set.

The formula (nth (state-idx s) fcaps) retrieves a letterbox from the array fcaps. The version of nth that we refer to here is the hypothetical operator that we discussed earlier, which in a single step delivers an element with the specified index from an array. We didn't say how that operator worked. It can be done. We leave it at that.

We want to figure out how well the idea of having one box for each letter in the alphabet works. There are 50 states and 26 letters, but the letters occur with different frequencies in state names. There are four states that start with A and eight that start with M but none that start with B or E. It would be better to put the same number of states in each array element. We could, for example, have 25 boxes and put two states in each box. Or, we could have 50 boxes and put one state in each box. We could even have 100 boxes and put one state in 50 of the boxes and no states in the other 50. Sounds stupid but it's not, depending on how the state-idx operator is defined, and we will talk about that shortly.

In any case, the lookup operation could be faster if the states were distributed more evenly among the boxes. To make that work, the definition state-idx would need to be more sophisticated than just going by the first letter in the name of the state. It would need to use more letters in the name to compute the index of the box containing the corresponding state/capital pair.

Choosing a way to distribute the states among the boxes and defining a special version of state-idx to compute the index of the right box is more art than science. This kind of problem comes up in many kinds of table-lookup problems. We are using state capitals as an example to discuss the issues, but the problem of looking up information associated

---

[79] We use the term *string* to mean a sequence of characters. Most programming languages, including ACL2, denote strings by putting double-quote marks around the characters: "This is a string."

with URLs is the same, only bigger and more complicated. For example, the first ten characters in the URL http://www.apple.com/ are the same as the first ten characters in http://www.cnn.com/. The last five characters are the same in both URLs, too. So, if we distribute all the URLs into an array of URL-boxes, defining an operator, say URL-idx, to compute the index of the box containing a given URL, then the operator URL-idx should take all of the characters in the URL into consideration. We have the same problem with state names but on a smaller scale, so let's get back to that discussion.

There are many ways to set up the letterboxes. They are not "letterboxes" anymore but rather boxes set up to hold state/capital pairs distributed into those boxes using more than just the first letter in the name of the state. We will refer to them as the *buckets* of the *hash table*. (Yes, we're going to call them buckets. It's hashing terminology.) And the definition of state-idx must be tailored to the way the states are distributed among the buckets. Some distributions of states and corresponding definitions of state-idx work better than others. The key is to look at more letters in the name of the state than just the first one. Choosing the details of the distribution of keys (state names in our example) and the definition of the index operator (state-idx in our example) is where the art comes in. The trick is to choose a distribution and operator definition that makes for a fast state-idx operation.

Let's discuss this problem in a more general context where the keys are strings of characters that we will call *words*, even though they might be odd-looking strings like URLs. A common solution is to use two steps. First, the word, which we refer to as the search key, is converted into a list of numbers by mapping each character to a number. For example, we could convert A to 1, B to 2, C to 3, and so on.[80] Using the A 1, B 2, C 3 scheme, the word "life" becomes the list [12 9 6 5] and the word "file" becomes [6 9 12 5].

The second step is to take the list of numbers corresponding to the letters in the search key and combine them into a single number called the *hash key*. The operator that computes the hash key is called the *hash operator*. Ideally, the hash operator associates a different hash key with each search key. One way to compute a hash key from the list of numbers corresponding to the letters in the search key is to add multiples of powers of a prime number, which is called the *hash base*.[81]

If we choose the prime number 31 as the hash base, the list [12 9 6 5] maps to the hash key $12 + 9 \times 31 + 6 \times 31^2 + 5 \times 31^3 = 155{,}012$. The numbers 12, 9, 6, and 5 are used as coefficients to powers of the hash base.[82] The list [6 9 12 5] maps to the hash key $6 + 9 \times 31 + 12 \times 31^2 + 5 \times 31^3 = 160{,}772$. That might be okay if there were hundreds of

---

[80] This is not the usual scheme, and we're conflating uppercase and lowercase letters, but other number/letter conversions will also work.

[81] Why choose a prime? There is a lot of number theory involved in these kinds of problems, but that's a story for a different book. We'll leave it as a mystery.

[82] This formula is similar to the ones discussed in chapter 6 for converting base 10 and base 2 numerals to numbers. It may happen that some of the digits in the numeral fall outside the usual range for base 31 digits, but

thousands of words that need to be associated with indexes, as there would be if the words were URLs, for example.

In the problem at hand, there are fifty states, so there are fifty search keys. Let's get back to that again. The fcapital operator uses the number returned from state-idx as an index to an element in the array containing the (short) lists of state/capital pairs, in which lookup will find the capital of the state. A hash table with 160,772 elements is overkill for a problem with fifty search keys. We had a hash table with 26 elements when we used the first letter of the state. Suppose we look for a solution with a hash table of one hundred elements. Then, we would need to compute a hash index between zero and ninety-nine from the hash key. One way to do that would be to consider only the last two digits of the hash key. Then, the hash index for the hash key 160,772 would be 72. The word "life" (not the name of a state, but you get the idea) would go into the bucket with index 72. There's nothing special about 100. If we wanted a fifty-bucket solution, we could compute the index from the hash key by taking the remainder in the division[83] by 50: (160,772 *mod* 50) = 22. The method works for a hash table with any number of buckets.

To pick a hash base and hash table size that works well is a matter of analyzing how the hash keys come out for the words we want to look up. If we want to extend this problem to take into account Canadian provinces or all the countries in the world, we could use an array with, say, a thousand elements. The hash operator could compute a hash key h from a search key, as before, then deliver (h *mod* 1,000) as the hash index.

We've discussed a lot of ideas in this chapter that are important and practical in computer science. Let's take a moment to recap these ideas. The general problem we set out to solve was how to quickly find a value associated with a name. In general, we think of having many key/value pairs, and what we'd like to do is to find the *value* associated with a given *key*. This general problem is so important that many computer languages offer ready-made solutions, which are called by various names: *associative arrays*, *dictionaries*, *maps*, *memories*, and others. We call them hash tables in this discussion, but they are all similar ideas.

The first solution we discussed was to keep all the key/value pairs in a single list and search for keys by going through the list one by one. This works, but it is too slow to scale up. On average, if search keys all have about the same likelihood of being the target of a search, the search will have to look at roughly half of the key/value pairs before it finds the one it's looking for. For many real-world problems, that is too long. A hash table alleviates this problem by splitting the list of key/value pairs into many smaller lists. Each of these lists goes into one of the buckets of the hash table. With this approach, the time required

---

the method in chapter 6 for computing numbers from numerals works for the hash-key computation, too. Not in the other direction, though. The hash coefficients can't be computed from the hash key.

[83] Box 6.3 (page 126) discusses division, remainders, and the mod operator.

to locate a key will be, at worst, proportional to the number of search keys in the biggest bucket, which is much smaller than the total number of search keys. There is a trade-off between the number of buckets and the size of the buckets. Smaller buckets, faster search, but more total space because, depending on how the hash index is computed, there can be a lot of empty buckets that take up space even though they are empty.

The big question is how to distribute the key/value pairs into the various buckets. That is determined by the hash operator and the computation of the hash index from the hash key. The goal in designing hash operators and hash tables is to put about the same number of search keys in each bucket. A popular hash operator for strings is to treat each character as a number, to use these numbers as the coefficients for a sum of powers of 31, and then to trim that number to a hash index by computing the remainder in the division by the number of buckets in the hash table. This overall process is illustrated in figure 12.3 (page 236).

In our state/capital example, the search keys were state names and the associated values were capital cities. Using a hash table with ten buckets and the hash operator we discussed, the key/value pairs would be associated with the buckets as seen in table 12.1 (page 235). As the table shows, the states are not distributed evenly in the buckets. Bucket 7 has eight states, but bucket 4 has only two. Keeping the same hash operator but using thirty buckets instead of ten does a better job of distributing states to buckets. Keeping the hash operator separate from the number of buckets makes it easy to change the size of the hash table, depending on how fast the search needs to be.[84]

Exercises

1. Consider the words "age," "cage," and "cape." Using the hash operator from this chapter, compute the hash key of each word and the bucket the word maps to, assuming there are ten buckets.

2. There are many websites on the internet that contain lists of words: most popular baby names, most popular English words, and so on.
    https://www.babycenter.com/top-baby-names-2017.htm
    https://www.englishclub.com/vocabulary/common-words-100.htm
   Pick a list of words and explore how the hash operator of this chapter works. Use 31 for the hash base and try different sizes for the hash table. In your project, try to answer the following questions:

   a) How many search keys are there in the biggest bucket?

   b) What is the average number of search keys in a bucket?

---

[84] For small problems like the state/capital problem, there are mechanized ways to design hash operators and hash arrays that lead to *perfect hashes* with no more than one key in each bucket. Sometimes there are empty buckets, but that doesn't affect the search time. It just takes up more space. Some of the methods of finding perfect hashes also provide ways to economize on the number of buckets. Hashing is an important and well-studied search method. You might find it interesting to look into the topic in more depth.

Exercises

Bucket	[key value] Pairs
0	["Washington" "Olympia"] ["Utah" "Salt Lake City"] ["Rhode Island" "Providence"] ["New Hampshire" "Concord"] ["Minnesota" "St Paul"] ["Kentucky" "Frankfort"]
1	["Nebraska" "Lincoln"] ["Louisiana" "Baton Rouge"] ["Hawaii" "Honolulu"] ["Alabama" "Montgomery"]
2	["Pennsylvania" "Harrisburg"] ["New York" "Albany"] ["New Mexico" "Santa Fe"] ["Maine" "Augusta"] ["Indiana" "Indianapolis"] ["Georgia" "Atlanta"]
3	["Missouri" "Jefferson City"] ["Colorado" "Denver"]
4	["Oregon" "Salem"] ["Michigan" "Lansing"] ["Arkansas" "Little Rock"] ["Arizona" "Phoenix"]
5	["Wisconsin" "Madison"] ["New Jersey" "Trenton"] ["Kansas" "Topeka"] ["Florida" "Tallahassee"] ["Alaska" "Juneau"]
6	["Wyoming" "Cheyenne"] ["Tennessee" "Nashville"] ["South Dakota" "Pierre"] ["Oklahoma" "Oklahoma City"]
7	["West Virginia" "Charleston"] ["Vermont" "Montpelier"] ["South Carolina" "Columbia"] ["Ohio" "Columbus"] ["Nevada" "Carson City"] ["Mississippi" "Jackson"] ["Idaho" "Boise"] ["Connecticut" "Hartford"]
8	["North Dakota" "Bismarck"] ["Montana" "Helena"] ["Massachusetts" "Boston"] ["Maryland" "Annapolis"] ["Iowa" "Des Moines"] ["California" "Sacramento"]
9	["Virginia" "Richmond"] ["Texas" "Austin"] ["North Carolina" "Raleigh"] ["Illinois" "Springfield"] ["Delaware" "Dover"]

Table 12.1
Ten-bucket hash table for state capitals.

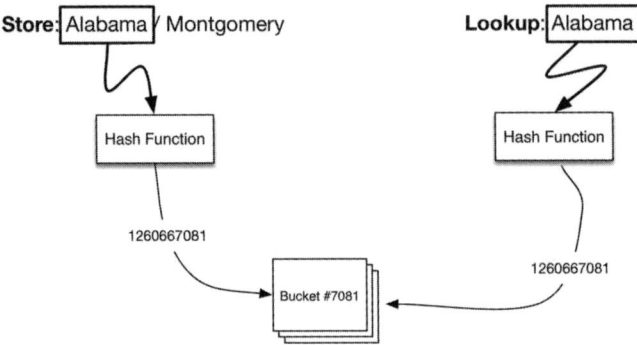

Figure 12.3
Hash table storage and retrieval.

    c) How many empty buckets are there?

    d) What is the average length of a nonempty bucket?

    e) Which one of these measures best estimates how well this hash works?

You can use the operators defined in figure 12.4 (page 237) to get started on the project. The operators use lists instead of arrays, so they are only prototypes for experimentation, not for practical use. The definitions invoke some ACL2 intrinsic operators that have not been discussed, such as coerce and char-code. The coerce operator converts a string to a list of characters, and the char-code operator converts a character to a number. The numbers are not the A 1, B 2, C 3, ..., but that is immaterial. Different characters are associated with different numbers, which is the important thing. If you are interested, you can find more information about these intrinsic operators in the ACL2 online documentation.

12.3 Some Applications

Hash tables provide an effective solution to the search problem, so it is no surprise that they are all around us. Hash tables are used behind the scenes in all sorts of applications. They are ubiquitous. Let's discuss some examples.

    Hashing facilitates finding the definitions of operators to run computer programs. When you run a computer program, the computer system needs to find the definitions of the operators the program uses. For example, the computer needs to know the definition of append to compute the value of (append [1 2] [3 4]). A lot of searches of this kind are required to run a program, so a fast lookup procedure is important.

    Hash tables offer a way to do it. Store the operator definitions in a hash table where the key is the name of the operator and the value is the definition. This is almost certainly the solution that your computer system uses when it runs your ACL2 programs. Another

12.3 Some Applications

```
(defun hash-op (hash-base chars)
  (if (consp chars)
      (+ (char-code (first chars))
         (* hash-base (hash-op hash-base (rest chars))))
      0))
(defun hash-key (hash-base wrd)
  (hash-op hash-base (coerce wrd 'list)))
(defun hash-idx (hash-base num-bkts wrd)
  (mod (hash-key hash-base wrd) num-bkts))
(defun rep (n x)
  (if (posp n)
      (cons x (rep (- n 1) x))  ; {rep1}
      nil))                      ; {rep0}
(defun bump-bkt (idx bkts)
  (if (posp idx)
      (cons (first bkts) (bump-bkt (- idx 1) (rest bkts)))
      (cons (+ 1 (first bkts)) (rest bkts))))
(defun fill-bkt-counts (hash-base num-bkts bkts wrds)
  (if (consp wrds)
      (let* ((idx (hash-idx hash-base num-bkts (first wrds)))
             (newbs (bump-bkt idx bkts)))
        (fill-bkt-counts hash-base num-bkts newbs (rest wrds)))
      bkts))
(defun hash-bucket-sizes (hash-base num-bkts wrds)
  (fill-bkt-counts hash-base num-bkts (rep num-bkts 0) wrds))
(defconst *example-tbl-25-most-common-English-words*
  (hash-bucket-sizes 31 10
    (list "the" "be" "to" "of" "and" "a" "in" "that" "have" "I"
          "it" "for" "not" "with" "he" "as" "you" "do" "at" "this"
          "but" "his" "by" "from" "they")))
```

Figure 12.4
Prototype hashing operators (slow: using lists, not arrays).

viable alternative is to use a binary search tree (chapter 11), but hash tables work better in this situation because there are a moderate number of search keys (operator names), and they don't change very often. With a binary search approach and, say, a few hundred keys, a binary search would have five or ten computational steps, while a good hash operator and corresponding hash table might find operator names several times faster.

The number of steps in a binary search is proportional to the logarithm of the number of keys. That's a lot less than the number of keys, but it does grow slowly as the number of keys increases. The amount of time it takes to perform a search using a hash table is proportional to the number of keys in the biggest bucket. Of course, the computation of the hash key and bucket index also has to be factored in, but whether there are hundreds of search keys or thousands, the time it takes to find a search key stays about the same.

Database systems, the original killer app of computing, are another common use of hashing. Some of the earliest databases were used to keep track of airline reservations, and now databases are used to store and process all kinds of data: student records at a university, transactions at every register in every Walmart, cast members for every movie produced in Hollywood, lifetime medical records for billions of people. Organizations use databases to store literally thousands of related records for each person involved in the system, which can easily amount to billions of records for an organization as a whole.

But what is a database system? At its core, a database system manages one or more tables, and each table is made up of one or more records consisting of several attributes. Think of a database table as a spreadsheet, where each record corresponds to a row in the spreadsheet and each column corresponds to an attribute. If the table records states and their capitals, each database record (or row in the spreadsheet) would have one attribute for the state name and another attribute for the capital.

It is common to design a database table in such a way that the data in a particular attribute is sufficient to single out a specific record in the database. For example, a table containing student records might use the attribute "Student ID" to identify each student, so there would be a column in the table containing student IDs. Given a student ID, the database system can find the record for that student in the table. Part of the magic of database systems is that they can retrieve records quickly using specialized data structures called *database indexes*. Database systems use indexes to retrieve specific records quickly, even when the table has millions of records that are stored in separate files on disk drives.

So, what is a database index? Database systems offer many different kinds of indexes, but there are two kinds that are the most common. One is based on hashing, the other on trees. In a sense, a database table is a key/value pair, where the key is an attribute that is sufficient to identify a particular record and the value is the record it identifies.

Hash-based and tree-based indexes in databases are, in essence, hash tables similar to those in this chapter and trees similar to those in chapter 11. The only substantial difference is that our trees and hash tables are ACL2 objects that reside in the fast-access memory of a computer, whereas database indexes are designed to retrieve records that can be stored in very large files.

However, database systems go beyond storing and retrieving records using a key. They also excel at finding information by combining records in different tables. For instance, one table may have information regarding a student's permanent address and another table may have information listing the students enrolled in any given course. A database query can combine these tables to find the zip codes that the students in any given course come from, perhaps showing that students in one region of the country are more likely than those in another region to enroll in history courses. That's the sort of insight that data scientists find valuable.

Combining separate tables in a database can be expensive. One way to proceed is to consider all possible pairs. In the zip code example, you could look at each course one at a time, and for each course consider all the students who could be enrolled in it. At a mid-sized university with 13,000 students and 80,000 enrollment records, that would require looking at over a billion combinations (1,040,000,000, actually).

Hash tables provide a simpler alternative. The student information table and the course enrollment table are connected through the student ID. So, before combining these tables, it is advantageous to hash both of them using the student ID. Once this is done, each bucket contains some entries from the student table and some from the course enrollment table. The entries in each bucket must be considered exhaustively, but there is less work to do than before. To see this, suppose that there are 1,000 buckets, each having an average of 13 students and 80 courses. Each bucket contains $13 \times 80 = 1{,}040$ combinations, for a total of 1,040,000 combinations. That makes retrieval a thousand times faster than the direct approach.

The reason that hash tables work so well in this context is that they take a big problem and turn it into many smaller problems. Instead of combining two tables of size 13,000 and 80,000, you can combine 1,000 tables of size 13 and 80. This savings is significant in itself, but it can be even more dramatic if the smaller problems can be performed on different computers. For example, if you have 1,000 computers, each one of them can be working on a different bucket, so the total time to find the answer is just the time to consider 1,040 combinations, which is a million times faster than the direct approach. This is the kind of thing that companies with very large databases do—Google, Facebook, and Amazon, for example. They use thousands of computers to process databases, and hash tables are the key to spreading the work evenly across the available computers.

Another important application of the hashing idea focuses on hash operators rather than hash tables. Suppose someone were to send you a very large file. After waiting a few hours for the file to transfer to your computer, how do you know that the file you received is the same as the one that was sent? It is possible that one of the characters in the file was transmitted incorrectly because of a network glitch. Or perhaps an intruder arranged to give you a false copy of the file.

Hash operators provide a way to address this problem. Before sending the file, the sender could use a hash operator to compute a hash value for the file. After you receive it, you would use the same hash operator to create a hash key, and then you could compare the keys to make sure they're identical. While it is *possible* that two files end up with the same hash key, a good hash operator makes it extremely unlikely. Depending on how important it is, the hash operator can be designed to set the bar at any level. The odds against ending up with the same hash key, given two different files, can be a thousand to one, or a million to one, or a billion to one. The required level of security determines what the odds against intrusion should be. Higher security costs more, but the sensitivity of the information can justify the cost. Variations of this idea are behind digital signatures and also behind procedures for keeping multiple copies of a file synchronized with one another.

Hash operators and hash tables are central to practical, large-scale computing. We'll consider more aspects of large-scale, practical computing in part IV of this book.

IV Computation in Practice

13 Sharding with Facebook

Facebook, with billions of users and a profound, worldwide impact on people, social interaction, commerce, and institutions, is a major force far beyond its beginnings as a way for college students to socialize online. Having so many users poses tremendous technical challenges, and Facebook's success is partly due to the ability of its engineers to deal with these problems. Many important technical innovations have made the Facebook website possible. Some of them are discussed in this chapter.

13.1 The Technical Challenge

To understand the challenges facing Facebook, let's consider just one of its main features. Facebook users can post frequent updates of their doings, and they can view the postings of their friends through online connections accessible from ubiquitous devices such as laptops, tablets, and cell phones. Making this possible requires two lists of information for each user: (1) a list of the user's friends and (2) a list of the user's status updates.

It's not unusual for these lists to have hundreds of items, so Facebook has to make hundreds of billions of items accessible, quickly, online to billions of people. That's this year. Next year, it may be trillions of items.

Traditionally, this data would be stored in a database using two tables, one for the status updates and one for the friends. Figure 13.1 (page 244) shows what these tables could look like. Database software inserts, deletes, and modifies rows on these tables as they evolve, and supports sophisticated queries that can retrieve information from the tables. For example, Facebook could use the following query to determine what status updates to display when a specific user logs in:

```
SELECT    s.User, s.Time, s.Status
FROM      Friends f, Statuses s
WHERE     f.User='John'
  AND     f.Friend = s.User
ORDER BY  s.Time DESC
```

FRIENDS		STATUSES		
User	Friend	User	Time	Status
John	Sally	John	Apr 21, 2011, 10:27 am	Checked into Starbucks in Norman.
John	Mary			
Mary	Sally			
Sally	David	Sally	Apr 21, 2011, 10:29 am	Saw *Battle: Los Angeles* last night. What a waste!
		Mary	Apr 21, 2011, 10:32 am	Is anybody going to the carnival this weekend?
		Sally	Apr 21, 2011, 10:33 am	Looks like the fires are getting closer to our house. Thinking about evacuating.

Figure 13.1
Tables for Facebook-style status updates.

This approach would work very well if Facebook had a small number of users, but it does not scale to billions of users. The problem is that a traditional database cannot process this query quickly enough to keep up with a continuous flood of online demand.

Facebook is not the only company that has this problem. For example, Amazon offers millions of items for sale. It encourages customers to write reviews for each item, and it keeps a history of purchases made by its customers so it can keep them informed about the status of orders, make suggestions for other purchases, and provide other services. If the information were stored in traditional database tables, retrieving it would take too long.

In fact, this is a problem faced by any Web 2.0 organization.[85] The web content that their users produce can take the form of passages created directly by users, such as Facebook updates or Amazon reviews, or it can be content created indirectly, such as Amazon purchase recommendations. When a Web 2.0 company is successful, its users produce much more content than traditional databases can handle.

---

[85] A Web 2.0 organization is one that allows large numbers of users to produce content for its website.

13.2 Stopgap Remedies

13.2.1 Caching

Web 2.0 companies cannot use traditional databases because, for example, it would take too long to retrieve information needed to display a user's welcome screen. One way to address this problem is to limit the number of queries that the database needs to carry out. For example, if many users check their home screens several times in a span of a few minutes, the computer can remember the results of previous inquiries instead of asking the database to retrieve the home screen every time a user wants to check it.

That's called *caching* the data. A cache is a key/value store in which data (values) are associated with keys that facilitate locating the data, like the search-tree keys of chapter 11 or the hash keys of chapter 12. When a query arrives, the database first checks the cache to see if the information is already there. If it is, it is simply reused. Otherwise, the database performs the query against its underlying store and puts the results in the cache so they can be reused later. A cache is much smaller than the underlying store so data in the cache is frequently flushed and replaced with other active data.

Caching is useful when three conditions are met. First, putting information in the cache and retrieving it must be faster than ordinary database operations. Much faster, so there is a significant gain when the results are in the cache to make up for the delay when they're not. Second, interactions with the database must frequently repeat the same transactions, so that results stored in the cache are often reused. Otherwise, caching data is a waste of time because it will have been flushed from the cache before a second or third request arrives. Third, retrieval rather than update must be the dominant type of database transaction. Updates can make the data in the cache inconsistent with information in the database, forcing it to be retrieved again from the underlying store, just as if there were no cache at all. If it's not retrievals but updates that are the dominant transaction, caching is a waste of time and effort.

Caching is used throughout the web. It is especially successful in storefront applications, where database queries often concern details about a particular product. Storefronts offer hundreds of thousands of products, but only a handful of them are really popular. So, there are frequent requests for the same information, which makes the cache effective.

Caching worked well for Amazon, at least before product reviews and recommendations became prevalent in their product pages. But caching cannot work well for Facebook. Users look at their welcome pages on an individual basis, and they make frequent postings. Retrieval of information does not dominate the pattern of transactions, and it is not the case that many incoming requests are for retrieval of the same data, over and over, the way it is with a storefront operation. Besides that, one update on Facebook can trigger changes in many pages of the website. This is typical in Web 2.0 applications, and this need for frequent updates of customized information for every user eliminates the advantages of caching. Another solution is needed.

13.2.2 Sharding

Sharding splits a database into many different databases. For example, John's Facebook friends and status updates may be stored in machine J, whereas Mary's friends and status updates are stored in machine M. Machines like J and M that store just a portion of the data are called *shards*. Because the database is stored on many different computers, no one computer has to shoulder the entire load. That's the upside. The downside is that it makes it harder to automate transactions with the database. Programmers have to specify how to distribute individual queries across all of the many computers that may be involved in resolving the query.

To see how this might work, suppose the computer needs to generate John's welcome page. The first step might be to find John's friends, which can be done by executing a query on machine J.

```
SELECT    f.Friend
FROM      Friends f
WHERE     f.user='John'
```

The query returns John's friends, Sally and Mary. The next step is to find Sally's and Mary's status updates. That leads to the following queries:

```
SELECT    s.Time, s.Status
FROM      Statuses s
WHERE     s.User='Sally'
ORDER BY  s.Time DESC

SELECT    s.Time, s.Status
FROM      Statuses s
WHERE     s.User='Mary'
ORDER BY  s.Time DESC
```

The first query should run on machine S, whereas the second query should run on machine M. The final step is to combine the results from the two queries and then merge them, keeping the combined list in reverse chronological order.

Queries like this show one of the shortcomings of sharding. Each query retrieves results from only one table because the related records are not necessarily stored in the same shard. In this particular example, the information needed to answer the query was distributed across shards J, S, and M. So the program had to collect the information from all the different sources and then combine it. That makes sharding more complicated than keeping all the information in one place, as in a traditional database.

Sharding also suffers from uneven distribution. What if, for example, one of the shards ends up with too many records? That shard would need to be split into pieces. For example, the system could split the shard M into two shards, Ma–Mp and Mq–Mz. That seems like a good idea, but in practice splitting shards is complicated because the software on all the computers that access the data needs to be modified to make it possible for the computers to find the shard that contains the information they're looking for.

FRIENDS		
Record ID	**Friend1**	**Friend2**
John	Sally	Mary
Mary	Sally	
Sally	David	

Figure 13.2
Storing friends lists in Cassandra.

13.3 The Cassandra Solution

Faced with these difficulties, Facebook engineers developed a solution that retained the benefits of sharding, but avoided some of the difficulties. The goal was to make it easy to split a shard into multiple pieces and to hide from the software the complexity of sharding.

Cassandra, the solution they devised, combines features from the Dynamo project at Amazon and the BigTable project at Google. From Dynamo, Cassandra borrows the idea of a replication ring, and from BigTable, a data model. Cassandra's data model groups records into different tables. Each record in a table is identified with a key. The key must be unique in a given table, but the same key may be used in different tables. Each record consists of one or more key/value pairs, and different records in a Cassandra table may have different keys.

For example, John's friends may be stored in a record like the one shown in figure 13.2. The important thing is that a program can retrieve all of John's friends by requesting the single record with ID John. Once John's friends are known, it is necessary to retrieve their status updates. This can be done by looking for the records in the status table that have the appropriate record IDs. Figure 13.3 (page 248) shows what the status table could look like.

The table structures we have presented assume that all fields will fit in a single record. That is, we assume that a single record can hold all of a user's friends or all of a user's status updates. Cassandra tables are designed to support thousands of fields. This will be enough for most users, but not for the heaviest users of Facebook. To deal with the heaviest users, Facebook can reuse the same idea with column names. To support an arbitrary number of friends or status updates, the values can be spread across multiple records, with IDs such as John1, John2, and so on.

The upshot is that the workflow for retrieving information from Cassandra is similar to the workflow for sharding but with a major difference. The queries that are generated do not need to be aware of which shard contains the information they need. In fact, Cassandra relies on sharding both for performance and for replication. The key innovation is that the shards are arranged in a ring. For simplicity, assume that the shards are labeled A, B, C, ..., Z. The ring arrangement means that each shard is connected to the next, and eventually

	STATUSES			
Record ID	**Time1**	**Time2**	**Status1**	**Status2**
Sally	Apr 21, 2011, 10:29 am	Apr 21, 2011, 10:33 am	Saw *Battle: Los Angeles* last night. What a waste!	Looks like the fires are getting closer to our house. Thinking about evacuating.

Figure 13.3
Storing status updates in Cassandra.

the last shard is connected to the first. For example, A could be connected to B, B to C, and so on, until Z is connected back to A.

Records are arranged in an order determined by a hash function that computes a hash value from the key of a record. The hash value is used to select a shard label (labels A through Z in the example). All hash values up to A are mapped to shard B, those between A and B to shard C, and so on. The hash function and mapping of hash values to shards is known to every shard, so a program can ask any of the machines to retrieve a given value. If the machine does not have the value, it can determine the shard containing the value and forward the request to that shard.

The ring makes it easier to rebalance the shards in case one of them becomes too large. Suppose, for example, shard B gets too large. To balance it, a new shard, say BM, is created and inserted between B and C. During the insertion process, B sends all of its records between BM and C to shard BM. When this process completes, all shards are notified of the new shard's existence and shard BM joins the ring. This can happen without the knowledge of any programs that are retrieving data from the system.

Cassandra also uses the ring for replication. Records that should be stored in A are also stored in B and C. This is important, because computers and disks can fail, but if shard A should fail, there are two more copies of its data. It can also serve to improve performance during spikes. If shard A becomes busy, shards B and C can take over some of the load.

Replication complicates the splitting of shards. When shard B is split into B and BM, for example, this also affects shards C and D because they store replicated records for shard B. Now this needs to be restructured, so that B's records are replicated in shards BM and C. Moreover, C and D should replicate the records for shard BM. This means that shards C and D need to participate in the insertion of BM into the ring. Shard C needs to know that some of the records it replicated for B are now associated with BM instead. Shard D needs to know that some of the records it was replicating on behalf of B can now be forgotten, and the rest need to be associated with BM. Finally, the new shard BM needs to receive not just its share of B's records but all of B's records so that it can replicate B's remaining data.

That's a lot of data movement, and it might be surprising that it helps. Part of the reason it works is that Facebook has the luxury of not having to get all the data right all the time. If people don't see all the latest postings of their friends until an hour or even several hours after they are posted, it's no big deal. Users may be a little out of sync for awhile, but the data is not time critical on a minute-by-minute basis.

13.4 Summary

Web 2.0 applications bring up many scaling challenges. These challenges go beyond what traditional database solutions can offer. So, leading Web 2.0 companies have developed custom solutions, and the idea of sharding plays a prominent role in some of those solutions. Cassandra, Facebook's solution, successfully met the challenge, and fortunately, Facebook decided to make Cassandra available to the programming community via an open-source process.

Programmers can download Cassandra (`http://cassandra.apache.org`) and use it to develop new software.

14 Parallel Computation with MapReduce

14.1 Vertical and Horizontal Scaling

Some important computer applications are so large that they would take unacceptably long to run on conventional computers. For example, a personal computer is more than powerful enough to balance your checkbook but not for a financial application that tracks credit card usage in real time to detect instances of fraud. The sheer number of credit card transactions make this application far too intense for a personal computer.

There are various ways to cope with problems of scale in large applications. The traditional approach is *vertical scaling*, which means running the application on a more powerful computer. This is the easiest solution in terms of software because the software does not need to change as the machine scales up. But what happens when the problem gets bigger than any single machine can handle? For example, maybe a big, fast computer could handle a billion credit card transactions an hour but not a hundred billion. At some point, the rate at which transactions take place will overwhelm the available computing technology.

Horizontal scaling offers an alternative. Instead of running the application on a single computer, the application is split into smaller chunks, and each chunk is run on a separate computer. Ideally, all of the computers involved are similar in most respects to personal computers, so that the cost of an additional machine is a small fraction of the overall system cost. This makes it economically feasible to scale the hardware platform as the computation requirements increase. Horizontal scaling has become the de facto solution for dealing with web services that have to deal with rapid growth, such as those provided by Facebook, Google, eBay, Amazon, Netflix, and many others.

The problem with horizontal scaling, however, is that it is not always easy to split an application into smaller chunks. In fact, this is particularly hard when the program is written using conventional programming languages, such as C++ or Java. In conventional languages, the program specifies a sequence of updates to records stored mostly in fast memory. That makes it difficult to manage multiple computers working on the computation at the same time because the cooperating computers need to coordinate their updates. The coordination problem can sink the effort to make the computation go faster. Getting the appropriate data and the right software components to the right computer at the right time presents a host of problems.

One advantage of an equation-based software model is that the different parts of the software are more decoupled than they are with the conventional approach. Everything is defined in terms of operators that, given operands with appropriate data, can produce results without interacting with other parts of the software. For that reason, it does not matter where or when the operations are performed. The problem of getting the right data to the right place remains, but the task of managing a great many small interactions between different parts of the software is greatly reduced.

The engineers at Google were faced with one of the largest scaling problems that computing had ever seen, namely, searching the entire, rapidly growing, worldwide web. To cope with the scale of this problem and its continual growth, the horizontal scaling approach was the only practical option. Even though they use mostly conventional programming methods for most of their software, they invented and adopted a way to manage large-scale components with a programming model called MapReduce that has a lot in common with equation-based software models. Since Google's introduction of MapReduce, it has been adopted in many other settings. For instance, the Apache Foundation implemented Hadoop, an open-source implementation of MapReduce that you can download for free to your computer. Hadoop is widely used in commercial applications and in research. We will describe the general nature of MapReduce systems like Hadoop and explain how the MapReduce framework simplifies the development of programs that can scale horizontally by focusing on just two operations: Map and Reduce.

14.2 The MapReduce Strategy

The MapReduce paradigm is applicable to problems that process data that can be represented as a sequence of key/value pairs. Not all problems lend themselves to this representation, but Google engineers recognized that many practical problems do fit in this category. The following are a few examples:

- *Counting Words in a Document.* The data in this case is the collection of words in a document. It can be organized as a list of word/count pairs, where the counts are initially set to one. Since any word may appear multiple times in the document, most words will, in the beginning, occur in more than one word/count pair. The objective is to produce a list of word/count pairs in which each word appears only once and the associated count is the number of occurrences of the word in the document.

- *Finding Words That Link to a Webpage.* The purpose of this operation is to find the words that are most commonly used to link to a particular URL. For example, your name may be the most common phrase used to link into your Facebook page. Google uses this kind of information to select which pages to display for a particular search. The MapReduce approach applies to this problem. The data is the collection of links on

14.2 The MapReduce Strategy

the internet.[86] Each link can be represented as a word/URL pair, where the same word may appear in many different pairs. To figure out which words are most commonly associated with a particular web address, this data needs to be reduced to a collection of URL/word pairs, where each URL will appear once, associated with the word that is used most often to link to that URL.

- *Finding Extreme Values.* Consider an application that finds the record high or low temperature for each of the fifty states of the US. The initial data consists of a list of city/temperature pairs. There would be one record for each recorded temperature in a city. It might be one per day, one per hour, one per minute, or a varying combination, depending on the city, and the records would extend over different periods of time for different cities, a hundred years for some cities, ten years for others, and so on. So, each city would occur in many different pairs. The desired outcome is a collection of city/temperature pairs where each city occurs only once, and the associated temperature is the highest (or lowest) one in the recorded data.

What all these applications have in common is that data processing can be split into three different parts, each involving a list of key/value pairs, and it is possible to process each individual data record or key/value pair without having to simultaneously examine all the other records. For example, consider the case of finding the highest temperature ever recorded in each of the fifty states. The computation could proceed in three stages:

1. **Input Data** pairs temperatures with associated sensors. Each pair could have a key identifying a specific temperature sensor and a value that consists of the city and state where the sensor was located, the date the sensor measured the temperature, and the temperature on that date. For example, the input data may contain the following records:

    - KLAR, Laramie, WY, 2009-05-13, 41
    - KLAR, Laramie, WY, 2009-05-14, 47
    - KOUN, Norman, OK, 2009-05-13, 76
    - KOUN, Norman, OK, 2009-05-14, 70
    - ⋮

    The first column is the key for each record (for example, KLAR), and the remaining columns comprise the value. Many records may have the same key in the input data because the sensor makes many measurements over time. The goal is to reduce that to records in which each key appears only once, associated with the maximum temperature measured by that sensor.

2. **Intermediate Data** is used to pass data from the Map operation to the Reduce operation. The Map operation processes the *input data* and produces key/value pairs that make up the *intermediate data*, and the Reduce operation processes the intermediate data to create *output data*. In this application, the goal is to find the highest temperature for each state, so the Map operation could extract the state and temperature from

---

[86] Nobody knows how many links there are on the internet, worldwide, but most estimates place it in the trillions, and some credible estimates place it in the hundreds of trillions.

each input data record. Then, the intermediate records would be state/temperature pairs.

- WY, 41
- WY, 47
- OK, 76
- OK, 70
⋮

In general, the Map operation may produce any number of intermediate data points for any given input data record, although in this case, precisely one intermediate record is generated for each input record.

3. **Output Data**, the final result of the MapReduce computation, is delivered by the Reduce operation. In the high-temperature example, this corresponds to the maximum temperature recorded for each state, so there would be exactly one record for each state that appeared in the input data.

- WY, 115
- OK, 120
⋮

As you can see, the data records are largely independent of one another, so the computer can process the May 13, 2009, temperature entry for Norman, OK, without considering the May 13, 2009, entry for Laramie, WY. This enables horizontal scaling because the different entries can be processed in different machines. However, this application also shows the need for combining the entries for a specific key at a later time. For example, to find the high-temperature record for Oklahoma, it will be necessary to consider all the entries for Oklahoma at some point.

The MapReduce paradigm applies well in these kinds of problems. The map operation, which receives each of the input key/value pairs, processes the pair and produces a large number of intermediate key/value pairs. These intermediate pairs use keys that may or may not be completely different than the input keys. In the word-counting example, the intermediate keys may be the same as the input keys, namely, the words that are being counted. On the other hand, when looking for words that are used to link to URLs, the input keys are the words, but the intermediate keys are the URLs. The MapReduce framework leaves this choice up to the software designer, which is one of the reasons that MapReduce is widely applicable.

The second step is the reduce operation, which combines all the entries for the intermediate keys and produces zero or more output key/value pairs. As before, the output key/value pairs may use the same keys as the intermediate or input key/value pairs, or they may use entirely different keys.

For example, consider the problem of counting words in a document. Assume that the document has already been read and that it has been broken up into key/value pairs, where

14.2 The MapReduce Strategy

each key is a word and the value is always one. These key/value pairs make up the input data for this problem. For instance, the Gettysburg address could be represented by a list of key/value pairs using the notation (*key* . *value*) to denote a pair.

- (four . 1)
- (score . 1)
- (and . 1)
- (seven . 1)
- (years . 1)
⋮
- (from . 1)
- (the . 1)
- (earth . 1)

The **map** operation takes in an input key and value and delivers a list of zero or more intermediate key/value pairs. For the word-count program, map could deliver a list with exactly one element, namely, the very same input key and value. In this case, the map part of MapReduce packages the result in a list, which is the form expected by the MapReduce system, but does not perform any additional computation.

$$map(k, v) = [(k . v)]$$

In a more elaborate example, map would perform a computation using the key/value pair supplied as its operand and deliver a list representing the results of that computation.

The **reduce** operation accepts an intermediate key and a list of all the values returned by any map operation for that key. It delivers a list of zero or more final key/value pairs. In the case of word count, reduce returns only one key/value pair, namely, the key and the sum of the counts in the list.

$$reduce(k, vs) = [(k . sumlist(vs))]$$

The operator **sumlist**, which adds all the elements of its input, would be defined by the software designer.

To make this discussion more concrete, consider the Gettysburg Address, which contains the word "nation" in four places. Because of this, the map operation will be called with the input key/value pair (**nation** . 1) four times, and each time it will return a list with the single intermediate key/value pair [(**nation** . 1)]. The MapReduce system collects all the values for each intermediate key and starts the reduce operation on those values. At some point, it will collect all of the four intermediate values for "nation" and invoke the reduce operation with the operands **nation** and [1 1 1 1]. The reduce operation will then return a list with the final key/value pair [(**nation** . 4)].

Much of the value of MapReduce is that the programmer only needs to define the map and reduce operations and does not need to deal with the details of carrying out the map and reduce operations in a collection of computers sharing the data across a network. It is the MapReduce framework that takes care of running the program in a single computer or in a cluster of hundreds or even hundreds of thousands of computers, depending on the size of the problem. It also takes care of things like sending the intermediate key/value pairs from the computers processing the map operation to the computers processing the reduce operation.

The MapReduce framework takes the input key/value pairs and splits them across many different machines. On each machine, MapReduce performs the map operation on each of the key/value pairs that is assigned to that machine. As it does this, it combines the intermediate key/value pairs returned by each map operation into a single list. The lists from all of the machines are then combined.

The intermediate lists must be combined because the reduce operation expects to see an intermediate key and all of the values associated with that key at the same time. For example, to find the maximum temperature for the key OK, the reduce task needs to see all the records associated with OK. Different intermediate keys, such as the ones for OK and WY, are independent, so they can be processed by different machines. That is, one machine can process OK temperatures at the same time that another machine is processing WY. But all of the OK records must go to the same machine running the reduce operation for OK. Therefore, the map operation collects all of the values for each of the intermediate keys, and then the reduce operation is called only once for each intermediate key.

Once all the values for a given intermediate key are collected, the MapReduce framework can call the reduce operation on that intermediate key. The result is a list of output key/value pairs. MapReduce collects all these results and returns them as the final result of the computation.

MapReduce does the work of distributing the program across multiple machines. That is one way it provides value. An engineer can develop a MapReduce program on a local computer and modify it until it behaves as required on a small data set. Then, the program can be submitted to a large MapReduce cluster to process a full-scale data set. The MapReduce system deals with the problem of scale automatically once the program has been developed.

14.3 Data Mining with MapReduce

Now that we have seen the basics of MapReduce, we can look at a larger example that illustrates how MapReduce is used in practice. The application we will discuss is a recommendation engine, a piece of code that is used to recommend new things to a person based on other things the person likes. For example, the product page that pops up on a typical visit to the Amazon website often has a section called "Customers Who Bought This

14.3 Data Mining with MapReduce

Item Also Bought" that recommends related items. Based on past purchases and browsing habits, the Amazon software builds a customized web page full of recommendations. How does Amazon do this?

Let's break the problem down into two components. First, Amazon needs to find *customers like you*. In the case of a single product, this means other customers who have bought that product. In the more general sense, it means other customers who have bought many of the same items that you have purchased in the past. Once the group of customers like you is identified, the purchases that people in the group have made can be examined to find the most popular items.

Finding the most popular items is a problem that is a lot like counting words in a document. Amazon keeps a history of all the purchases that each customer has made. To process this list with MapReduce, think of it as consisting of entries of the form customer/item, meaning that at some time the specified customer bought that particular item. The map operation produces intermediate entries of the type item/1, meaning the given item was bought (once), which is similar to what the word-count program did. Such an entry should be generated for each purchase made by a customer in the reference group. That is, the map operation filters out the purchases made by customers who are not like you. The reduce operation is identical to that of word-count, but this time it counts the number of purchases for each item. In the end, the computation must consider the results of the reduce operation and select the items that were purchased most often.

Unfortunately, that leaves the first problem unsolved: finding the group of customers who are most like you. This is the most critical aspect for generating useful recommendations. For instance, if all of your purchases from Amazon have been gardening books, you are likely to ignore a recommendation engine that alerts you to the latest novel in a long-running vampire series. Worse, you may start thinking of the recommendations as unwarranted spam.

How can Amazon find other customers like you? Imagine that you have rated all the purchases you have made, giving each item a grade between 0 (hated it) and 5 (loved it). To keep things simple, imagine that Amazon sells only two items. Then your ratings for these items can be expressed as a pair of numbers, say (2, 0). Now suppose that other customers have also rated the items. The customers who are most like you are the ones whose ratings are close to yours. One way to measure how close other ratings are to the pair (2, 0) is to view the pair as the coordinates of a point in a two-dimensional plane. Using this notion of closeness, the customers who are most like you have purchase histories that correspond to nearby points in that plane.[87]

---

[87] To determine what points are "nearby," you will need some notion of distance in the plane. You could use the standard Euclidean distance or, instead, a metric customized to determining which customers are most like you. This is just one of the complex issues that come up in fashioning methods for finding clusters in data.

Of course, Amazon sells many more than two items, and neither you nor any of Amazon's other customers are likely to have rated even a fraction of them. But the principle stays the same. Instead of using pairs to represent ratings, many more coordinates would be needed, as many coordinates as the number of different products Amazon sells. The coordinates would still specify a point in space, but it would be a space with many dimensions. The customers most like you will still be represented by points close to yours in that multidimensional space.

A remaining complication is that customers do not always explicitly rate the items they like, but this can be resolved by using implicit ratings. For example, if you buy an item, we can give it a rating of 4 unless you explicitly change it. And any product that you have never looked at can be given a rating of 0.

The problem is to find which points in this huge domain with many dimensions are near your own. Or, viewed another way, the problem is to find groups of points that are clustered together. One cluster, for example, may consist of avid gardeners, whereas another is made up of fans of vampiric fiction.

In general, finding clusters in a large data set is a difficult problem that has been studied extensively by scientists and mathematicians for a long time. There are many useful approaches. One that, on the surface at least, is straightforward to describe starts with guessing some cluster locations, then gradually refining the guesses by making computations based on the data. The computation could proceed as follows:

1. Initially, guess a location for each cluster. The guess can be taken as the estimated center point of each alleged cluster.
2. For each point in the data set, decide which cluster center is nearest to the point. Split the points into clusters so that each point is in the cluster determined by the nearest center point.
3. Recalculate each cluster's center point by averaging all the points that were assigned to that cluster.
4. Repeat the previous two steps until things settle down or meet some other predetermined conditions.

The middle two steps can be implemented using MapReduce. The map operation can assign points to clusters, and the reduce operation can compute the new center point of each cluster. The distances between data points and estimated cluster centers determines which cluster the data point belongs to, and the cluster determines the products of interest to particular customers.

The map and reduce operations must conform to the expectations of the MapReduce framework, which requires the map operation to be based on key/value pairs. In this application, the programmer defines map operation to figure out, for individual customers, which cluster center is closest to the customer's purchase history. The MapReduce framework then distributes the map operation across all customers.

14.3 Data Mining with MapReduce

Let's say that the purchase history for the customer is the first operand of the map operation and the list of cluster centers is the second operand. Using this information, the map operation selects the center that is closest to the customer's purchase history. The MapReduce framework requires the map operation to deliver its result in the form of a list of key/value pairs, and we can meet this requirement by specifying that the map operation delivers a list of just one element, which is a pair consisting of the selected center (the key) and the purchase history (the value).

$$map(hist, centers) = [(closest\_center(hist, centers) \,.\, hist)]$$

The MapReduce framework applies the map operation to all the customers, gathers up the results, and packages them in the form of intermediate key/value pairs, one for each cluster center (the key). The value that the framework associates with the key is the list of points in the purchase histories associated with the cluster center that the key represents. In the equation that defines the map operation, the center delivered as the key is computed by an operator called *closest_center*, which does the work of selecting one of the centers from the list supplied as its second operand. The one it selects is, of course, the one that is closest to the customer history supplied as its first operand.

The *closest_center* operation would need to be defined, and the details depend on how distances between points are measured. We have deferred deciding those details, so we will leave that part of the computation out of the discussion. In any case, an intermediate key/value pair is a cluster center (the key) and a list of nearby purchase histories (the value). Each purchase history is a point for which the key is the nearest cluster center to that point. The MapReduce framework sends the intermediate key/value pairs to the reduce stage of the process. Now we need to turn our attention to the reduce operation.

The reduce operation receives an intermediate key/value pair from the MapReduce framework. The key is the center of a cluster, which is treated by the reduce operation as an identifier for the cluster. The value is a list of the points in the cluster. The job of the reduce operation is to compute a new center for the cluster (the key) by averaging the list of points (the value). The MapReduce framework requires the reduce operation to deliver a list of key/value pairs, so we package its result as a list consisting of exactly one key/value pair, namely, the cluster identifier (the key) and the new center (the value).

$$
\begin{aligned}
reduce(cluster, points) &= [(cluster \,.\, average(points))] \\
average(points) &= avg(points, (0,0), 0) \\
avg(points, sum, count) &= \begin{cases} sum/count & \text{if } points = [\,] \\ avg(rest(points), \\ \quad sum + first(points), \\ \quad count + 1) & \text{otherwise} \end{cases}
\end{aligned}
$$

We need to explain some of the operators and terms in the equations that define the reduce operation because they are not, in all cases, what might be expected. The *points* variable refers to a list of points, and each point is a pair of numbers representing product ratings by a customer. The operand (0, 0) in the equation that defines *average* is also a pair of numbers, both zeros in this case. These zeros serve as a starting point for adding up the product ratings to make it possible to compute averages.[88]

The *sum* variable is also a pair of numbers. Therefore, the addition (+) and division (/) operators with *sum* as a left-hand operand are not the usual arithmetic operators. The formula *sum* + *point* denotes $(s_1, s_2) + (r_1, r_2)$, which stands for the pair $(s_1 + r_1, s_2 + r_2)$. Similarly, *sum*/*count* denotes $(s_1, s_2)/count$, which stands for $(s_1/count, s_2/count)$. The definitions also refer to the operators *first* and *rest*. The operator *first* delivers the first point in the list of points, and the operator *rest* delivers all the points in the list after the first one.

Finally, the equation defining *avg* selects one of two formulas, depending on the value of *points*, which is a list. If the list is empty (that is, if *points* = [ ]), the formula *sum*/*count* is selected as the value of *avg*. If *points* is not the empty list, the definition selects a more complicated formula as the value of *avg*.

There is an important subtlety in the way the equations define the operator *average* that computes a new center point of a cluster from points assigned to that cluster in the map stage of the MapReduce computation. The work of computing the average is done by the *avg* operator. The definition of *avg* is inductive and employs a common trick known as tail recursion that makes computations like averaging much faster. We think it is worth complicating the discussion with this technique because the main point of MapReduce is to be able to compute things quickly. The tail recursion trick avoids supplying, as an operand to another operator, the value that *avg* delivers. This makes the computation faster because the computer is able to avoid a lot of bookkeeping that is required when an inductive invocation is nested in an invocation of another operator.[89]

The result of the reduce operation is a new list of center points in the same format as the initial guess. That makes the output of the reduce operation suitable as the initial guess for additional map computation steps, which makes it possible for the computer to perform as many map/reduce cycles as necessary to find clusters. Figure 14.1 (page 261) diagrams the MapReduce process that we applied to implement the four-part clustering procedure discussed earlier (page 258).

---

[88] There are only two product ratings to simplify the example. In practice, there would be many product ratings, so the data item representing a customer's ratings (and the initial item with the zeros that start the calculation) would have many components, each processed in a manner analogous to the two-component ratings in the example.

[89] Section 5.4 discusses an example illustrating the effectiveness of tail recursion.

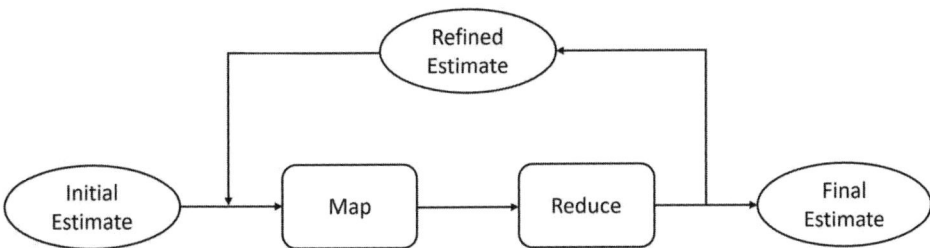

Figure 14.1
Iterative MapReduce operation.

14.4 Summary

Some problems are too large to solve on a single, conventional machine. That leaves two options. Get a large machine (a supercomputer, for example), or break the problem down into many tasks and perform each task on a separate computer. The first approach, vertical scaling, is limited by the size and speed of the most powerful computer available. In addition, it is difficult to use in practice because as the problem grows, several scaling steps may be necessary, and each step requires an expensive migration to a more capable computer. If we were trying to use vertical scaling to deal with the massive, rapid increases in the volume of traffic on websites like those of Google and Amazon, we might need an upgrade every few days, which would be impossible with vertical scaling.

An alternate approach, horizontal scaling, offers the promise of virtually unlimited scalability and incremental increase in the cost of the solution as the problem size grows. However, it is much more difficult to write programs that are split across multiple machines, and this solution works only for big computations that can be carried out in the form of many smaller computations. Such problems usually involve a lot of data, and the data needs to be in clumps that can be processed independently.

MapReduce is a framework related to the equation-based software model that facilitates writing programs that can be distributed across a large number of computers. MapReduce programs are organized around two operations. The map operation applies to each of the input values, and it generates an intermediate result. The intermediate results are subsequently combined using the reduce operation. If the results generated by the reduce operation are in the same format as the input to the map operation, multiple MapReduce passes can be performed. This is useful in programs that estimate optimal results by refining an initial estimate, such as programs that find clusters in a large data set.

15 Generating Art with Computers

This book has explored deep connections between logic and computing, including the use of equations and logic to design and analyze computer programs. In this chapter, we're going to turn our attention to artistic creativity. Can logic and equations be useful in creating works of art?

15.1 Representing Images in a Computer

To create visual art with a computer, we need some way to represent images in computers. How does a computer store a picture? The answer, it turns out, is surprisingly pedestrian. Ignoring color for now, you can think of a computer display, such as the screen on your computer or phone, as a collection of millions of tiny dots arranged in a rectangular grid with a fixed number of rows and columns. Different displays have different dimensions. For reference, let's say there are M rows and N columns in the grid. Each dot is called a *pixel*, a term in common use now that at one time was short for "picture element."

A laptop computer might have a 1440×900 display: 900 rows and 1440 columns. Continuing to ignore color for the moment, each pixel can be either emitting light (turned on) or not emitting light (turned off), so a picture in a computer can be represented by specifying which pixels are turned on and which are turned off. A straightforward way to do this is to use a list of numbers in which each number corresponds to a pixel. Even better, we can use a list of rows, each of which is a list of N 0/1 elements (one for each pixel in the row). The off/on status of the pixels in the row would match the 0/1 elements in the list, and there would be M such lists, one for each row on the display. This list of lists forms a kind of matrix.

Figure 15.1 (page 264) illustrates this idea with a 4×4 grid of squares representing a 4×4 section of pixels in a computer display. The pixel in row i and column j is "on" (indicated in the figure by a black square in the diagram) when the jth entry in the ith list is a 1. When the entry is a 0, the pixel is "off" (indicated by a white square).

Using lists to represent images is an important concept. The first chapter of the book talked about boundaries and interfaces between software and hardware, and these ideas are

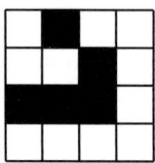

```
[[0 1 0 0]
 [0 0 1 0]
 [1 1 1 0]
 [0 0 0 0]]
```

Figure 15.1
A picture encoded as a matrix of pixels.

well illustrated in this representation of images. From the software perspective, generating an image amounts to generating a list of numbers. It is up to the hardware to interpret the list of numbers as an image and display it.

This kind of layering is a common strategy for dealing with complex problems in computer science. Each layer in the solution provides a service to the layers next to it. In this example, the software layer generates lists of zeros and ones for the hardware layer, and the hardware layer turns lists of zeros and ones into physical images. Both layers represent images but in different forms. The software generates an abstraction of an image in the form of zeros and ones, and the hardware layer turns that abstraction into an image on a screen. That is, the hardware builds a concrete version of the abstraction specified by software. So far in this example, there are only two layers, but in complex systems, there can be dozens of layers.

So far so good with black-and-white images, but what about color images? The human eye perceives color using specialized cells called *cones*, which are located in the central part of the retina. There are three types of cones, each one sensitive to a different range of colors. There is some overlap in the range of colors that each type of cone can detect, but it is mostly accurate to think in terms of cones that are sensitive to shades of red, green, and blue. The brain forms an image by interpreting the measurements of colored-light intensities that the cones make.

In an analogous way, color images can be created by using three dots on a screen for each pixel, one for red, one for green, and one for blue. If the red dot in a pixel is on and the green and blue dots are off, the pixel looks red. Other colors come from combinations of red, green, and blue dots. The following list represents the first row of pixels in the black-and-white image shown in figure 15.1: [0 1 0 0].

In a color image, the row of pixels could be represented by a list of three-element lists: [[0 0 0] [1 0 0] [0 0 0] [0 0 0]]. In this representation, triples of 0/1 values specify the off/on status of color dots in the corresponding pixel. The first element of the triple is for the red dot, the second for the green dot, the third for the blue dot. That is the conventional order: red-green-blue (RGB). In this scheme, [1 0 0] represents pure red, so the list, as shown, represents a red dot in the second pixel of the row.

With this convention, there would be eight possible colors because each element in the triple can be either zero or one, which leads to 2^3 (that is, 8) combinations: black [0 0 0], white [1 1 1], red [1 0 0], green [0 1 0], blue [0 0 1], cyan [0 1 1], magenta [1 0 1], and yellow [1 1 0]. That might be an improvement over black-and-white, but it's a crude representation of color compared to the colors the eye can perceive.

Getting better color requires a more subtle blending of red, green, and blue. For example, a pixel with red and green on and blue off will appear to be yellow, but by using a range of color intensities, not just on and off, a blend of red and green can produce other colors, such as orange. Most computer screens can produce 256 different intensity levels for each colored dot.[90] Software represents a color intensity as a number in the range 0 to 255. Using this scheme, the list [255 140 0] produces a dark shade of orange.

A matrix of red-green-blue intensities is an abstraction of an image with realistic color, and hardware converts this matrix to a display with colored dots whose intensities reflect the numbers in the matrix. Additional layers between the image abstraction in the software and physical image on the screen can be helpful. Imagine, for example, an operator called line that paints a line on an image.

The line operator would figure out which pixels are on the line and build a matrix with the appropriate triple for the chosen color at each of those pixels: (line *image x_1 y_1 x_2 y_2 color*). The *image* operand is a matrix of pixels (red-green-blue triples) representing an image, the x and y values specify the row/column coordinates of the endpoints of the line, and *color* is a red-green-blue triple. The operator delivers a new matrix of color intensities like the *image* operand but with the specified color at each pixel on the line.

There are, of course, many other useful operators in the same vein as the operator line that can be used to draw triangles, squares, rectangles, polygons, circles, ellipses, and the like. The important thing to realize is that all of these operators work on an image represented as a matrix of color intensities. The operators add to the system a layer that understands basic geometric shapes, and the geometry abstraction interacts with the layer that represents an image as a matrix of pixels.

Operators that create shapes are not the whole story. Operators of another kind blur images in useful ways. For example, suppose that an image has a very dark line on a bright background. This creates a sharp boundary between the line and the background, but it can be smoothed out by averaging the colors of neighboring pixels. The bright pixels on one side of the boundary become a little darker and the dark pixels on the other side become a little lighter, resulting in a softer boundary. This is the basic idea behind operators that manipulate images at the level of individual pixels and their neighbors. Image-processing operators of this kind form a class known as *image filters*. There are a lot of filters in image-

---

[90] The human eye can detect more than 256 shades of each color, but 256 shades are enough to produce realistic color on computer screens.

processing software, and they provide another abstraction layer that facilitates creating representations of images with software.

Box 15.1
Bresenham Line-Drawing Algorithm

Drawing a line from one point to another on a grid of points is tricky. There are many subtle cases to consider, and if you don't make good choices, the result does not look like a line at all. If the starting point has coordinates $(1, 1)$ and the end point $(10, 10)$, then the line consists of the points $(1, 1)$, $(2, 2)$, $(3, 3)$, and so on up to $(10, 10)$. No problem.

However, if the end point is $(2, 10)$ instead of $(10, 10)$, then choosing the right points for the line is more difficult. Sketch a 10×10 grid and try drawing a line from $(1, 1)$ to $(2, 10)$. This is an extreme case, but many lines lead to similar issues. Geometry on a grid, digital geometry, presents a lot of problems like this one that are hard to solve.

Another problem is speed. In many graphics applications, the computer needs to draw millions or billions of lines per second, so speed is important. In the 1960s, the computer scientist Jack Bresenham invented a fast way to compute the list of grid points closest to the line between any two given endpoints. Today, many decades later, most of the line-drawing in digital images still makes use of some version of the Bresenham algorithm.

15.2 Generating Images Randomly

One way to generate artistic images is to insert a layer of randomness on top of the layers that create geometric shapes. An artist may layer paints on a canvas with a brush that has a certain texture, and the brush strokes will create a complex pattern that is more than just a line. Figure 15.2 (page 267) shows a pair of images. One is a straight line and the other shows how adding a lot of nearby, short lines in more or less random orientations creates an image similar to a line but with more texture, like a brush stroke, perhaps. Similar operations can produce other geometric shapes with a degree of randomness, such as in noisy rectangles or circles that are mostly but not exactly round.

But what do we mean by random? None of the digital circuits that we have studied exhibit random behavior. Instead, they conform to precise equations and deliver the same results every time they are supplied with the same data. The following sequence of numbers appears to be random:

$$0, 1, 5, 3, 4, 8, 6, 7, 2, 0$$

15.2 Generating Images Randomly

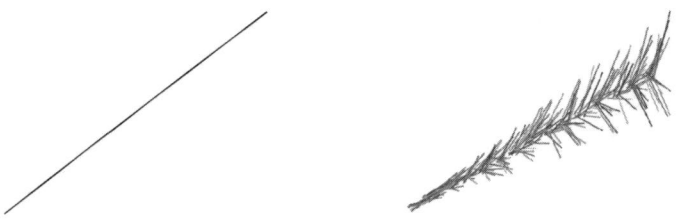

Figure 15.2
Straight line vs. a line made up of random segments.

However, it is not random. It is produced by an operator r defined by the following equations, which are presented in both traditional, algebraic form and in ACL2 notation:[91]

$$
\begin{array}{ll}
\text{Axioms for } r \\
\hline
r(n + 1) = (4 \cdot r(n) + 1) \bmod 9 & \{\text{r1}\} \\
r(0) = 0 & \{\text{r0}\}
\end{array}
$$

```
(defun r (n)
  (if (zp n)
      0                              ; {r0}
      (mod (+ (* 4 (r (- n 1))) 1) 9)))  ; {r1}
```

As you can easily check if you recall that (x mod d) is the remainder in the division $x \div d$ (box 6.3, page 126), $r(0) = 0$, $r(1) = 1$, $r(2) = 5$, $r(3) = 3$, and so on. So, r is an operator that appears to produce random numbers, but the process is completely deterministic. The sequences produced by such operators are called *pseudo-random* sequences, and they provide a way to introduce an aspect of randomness into images.

There are many ways to produce pseudo-random sequences. It is a well-studied problem with a long history and lots of fascinating ramifications. It happens that the operator r belongs to the most commonly used family of pseudo-random number generators: the linear congruential generators. A linear congruential generator has a *seed*, which is the value it delivers as the first number in the sequence. It also has a *multiplier*, an *increment*, and a *modulus*. The $(n + 1)^{\text{th}}$ number in the sequence is produced by multiplying the nth number by the multiplier, adding the increment, and delivering the remainder when that sum is divided by the modulus. The multiplier, increment, and modulus for the operator r are 4, 1, and 9, respectively. One of the effects of the modulus is to keep the numbers within a certain range, which is the range 0, 1, . . . 8 in the case of the operator r.

---

[91] You can ignore the ACL2 if you have not yet studied the chapters that introduce it. The ACL2 versions of the equations don't add any information necessary to understanding the concepts presented in this chapter.

As we have defined the operator r, $r(n)$ is the nth number in the sequence. This is not the usual way to produce pseudo-random sequences. The usual method computes each new number in the sequence from the previous one. That is, the invocation of the operator supplies the value delivered by the previous invocation as the operand. This procedure is much faster because the definition is no longer inductive. The following definition of rs employs this noninductive strategy:[92]

$$\frac{\text{Axiom for } rs}{rs(s) = (4s + 1) \bmod 9 \quad \{rs\}}$$

```
(defun rs (s)
  (mod (+ (* 4 s) 1) 9))   ; {rs}
```

Because the modulus is 9 in the definitions of the operators r and rs, they always deliver a natural number between 0 and 8. All linear congruential random number generators have a fixed range (although it is usually a much bigger range than 0 to 8), so they always deliver a natural number between 0 and $D - 1$, where D is the modulus.[93]

The operator defined by the equations in figure 15.3 (page 269) draws a ragged line by drawing short line segments at randomized angles from starting points along the line from (x_1, y_1) to (x_2, y_2). The starting points of the short line segments are spaced one unit apart in the x direction. The x coordinate of the endpoint of a short segment is ten more than the x coordinate of the starting point (that is, $x_1 + 10$). The y coordinate of that endpoint is computed by moving in the general direction of the target (x_2, y_2) but with the direction randomly adjusted by a small amount. The adjustment is made by adding a random number between $-1/2$ and $1/2$ to the *slope*, $(y_2 - y_1)/(x_2 - x_1)$, then multiplying that number by ten (the difference between the starting x coordinates of the endpoint and starting point of the short segment, in the usual manner of computing coordinates along a straight line, algebraically).

The slope used in the computation is a ratio of two integers but it is (usually) not an integer. Since grid-points have integer coordinates, the numbers computed using the slope are rounded to the nearest integer by an operator called **round**. That way, all of the coordinates supplied as operands to the **line** operator and the operator **ragged** are pairs of integers representing grid points. The **line** operator won't draw any lines that extend outside the boundaries of the image, so if any of the computed coordinates are out of range, the line segments that are drawn will be truncated at the image boundary.

The short segments are drawn in this way along the line between (x_1, y_1) and (x_2, y_2), with starting points whose x coordinates take all the integer values between x_1 and x_2.

---

[92] Again, you won't miss the point if you ignore the ACL2 version of the equation.

[93] It is often more convenient to work with random numbers between 0 and 1 instead of integers between zero and the modulus. That is easily accomplished by dividing the generated number by the modulus.

15.2 Generating Images Randomly

Axioms for ragged

$ragged(image, x_1, y_1, x_2, y_2, seed) =$ {rg1}
    $ragged(line(image, x_1, y_1, rx, ry, black), x, y, x_2, y_2, s)$ if $x_1 < x_2$
   where
   $s = rs(seed)$
   $jostle = s/9 - 1/2$
   $slope = (y_2 - y_1)/(x_2 - x_1)$
   $rx = x_1 + 10$
   $ry = y_1 + round(10 \cdot (slope + jostle))$
   $x = x_1 + 1$
   $y = y_1 + round(slope)$
   $b = [0\ 0\ 0]$
$ragged(image, x_1, x_2, y_1, y_2, seed) = image$ if $x_1 \geq x_2$ {rg0}

```
(defun ragged (image x1 y1 x2 y2 seed)
 (if (< x1 x2)
     (let* ((s (rs seed))
            (jostle (- (/ s 9) 1/2))
            (slope (/ (- y2 y2) (- x2 x1)))
            (rx (+ x1 10))
            (ry (+ y1 (round(* 10 (+ slope jostle)))))
            (x (+ x1 1))
            (y (+ y1 (round slope)))
            (black (list 0 0 0)))
       (ragged (line image x1 y1 rx ry black) x y x2 y2 s))  ; {rg1}
     image))                                                  ; {rg0}
```

Figure 15.3
Operator to draw a ragged line using random numbers.

This makes the starting points close together along the points on the grid that are closest to the line between (x_1, y_1) and (x_2, y_2). The effect is a ragged line that looks more organic than a simple, straight line. This is the idea behind the line made of random segments in figure 15.2 (page 267).

In the inductive equation {rg1} in the definition of the operator **ragged**, the *seed* operand is supplied to the pseudo-random number generator *rs*, and that operator uses it to produce a number *s* between 0 and 8. That number is supplied as the last operand in the invocation of **ragged** in the inductive equation {rg1} (so it can be used to generate the next random number), and it is also used to jostle the slope of the short line segment drawn by the operator **line**. The operator **line** produces a new image, which is the first operand in the invocation of **ragged** in equation {rg1}. In this way, the image accumulates the line segments that the operator **line** draws at each stage.

15.3 Generating Purposeful Images

Random images can be nice to look at, but should they be called art? Can a computer program produce real works of art? That is a philosophical question with lots of answers. But, regardless of the answer, there are a handful of programs that, it has been reasonably argued, are artists. Let's look at one of them.

The program AARON, written over a span of decades by artist Harold Cohen, was designed in part as an exploration of artmaking. Cohen's initial goal was to determine when a group of abstract marks can be recognized as a coherent image. The earliest versions of his program could do very little, and those early versions knew very little about art. Those versions knew the difference between a figure and the ground, and also the difference between an open figure and a closed figure, but not much more.

What made the early versions of AARON effective was a layer that stood on top of the low-level drawing layer. This layer was essentially a list, but instead of pixel colors, this list contained graphical objects and their locations. Adding a new figure to the canvas was accomplished by adding a new object to the master list. The real breakthrough was that this list could be examined during the process of adding a new figure, so AARON could effectively reflect on the work it had done earlier, then proceed with additional work. This allowed the software to make decisions based on artistic principles, such as balance and proportion. This breakthrough proved so successful that it carried over into all subsequent versions of AARON.

An obvious way to enhance AARON would have been to add more graphical primitives, such as new figures that it could draw. But the next step in AARON's evolution was something more profound. What Cohen did was to enhance the program so that it could "scribble" around "core figures." This idea came from observing the way children scribble on paper, and in particular the key moment when the children seemed to realize that a figure they scribbled actually represented a realistic object in some sense. From AARON's

15.3 Generating Purposeful Images 271

perspective, the main improvement was a two-step strategy, where core figures were placed in a virtual world and then the picture was allowed to evolve by tracing a kind of path around the core figures. This strategy resulted in paintings with much more complexity and a sense of realism, in that the end product showed realistic-looking objects that seemed to be inspired by the real world. Certainly, if a human artist had drawn these shapes, they would have been considered to be reflections of the real world.

From this point on, AARON became more and more *representational* in the works it produced. This came from three main sources of improvement. The first was a database of core figures that combined to represent visually some objects from the real world. Each core figure was represented as a list of key points that provided essentially an outline, and the figures were connected in space and orientation. For example, a plant could be represented by a trunk, some branches, and some leaves. Each of those may be a core figure, and the figures would be related to each other geometrically, which is to say that the branches are connected to the trunk and the leaves to the branches. This basic strategy worked to model even complex objects, such as a basic human form and a recognizable image of the Statue of Liberty.

The second improvement was a series of algorithms that could create reasonable objects made up of core figures. It is not feasible to model all possible objects using core figures, and so Cohen wrote some algorithms that allowed AARON to imagine plants, or as he called them *quasi-plants*. It probably won't surprise you to learn that these algorithms made extensive use of pseudo-random numbers. At this point, AARON's paintings featured recognizable human shapes in an environment that included many plant-like shapes.

Finally, the third improvement was the development of a set of rules that let AARON render an image from a world described by core figures. Essentially, these rules amount to expertise in drawing. For example, the core figures occupy a space in three dimensions, and the rules determine how AARON is to deal with geometric problems such as perspective and occlusion. That is, figures farther away are smaller in the final image, and a core figure near the front may partially hide a figure that is farther away. AARON is truly becoming an expert painter, which is one aspect of being an artist.

AARON's paintings from this time represent a high point in its work, and they have been exhibited in museums. Nevertheless, there was a hidden limitation, in that the core figures used to build AARON's model of the world were two-dimensional. The figures could be placed in a three-dimensional world, but the realism in AARON's work was partially due to the limitations in the scale of its paintings. Cohen wanted to create larger and more complex paintings, and he felt that this increase in scale would require a more detailed model of the real world in AARON's data structures.

So, Cohen embarked on a new modeling phase with more detailed and fully three-dimensional models of the objects that AARON would draw, such as the human body. These models were made of groups of three-dimensional points that were related to other similar groups in a way that is clearly derived from the earlier core figures. In a tradition dating back to da Vinci and Michelangelo, the models for the human body were taken directly from anatomical studies. This shift created many complications because AARON's model was now three-dimensional, so it had to create two-dimensional projections of core images before rendering a painting.

The complications are significant, and you might find it interesting to pursue it further by reading Cohen's own description of his work or Pamela McCorduck's very readable account of AARON's evolution as an artist—even to the point of mixing its own pigments and painting its own artworks on a physical canvas.

What we want to leave you with is an appreciation that simple principles can indeed lead to complex behavior. AARON's knowledge of the real world is encoded as a set of points, and its expertise in drawing is encoded as a set of rules. This is analogous to the way we used equations and logic to write programs, create models of digital circuits, and reason about properties of those abstractions. AARON is an impressive program, and it can be plausibly argued that it exhibits artistic creativity. But at its core, it is composed of simple principles that, taken as a whole, capture a way of painting.

Index

AARON software, 270
above the line (*see* proof)
absorption, 25–27
abstraction, 25
absurdity, equation, 30
ACL2, 112, 113
  adder-ok, 146
  admit, 107, 128
  circuit model, 141, 142, 144, 145
  equation, 93
  helping, 107
ACL2 book (*see* book)
ACL2r, 13
Adams, Douglas, 194
add, add-1, add-c (*see* bignum), 159, 160
adder (*see* circuit)
adder-ok theorem (*see* theorem)
adder2 (*see* circuit)
adder2 theorem (*see* theorem)
addition
  bignum, 157, 159, 160
  binary numeral, 140–146, 149, 160
  carry, 139–142, 144, 145, 160
  circuit, 141, 142, 144, 145
  decimal numeral, 140
additive law of concatenation, 90
Adelson-Velskii, Georgy, 203
admit, ACL2, 107, 128
Agda, 112
algebra, Boolean (*see* Boolean, algebra)
algorithm analysis (*see* computation steps)
Amazon, 244, 257
and (logic gate), 34

and-gate (*see* operator)
Apache Foundation, 252
append
  additive law, 90
  associative, 104
  operator, 73, 86, 93
  prefix theorem, 98, 106
  suffix theorem, 97, 103, 104
argument, 6
arithmetic
  bignum addition, 159, 160
  bignum multiplication, 164, 165
  binary numeral, 140–142, 144–146, 149
  clock, 74, 126
  decimal numeral, 140
  modular, 74, 126
  negative numeral, 151, 152, 154
  paper and pencil, 139
array
  search, 203, 228, 230
associative, 29, 104
Assume (natural deduction), 39–42, 44, 45, 47, 48, 50
discharge, 23, 39, 43–45, 47, 50
atomic formula, 27
AVL tree, 203
  balance, 205, 209, 210, 214, 215
  balance factor, 223
  build, 224
  dat operator, 207
  diagram, 206
  empty, 206
  height definition, 205

height operator, 206, 207, 210
hook (insert key), 211
ht (height, fast), 223, 224
ht (height, fast) test, 225
ins (insert key) examples, 211
ins (insert key) operator, 222
iskeyp predicate, 207
key operator, 207
keyp predicate, 207
lft operator, 207
mktr operator, 207
mktr, fast, 223
node, 206
order, 205, 208
ordp predicate, 208
rebalance, 216, 218–221
representation, 206
representation, formal, 207
rgt operator, 207
root, 206
rotate, 221
rotate, double, 221
rotate, fast, 223
size operator, 207
subtree, 206
subtree, extract, 207
treep predicate, 207
treep predicate, fast, 223
unbalanced, 209
unique keys theorem, 208
zig, zag operators, 218
axiom
  append, 86, 93
  associative, 19, 29
  Boolean algebra, 19, 29
  commutative, 19, 29
  DeMorgan, 19, 29
  distributive, 29
  exponent, 92
  Fibonacci, 108
  grammar, 29
  idempotent, 19
  if-true, if-false, 85
  member-equal, 99
  numeric algebra, 16
  parentheses, 19, 29
  software, 93

axiom, by name
  {2s+}, {2s−}, 151
  {∨ associative}, 19, 29
  {∨ commutative}, 19, 29
  {∨ deMorgan}, 19, 29
  {∨ distributive}, 19, 29
  {∨ idempotent}, 19, 29
  {∨ identity}, 19, 29
  {∨ null}, 19, 29
  {add01}, {add10}, 160
  {add0y}, {addx0}, {addxy}, 160
  {addc0}, {addc1}, 160
  {app0}, {app1}, 86, 93
  {atomic release}, 19, 29
  {bits0}, {bits1}, 135
  {cap0}, {cap1}, {cap2}, 228
  {cons}, 82
  {consp}, 81
  {d1}, {d2} (steps in dmx), 189
  {dgts0}, {dgts1}, 128
  {dmx0}, {dmx1}, {dmx2}, 173, 189
  {dmx2-0x}, {dmx2-1x}, 175
  {double negation}, 19, 29
  {expt0}, {expt1}, 92
  {f0}, {f1}, {f2} (Fibonacci), 108
  {fcap}, 230
  {fib1}, {fib2}, 108
  {fin1}, {fin2}, 136
  {fst}, {fst0}, 82
  {gib0}, {gib1}, {gib2} (Fibonacci), 110
  {hook0}, {hook<}, {hook>}, {hook=}, 211
  {idempotent}, 29
  {identity}, 29
  {if-true}, {if-false}, 85
  {implication}, 19, 29
  {ins0}, {ins1}, {ins2}, 178, 179
  {isrt0}, {isrt1}, {isrt2}, 179
  {look0}, {look1}, {look2}, 230
  {lst}, 102
  {mem0}, {mem1}, 99
  {mg0}, {mg1}, {mgx}, {mgy}, 182, 183
  {msrt0}, {msrt1}, {msrt2}, 184
  {mul0y}, {mul0xy}, {mul1xy}, 164
  {mulx0}, {mulxy}, 164
  {mux0x}, {mux0y}, {mux1y}, {mux11}, 170
  {mux2-0x}, {mux2-1x}, 172
  {n10.0}, {n10.1}, 130

Index 275

{natp0}, {natp1}, 91
{natp}, 75
{nlst}, 81
{nth0}, {nth1}, 229
{numb0}, {2numb}, {2numb+1}, 135
{ord}, 208
{p0}, {p1}, 116
{pad+}, {pad−}, 136
{pfx0}, {pfx1}, 98
{r0}, {r1}, 267
{redundant grouping}, 29
{rep0}, {rep1}, 99
{rg0}, {rg1}, 269
{rs}, 268
{rst}, {rst0}, 82
{s1}, {s2} (steps in msort), 194
{self-implication}, 19, 29
{sfx0}, {sfx1}, 95
{snd}, 142

baffled?, 16, 37
balance
　after AVL insert, 212
　AVL rebalance, 216, 218–221
　AVL tree, 205, 209, 210, 214, 215
balance factor (AVL), 223
balp, AVL tree (*see* predicate)
base case (*see* induction)
below the line (*see* proof)
bignum
　addition (add, add-1, add-c), 157, 159, 160
　multiplication (mul, mxy), 164, 165
binary numeral, 135
　addition, 140–146, 149, 160
　multiplication, 164
binary search, 202, 204
binary tree, 204
　AVL, 203
　height, 210
bit (binary digit), 134
bits (*see* operator)
black-and-white, 264
book
　ACL2, 104
　arithmetic-3/floor-mod, 128
　arithmetic-3/top, 104, 174

certified, 104
　directory (:dir), 71, 72, 104, 128, 174
　doublecheck, 72
　testing, 71, 72
Boolean
　algebra, 19, 29
　axiom, 19
　circuit, 34
　equivalence (↔, iff), 32, 171, 180, 185
　formula, 28
　gate, 33, 34
　grammar, 28
　model, 68
　signals, 33
　variable, 63
bound variable, 52, 54
　capture, 58
Boyer, Robert, 112
brackets
　ceiling, 77, 202
　floor, 77
　round, 79
　square, 79, 80, 84
Bresenham, Jack, 266
bucket, hash, 232

C++, 113
cache, 245
capital (*see* operator)
capitals of states, 230
capture, variable, 58
carry, addition, 139
Cassandra, 247, 249
ceiling (*see* operator)
ceiling brackets, 77, 202
Celsius vs Fahrenheit, 188
certified book (*see* book)
check-expect test, 71
chess, 12
Church, Alonzo, 113
Church–Turing hypothesis, 114
circuit
　ACL2 model, 141, 142, 144, 145, 148
　adder (ripple-carry), 145
　adder2, 144
　addition, 141, 142, 144, 145

Boolean, 34
  diagram, 34, 36
  full-adder, 142
  half-adder, 141
  numeric order, 153
  subtraction, 153
  two-bit adder, 144
  two's-complement negation, 153
  wiring, 34
circular definition (*see* definition)
citation
  equation, 18
  inference rule, 38, 40
clock arithmetic, 74, 126
closed-form formula, 190
clusters, finding, 258
coefficient, polynomial, 130
Cohen, Harold, 270
color, 264
  eight, 265
  intensity, 265
  pixel, 264
common notions, 18
commutative, 18, 19, 29, 30, 37, 42, 45, 49
compare
  ACL2 values (equal), 86
  circuit, numeric, 153
  numbers ($<, <=, =, >=, >$), 72, 75, 76, 86, 180, 196
computation models, 5
computation steps
  ACL2, 186–188
  counting, 186–188
  dmx operation, 189, 190
  insertion-sort (isort), 198
  merge-sort (msort), 195
  mrg (merge) operation, 191, 192
  msort vs isort, 197
  one-step operators, 187
computation time, 186
  AVL build, 224
computer word, 150
concatenate, 73, 86
  additive law, 90
conclusion, 38, 45
  implication ($\rightarrow$), 23
  theorem, 37

cons, 189
cons (*see* operator)
consp (*see* predicate)
contradiction
  inference rule, 39
  proof by, 48
contrapositive, 30
copy-and-paste programming, 25
Coq, 112
Coquand, Thierry, 112
core figures, 271
correctness property, 98

dat (AVL) (*see* operator)
data mining, 256
data, random test, 72–76, 83, 86, 102, 127
database, 243
deduction
  natural, 38, 39
  proof by, 37
  rule of thumb, 49
deductive proof, 39
deductive reasoning, 37, 38
Deep Blue, 9, 12
definition
  defun, 92
  defun hint, 172, 183
  equation ($\equiv$), 53, 80
  inductive (circular), 10, 91, 128
  operator, 92
  property, 72–76, 83, 86, 102, 127
  tail recursive, 109
  theorem, ACL2, 103
  why inductive?, 92
defproperty, 72–76, 83, 86, 102, 127
defthm, 103
defun, 92
  induction hint, 172, 183, 184
  lemma hint, 184
del (delete list element) (*see* operator)
DeMorgan, 30, 49
derivation, 19
dgts (*see* operator)
diagram
  AVL tree, 206
  circuit, 34, 36

Index

full-adder circuit, 142
half-adder circuit, 141
ripple-carry circuit, 145
two-bit adder circuit, 144
digit, 124
digital gate (*see* gate)
digital geometry, 266
directive
  include-book, 72, 104, 128, 174
  induction hint, 172, 183, 184
  lemma hint, 184
directory (:dir)
  :system, 104, 128, 174
  :teachpacks, 71, 72
discharge assumption, 23, 39, 43–45, 47, 50
disheartened?, 16, 37
disjunctive syllogism, 50
distributive, 18, 19, 29, 30
dividend, 74
division
  floor (round down), 126
  floor, mod, 74, 77, 128, 129
  long division, 74, 126
  mod (remainder), 74, 126
  third grade, 74, 126
divisor, 74
dmx theorems (*see* theorem)
dmx, dmx2 (demultiplexer) (*see* operator)
double rotation, AVL, 221
DoubleCheck, 72
Dr. Seuss, 27
duplicate key, none (AVL), 208

elimination rule (*see* inference rule)
empty list (nil), 79, 80
empty tree, 204, 206
emptyp (AVL) (*see* predicate)
engineering, 37
equal (*see* predicate)
equal, three-line ($\equiv$), 53, 80
equal, vs =, 86
equation
  absorption, 26, 27, 30
  ACL2, 93
  append, 86
  as axiom, 19

as theorem, 19
associative, 19, 29, 30
Boolean algebra, 19, 29
citation, 18
commutative, 19, 29, 30
defining ($\equiv$), 53, 80
DeMorgan, 19, 29, 30
distributive, 19, 29, 30
excluded middle, 21
grammar, 19, 29, 31
matching, 26, 31
numeric algebra, 16
parentheses, 19, 28, 29
proof by, 18, 20–22, 26
quantifier, 58
software, 91, 93
equation, by name
$\{2s+\}, \{2s-\}$, 151
$\{\exists\vee\}$, 58
$\{\exists\wedge\}$, 58
$\{\exists\rightarrow\}$, 58
$\{\rightarrow\exists\}$, 58
$\{R\exists\}$, 58
$\{\forall\vee\}$, 58
$\{\forall\wedge\}$, 58
$\{\forall\rightarrow\}$, 58
$\{\rightarrow\forall\}$, 58
$\{R\forall\}$, 58
$\{+ \text{ associative}\}$, 16
$\{+ \text{ complement}\}$, 16
$\{+ \text{ identity}\}$, 16
$\{+ \text{ commutative}\}$, 16
$\{\times \text{ commutative}\}$, 16
$\{\times \text{ identity}\}$, 16
$\{\times \text{ null}\}$, 16
$\{\neg \text{ as } \rightarrow\}$, 30
$\{\neg False\}$, 30
$\{\neg True\}$, 30
$\{\vee \text{ absorption}\}$, 27, 30
$\{\vee \text{ associative}\}$, 19, 29
$\{\vee \text{ commutative}\}$, 19, 29
$\{\vee \text{ complement}\}$, 21, 30
$\{\vee \text{ deMorgan}\}$, 19, 29
$\{\vee \text{ distributive}\}$, 19, 29
$\{\vee \text{ idempotent}\}$, 19, 29
$\{\vee \text{ identity}\}$, 19, 29
$\{\vee \text{ null}\}$, 19, 29

{∧ absorption}, 26, 27
{∧ associative}, 30
{∧ commutative}, 30
{∧ complement}, 30
{∧ deMorgan}, 30
{∧ distributive}, 30
{∧ idempotent}, 30
{∧ identity}, 30
{∧ implication}, 30
{∧ null}, 25
{→ identity}, 30
{absurdity}, 30
{add01}, {add10}, 160
{add0}, {add1}, {add.bit0}, {add.bits}, 145
{add0y}, {addx0}, {addxy}, 160
{add10}, {add11}, 159
{addc0}, {addc1}, 160
{app0}, {app1}, 86, 93
{atomic release}, 29
{bits0}, {bits1}, 135
{cap0}, {cap1}, {cap2}, 228
{cons}, 82
{contradiction}, 30
{contrapositive}, 30
{currying}, 30
{d1}, {d2} (steps in dmx), 189
{dgts0}, {dgts1}, 128
{dmx0}, {dmx1}, {dmx2}, 173, 189
{dmx2-0x}, {dmx2-1x}, 175
{double negation}, 19, 29
{expt0}, {expt1}, 92
{f0}, {f1}, {f2} (Fibonacci), 108
{fcap}, 230
{fib1}, {fib2}, 108
{fin1}, {fin2}, 136
{fst}, {fst0}, 82
{gib0}, {gib1}, {gib2} (Fibonacci), 110
{h0}, {h1}, 10
{hook0}, {hook<}, {hook>}, {hook=}, 211
{ht0}, {ht1}, 207
{idempotent}, 19, 29, 30
{identity}, 29
{if-true}, {if-false}, 85
{implication}, 19, 29
{ins0}, {ins1}, {ins2}, 178, 179, 196
{isrt0}, {isrt1}, {isrt2}, 179, 196
{iter0}, {iter1}, 119

{look0}, {look1}, {look2}, 230
{mem0}, {mem1}, 99
{member-equal}, 99
{mg0}, {mg1}, {mgx}, {mgy}, 182, 183
{msrt0}, {msrt1}, {msrt2}, 184
{mul0y}, {mul0xy}, {mul1xy}, 164
{mulx0}, {mulxy}, 164
{mux0x}, {mux0y}, {mux1y}, {mux11}, 170
{mux2-0x}, {mux2-1x}, 172
{n10.0}, {n10.1}, 130
{natp0}, {natp1}, 91
{natp}, 75
{nmb1}, 136
{nth0}, {nth1}, 229
{numb0}, {2numb}, {2numb+1}, 135
{ord}, 208
{p0}, {p1}, 116
{pad+}, {pad−}, 136
{pfx0}, {pfx1}, 98
{r0}, {r1}, 267
{redundant grouping}, 29
{rep0}, {rep1}, 99
{rg0}, {rg1}, 269
{rs}, 268
{rst}, {rst0}, 82
{s1}, {s2} (steps in msort), 194
{self-implication}, 19, 29
{sfx0}, {sfx1}, 95
{snd}, 82
{sz0}, {sz1}, 207
equivalence
  Boolean (↔, iff), 32, 171, 180, 185
  by definition (≡), 53, 80
  of operators, 177
Euclid, 18
excluded middle, 21
exclusive or, 27, 32, 34
existential quantifier (∃), 52
  empty universe, 52
  negation, 59
exponents, law of, 111
expt (*see* operator)
extensional, 51
eye, human, 264

F-from-C (*see* operator)

Index 279

Facebook, 243
Fahrenheit vs Celsius, 188
fcapital (*see* operator)
feasibility, 23
fib, fib-fast (*see* operator)
Fibonacci
  definition, 109
  fast, 110
  numbers, 108
  operator, 108, 110
filter, image, 265
fin (*see* operator)
first (*see* operator)
floor (*see* operator)
floor brackets, 77
forall (∀), 51
  empty universe, 52
  negation, 59
formalism, 16, 18, 38
formula
  atomic, 27
  closed form, 190
  grammar, 28, 29, 31
  let∗, 75
  logic, 34
  matching, 26, 31
  prefix notation, 71
  substitution, 15
  variables, 27, 43
free variable, 52, 54
  capture, 58
frustrated?, 16, 37
full-adder (*see* circuit)
full-adder theorem (*see* theorem)
function, 6
functionally complete, 33
funk, in a?, 16, 37

game, rock-paper-scissors, 7
gate
  and, 34
  functionally complete, 33
  inverter, 34, 35
  logic, 33–35
  nand, 34, 35
  negation, 34

nor, 34
not, 34
or, 34
universal, 33, 35, 36
xnor, 34
xor, 34
Gentzen, Gerhard, 38
geometry, digital, 266
gib (*see also* operator), 110
  lemmas (base, inductive), 110
Gödel, Kurt, 5
Gordon, Mike, 112
grammar (*see* Boolean grammar, defun,
    defproperty, defthm, let∗)
greater or equal (>=) (*see* predicate)
greater than (>) (*see* predicate)
grid, image, 263

Hadoop, 252
half-adder (*see* circuit)
halting problem, 113–115, 117, 118
hardware, 3
harmonic series, 10, 13
Harper, Robert, 112
hash, 227
  base, 232
  bucket, 232
  key, 232
  operator, 232
  perfect, 234
  search, 232
  table, 232
height
  AVL tree, 210
  maximum change in, 212
  test ht operator, 225
  tree, 205–207
height (AVL) (*see* operator)
HOL, 112
Honsell, Furio, 112
hook (AVL insertion) (*see* operator)
horizontal scaling, 251, 261
Horner, 135
Horner's rule, 130
ht (AVL height, fast) (*see* operator)
Huet, Gérard, 112

hypothesis, 38, 44
  Assume (natural deduction), 39, 42, 44, 45, 47, 48, 50
  Church–Turing, 114
  discharge, 23, 39, 43–45, 47, 50
  implication (→), 23
  induction, 89, 133, 134
  theorem, 37

I (isort recurrence), 198
idempotent, 29
identity, 39
if (*see* operator)
iff (*see also* operator), 171, 180, 185
image, 263
  black-and-white, 264
  blur, 265
  color, 264
  line, 266
image filter, 265
implication (→), 23
  conclusion, 23
  hypothesis, 23
  introduction, 45
import (*see* book)
include-book (*see* directive)
incompleteness, 5
incomputable (*see* uncomputable)
index, by name
  {add10}, {add11}, 159
induction
  base case, 89
  double, 193
  hypothesis, 89, 133, 134
  inductive case, 89
  inference rule, 89, 133, 134
  proof by, 88, 89, 132–134
  rationale, 88, 132
  strong, 132–134
inductive case (*see* induction)
inductive definition (*see* definition)
inductive, why?, 92
infeasibility, 23, 68
inference rule
  citation, 38, 40
  discharge assumption, 39, 43–45, 47, 50

  matching, 44
  scope of citation, 40
  table of, 39
  theorem as, 46
inference rule, by name
  {¬ elimination}, 39, 47, 48, 50
  {¬ introduction}, 39, 47
  {→ introduction}, 39, 44, 45, 47, 50
  {∨ elimination}, 39, 45, 50
  {∨ introduction 1}, 39, 45
  {∨ introduction 2}, 39, 45
  {∧ elimination 1}, 39, 42
  {∧ elimination 2}, 39, 42
  {∧ introduction}, 39, 42
  {contradiction}, 39, 50
  {identity}, 39, 44, 50
  {induction}, 89
  {modus ponens}, 39, 40, 47, 50
  {reductio ad absurdum}, 39, 48
  {strong induction}, 133
infix notation, 71
ins (AVL insert) (*see* operator)
insert, in order (*see* operator)
insertion-sort, 179, 196
insertion-sort (isort), 196
inside cases
  AVL rotate, 218, 219
  double rotate, 221
instruction set, 3
intensional, 51
introduction rule (*see* inference rule)
inverter, gate, 34, 35
invoke (invocation), 10, 79
Isabelle, 112
iskeyp (AVL) (*see* predicate)
isort (*see* operator)
iteration, 119

Java, 113
jostle, 270

Kansas, 230
Kaufmann, Matt, 112
key
  AVL tree, 206
  duplicates, none, 208

Index 281

unique (theorem), 208
key (AVL) (*see* operator)
key/value pair, 252
keyp (AVL) (*see* predicate)

LabView, 4
lambda calculus, 5
Landis, Evgenii, 203
law of the excluded middle, 21
layering, 264
LCF, 112
leading zero, 125, 136
left subtree (AVL), 206
Leibniz, Gottfried, 13
lemma, 184
len (*see* operator)
less or equal (<=) (*see* predicate)
less than (<) (*see* predicate)
let∗, local name, 75
letterbox array, 230
LF, 112
lft (AVL) (*see* operator)
line
  draw, 266
  ragged, 268
line (natural deduction)
  above/below, 40
  dashed, 41
linear congruential, 267
list, 73, 79
  cons (*see also* operator), 79, 82
  ellipsis, 81
  first (*see also* operator), 82
  for numeral, 125
  nonempty, 80
  numbered, 81
  operator (*see also* operator), 73
  properties, 83
  rest (*see also* operator), 82
  square bracket notation, 79, 80, 84
  true list, 105
local names (let∗), 75
log2-ceiling (*see* operator)
logarithm, 202
logic
  formula, 34

gate, 33, 35
in action, 5, 35
mechanized, 101, 112
real world, 63
signal (0 & 1), 33
symbolic, 15
logic operator (*see* operator, logic)
long division (*see* division)
lookup (*see* operator)
loop (*see* operator)

MapReduce, 252, 261
matching
  equation, 26, 31
  inference rule, 44
mathematical induction (*see* induction)
matrix, 263
max (*see* operator)
McCorduck, Pamela, 272
mechanized logic, 101
member-equal (*see* predicate)
memristor, 33
merge, ordered (mrg), 183
merge-sort, 184
metavariable, 16, 43
Milner, Robert, 112
mining, data, 256
MIT Media Lab, 4
mktr (AVL) (*see* operator)
mod (*see* operator)
model, 63
  Boolean, 67, 68
  of circuit, 148
  of computation, 5
  predicate, 70
  propositional, 69
modular arithmetic, 74, 126
modulus (modular arithmetic), 74
modus ponens, 39–41, 47
modus tollens, 47
Moore, J Strother, 104, 112
mrg (ordered merge) (*see* operator)
msort (merge-sort) (*see* operator)
mul, mxy (*see* bignum), 164
multiplication
  bignum, 164, 165

binary numerals, 164
mux theorems (*see* theorem)
mux, mux2 (multiplexer) (*see* operator)

name, local (let∗), 75
nand
  commutes, 49
  is all you need, 36
  logic gate, 33–36
Natarajan, Shankar, 112
natp (*see* predicate)
natural deduction, 38, 39, 49
  tree diagram, 41
  where to start, 49
natural number, 73
negation (not, logic gate), 34, 35
negative × negative, 18
negative numeral, 150–152, 154
nested, vs top level, 109
Newton, Isaac, 13
nil, 79, 80
Nim, 63
nmb10 (*see* operator)
node, in AVL tree, 206
nonstandard analysis, 13
nor
  commutes, 49
  elimination, 49
  logic gate, 33
Norell, Ulf, 112
not (logic gate), 34, 35
nth (*see* operator)
nthcdr (*see* operator)
numb (*see* operator)
number
  comparison circuit, 153
  from digits, 130
  from numeral, 130
  negative, 150–152, 154
  order (<, <=, =, >=, >), 72, 75, 76, 86, 180, 196
  vs numeral, 123
numbered list notation, 81
numeral
  addition, 140
  bignum addition, 160

binary, 135
binary addition, 140–146, 149, 160
binary multiplication, 164
decimal, 123
from number, 128
hexadecimal, 123
leading zero, 125, 136
length, 136
list for, 125
list representation, 135
negative, 150–152, 154
Roman, 123
sequence for, 125
subtraction, 153
two's-complement, 150–152, 154
vs number, 123

occurs in search tree, 204
  keyp (*see* predicate), 207
occurs-in list (*see* predicate)
ones & zeros, 33, 134, 146
operand, 6
operator, 6
  defun, 92
  functionally complete, 33
  infix notation, 71
  numeric order (<, <=, =, >=, >), 72, 75, 76, 86, 180, 196
  prefix notation, 71
operator, by name
  add, add-1, add-c (*see* bignum), 159
  adder, 145
  adder2, 144
  and-gate, 141
  append, 73, 86, 93
  balp (predicate), 210
  bits (numeral from number), 135
  capital, 228
  ceiling (divide, round up), 77
  cons (insert at front), 79, 82
  consp (*see* predicate), 81
  dat (AVL, root data), 207
  del (delete list element), 181
  dgts (digits from number), 125, 128
  dmx (demultiplexer), 173, 189
  dmx2 (demultiplexer), 175

emptyp (AVL, *see* predicate), 207
equal (*see* predicate), 86
expt (x^n), 92
F-from-C, 188
fcapital, 230
fib, fib-fast (Fibonacci), 108, 110
fin (extract last element), 136
first (extract first element), 82
floor (divide, round down), 77, 126
full-adder, 142
gib (iterative Fibonacci), 110
half-adder, 141
hash-op, hash-key, hash-idx, 237
height (AVL tree), 207
hook (AVL insert key), 211
ht (AVL height, fast), 223, 224
if (select formula), 85
iff (Boolean equivalence), 180, 185
ins (AVL insert key), 222
insert (in order), 179, 196
iskeyp (AVL, *see* predicate), 207
isort (insertion-sort), 179, 196
key (AVL, root key), 207
keyp (AVL, *see* predicate), 207
len (length of list), 73
lft (AVL, left subtree), 207
list, 73
log2-ceiling, 225
lookup, 230
loop, 116, 119
max (of two operands), 76
member-equal (*see* predicate), 99
mktr (AVL, make tree), 207
mktr, fast, 223
mod (remainder), 73, 74, 126
mrg (ordered merge), 183
msort (merge-sort), 184
mul, mxy (*see* bignum), 164
mux (multiplexer), 169, 170, 172
mux2 (multiplexer), 172
nmb10 (number from digits), 130
nth, 229
nthcdr (suffix of list), 95
numb (number from bits), 135
occurs-in (*see* predicate), 171, 173
or-gate, 141
p (paradox), 116

pad (append padding), 136
permp (*see* predicate), 181
posp (*see* predicate), 94
prefix (of list), 98
r (pseudo-random), 267
ragged (line), 269
rep (list of duplicates), 99
rest (drop first element), 82
reverse (elements in list), 77
rgt (AVL, right subtree), 207
rot+, rot- (AVL rotate), 221–224
    fast, 223
rs (pseudo-random sequence), 268
second (extract second element), 142
size (AVL tree), 207
state-idx, 230
swap2, 188
treep (AVL, *see* predicate), 207
    fast (*see* predicate), 223
twos (two's-complment), 151
up (*see* predicate), 180
xor-gate, 141
zig, zag (AVL rotate), 218
zp (*see* predicate), 94
operator, logic
    and ($\wedge$), 19, 29, 34
    Boolean equivalence ($\leftrightarrow$, iff), 32, 171, 180, 185
    exclusive nor, 34
    exclusive or, 34
    functionally complete, 33
    implication ($\rightarrow$), 19, 29
    nand, 34, 35
    nor, 34
    not ($\neg$), 19, 29, 34
    or ($\vee$), 19, 29, 34
operators, equivalence of, 177
or (logic gate), 34
or, exclusive, 27, 32, 34
or-gate (*see* operator)
orange, 265
order
    AVL insert, 212
    AVL tree, 205, 208
    numeric (<, <=, =, >=, >), 72, 75, 76, 86, 180, 196
    total, 204

ordp (AVL tree) (*see* predicate)
outside cases, AVL rotate, 218
Owre, Sam, 112

p (paradox) (*see* operator)
pad (*see* operator)
painting, 266
panic, don't, 16, 37
paradox, halting problem, 115
parameter, 6
parentheses, 28, 29, 79
Paulson, Lawrence, 112
perfect hash, 234
perfect shuffle, 169
permp (*see* predicate)
Pfenning, Frank, 112
picture, 263
pile of socks, 202
pixel, 263
  color, 264
plants, quasi, 271
Plotkin, Gordon, 112
posp (*see* predicate)
Prawitz, Dag, 38
predicate, 51, 53
  multi-index, 53
  numeric order (<, <=, =, >=, >), 72, 75, 76, 86, 180, 196
  true-listp, 105
  universe of discourse, 51
  vs proposition, 51
predicate, by name
  balp (AVL balance), 210
  consp, 81
  consp (nonempty list), 81
  emptyp (AVL tree), 207
  equal, 86
  H (termination), 114
  iskeyp (AVL tree), 207
  keyp (AVL, occurs in), 207
  member-equal (element of list), 99
  natp, 75
  occurs-in (list), 171, 173
  ordp (AVL tree), 208
  permp (permutation), 181
  posp (positive integer), 94

treep (AVL), 207
treep, fast (AVL), 223
up (increasing order), 180
zp (natural number zero), 94
prefix notation, 71
prefix, of list (*see* operator)
programming language, 113
programs, as axioms, 93
proof, 19, 38, 45
  above/below the line, 40
  Assume (natural deduction), 39, 42, 44, 45, 47, 48, 50
  assumption, 41
  by contradiction, 48
  by induction, 89, 132–134
  dashed line, 41
  deductive, 39, 40, 42, 46, 49
  discharge assumption, 43, 44
  reductio ad absurdum, 48
  tree diagram, 41
  with equations, 18, 20–22, 25, 26
proof bar (Proof Pad), 103
Proof Pad, 71
property
  correctness, 98
  defproperty, 72
  doublecheck, 72
  of lists, 83
  round-trip, 174
proposition, 51
  vs predicate, 51
pseudo-random, 267
PVS, 112

quantifier, 52
  $\exists$, there exists, 52
  $\forall$, forall, 51
  empty universe, 52
  equation, 58
  existential ($\exists$), 52
  negation, 59
  reasoning with, 56, 58
  universal ($\forall$), 51, 59
quasi-plants, 271
query, 243
quote mark, single, 84

Index 285

quotient, 74

r (pseudo-random) (*see* operator)
ragged (line) (*see* operator)
ragged line, 268
random, 266
 number generator, 267
 pseudo, 267
random data, 72–76, 83, 86, 102, 127
reasoning
 about tail recursion, 109
 deductive, 37, 38
 with equations, 15, 16, 18, 19, 58
 with quantifiers, 56, 58
reciprocal, 10
record, database, 246
recurrence equations
 deriving, 199
 dmx, 189
 isort (insertion-sort), 198
 merge-sort (msort), 193, 194
 mrg (merge) operation, 192
 solving, 189, 199
recursive, 10, 109, 260
red-green-blue, 264, 265
reduce, map, 252, 254, 261
reductio ad absurdum, 39, 48
remainder, 74
rep (*see* operator)
replication, database, 248
representation
 AVL tree, 206, 207
 binary numeral, 135
 decimal numeral, 125, 128
rest (*see* operator)
retrieval, database, 245
reverse (*see* operator)
RGB, 264
rgt (AVL) (*see* operator)
right subtree (AVL), 206
ripple-carry adder (*see* circuit)
ripple-carry theorem (*see* theorem)
Robinson, Abraham, 13
rock-paper-scissors, 7
root node, AVL tree, 206
rot+, rot- (*see* operator)

rotate AVL
 fast, 223
 formal, 221
 inside cases, 218, 219
 inside, double, 221
 outside cases, 218
 unbalance, 220
 zig, zag, 218
round-trip property, 174
rs (pseudo-random sequence) (*see* operator)
rule of inference (*see* inference rule)
rule of thumb
 natural deduction, 49
Rushby, John, 112

S recurrence (steps in msort), 194
scaling
 horizontal, 251, 261
 vertical, 251, 261
scared?, 16, 37
Schürmann, Carsten, 112
scope
 inference rule citation, 40
 let∗ name, 75
Scratch, 4
search
 array, 203, 228, 230
 binary, 202
 binary vs linear, 203
 hash, 232
search tree, 204
 balance, 205, 209, 210, 212, 214, 215
 building, 204
 diagram, 206
 empty, 204, 206
 height, 205–207
 key occurs in, 204
 node, 206
 order, 205, 208
 ordered, 204
 rebalance, 216, 218–221
 representation, 206, 207
 root, 206
 sibling, 204
 subtree, 206
second (*see* operator)

sequence, 73, 79
  for numeral, 125
  pseudo-random, 267
series, harmonic, 10, 13
set, 51
Seuss, Dr., 27
sharding, 246
shuffle, perfect, 169
sibling, in tree, 204
signals, Boolean, 33
single-quote mark, 84
size (AVL) (*see* operator)
Skolemization, 56, 58, 62
social media, 243
socks, pile of, 202
Socrates syllogism, 40
software, 3, 5
  as axioms, 93
  as equations, 91
sorting
  bubble sort vs quicksort, 177
  insert (in order), 179, 196
  insertion-sort (isort), 179, 180, 198
  merge, ordered (mrg), 183
  merge-sort (msort), 184, 192
  msort theorems, 185
  msort vs isort, 186, 197
square brackets, 79, 80, 84
state capitals, 230
state-idx (*see* operator)
status update, database, 244
stock number search, 201
store front, 244
string, 231
strong induction, 132–134
struggling?, 37
substitution, 15
subtraction, binary numeral, 153
subtree
  AVL tree, 206
subtree, definition, 204
swap2 (*see* operator)
syllogism
  disjunctive, 50
  Socrates, 40
symbol, for number, 124
symbolic logic, 15

syntax (*see* Boolean grammar, defun, defproperty, defthm, let∗)
system, :dir, 104, 128, 174

tail recursion, 109, 260
  reasoning about, 109
teachpacks
  :dir, 71, 72
  doublecheck, 72
  testing, 71, 72
termination predicate (H), 114
testing
  check-expect, 71
  doublecheck, 72
  teachpack, 71, 72
  with random data, 72, 74
theorem, 19, 38
  algebra, ACL2, 104, 128
  append associative, 104
  append-prefix, 106
  append-suffix, 103, 104
  as inference rule, 46
  bignum addition, 160
  bignum multiplication, 165
  constraints, 106
  defthm, 103
  DeMorgan, 49
  Fibonacci fast, 110
  full-adder, 143
  gib lemmas (base, inductive), 110
  halting problem, 114, 115
  Horner, 130, 135
  implication, constraint, 106
  import (*see also* book), 104
  insertion-sort (isort), 180
  merge, ordered, 183
  merge-sort (msort), 185
  paradox, 115, 118
  ripple-carry adder, 146, 149
  Socrates, 40
theorem, by name
  $\{\neg \text{ as } \rightarrow\}$, 30
  $\{\neg \text{ truth table}\}$, 22
  $\{\neg False\}, \{\neg True\}$, 30
  $\{\neg\neg \text{ forward}\}$, 48
  $\{\rightarrow \text{ chain}\}$, 47

Index

{→ identity}, 30
{→ truth table}, 24
{∨ absorption}, 27, 30
{∨ commutes}, 45
{∨ complement}, 21, 30, 49
{∨ truth table}, 20
{∧ absorption}, 26
{∧ associative}, 30
{∧ commutative}, 30
{∧ commutes}, 37, 42
{∧ complement}, 30, 49
{∧ deMorgan}, 30
{∧ distributive}, 30
{∧ idempotent}, 30
{∧ identity}, 30
{∧ implication}, 30
{∧ null}, 25
{absurdity}, 30
{adder-ok}, 146, 149
{adder2-ok}, 144
{additive law of concatenation}, 90
{app-assoc}, 98, 104
{app-nil}, 99
{append-prefix}, 98, 106
{append-suffix}, 97, 98, 103, 104
{bal-ht2}, AVL balanced if ht ≤ 2, 212
{bignum-add-ok}, 160
{bignum-mul-ok}, 165
{bits-ok}, 135
{consp = ($len > 0$)}, 84
{contradiction}, 30
{contrapositive}, 30
{currying}, 30
{DeMorgan ∨ backward}, 49
{DeMorgan ∨ forward}, 49
{dgts-ok}, 130, 132
{disjunctive syllogism}, 50
{dmx computation steps}, 190
{dmx-inverts-mux}, 174
{dmx-len-first}, {dmx-len-second}, 174
{dmx-length}, 175
{dmx-shortens-list}, 185
{dmx-val}, 175
{expt}, 92
{fib=fib-fast}, 110
{full-adder-ok}, 143
{gib-base-equation}, 110

{gib-inductive-equation}, 110
{H+}, {H−}, 117, 118
{h0}, {h1}, 117
{halting problem}, 114, 115, 118
{hF}, {hT}, 116
{hi-1}, 136
{Horner 2}, 135
{Horner 10}, 130
{ht-emp}, 207
{i-ht}, AVL height after insert, 212
{i-ord}, AVL ordered after insert, 212
{idempotent}, 30
{isort-len}, {isort-ord}, {isort-val}, 180
{keys unique}, 208
{leading-0}, {leading-0s}, 136
{len-bits}, {len-bits≤}, 136
{len-nat}, 84
{len-pad}, 136
{log-bits}, 136
{minus-sign}, 153
{mod-div}, 128
{modus tollens}, 47
{mrg computation steps}, 192
{mrg-length}, {mrg-ord}, {mrg-val}, 183
{msort $n\ log(n)$}, 195
{msort-len}, {msort-ord}, {msort-val}, 185
{mux-inverts-dmx}, 174
{mux-length}, 170, 172
{mux-val}, 171, 173
{nand commutes}, 49
{nmb1}, 136
{nor commutes}, 49
{nor elimination 1}, 49
{p0}, {p1}, 117
{paradox}, 115, 118
{pF}, {pT}, 116
{pfx-mod}, 137
{plus-sign}, 153
{rep-len}, 99, 107
{rst1}, 84
{self-implication}, 44
there exists (∃), 52
  empty universe, 52
  negation, 59
third-grade long division (*see* division)
three C's, 91, 128
three-line equal (≡), 53, 80

time of computation (*see* computation steps)
timing AVL build, 224
top level, vs nested, 109
total ordering, 204
tree
 AVL, 203
 balance, 212, 214, 215
 binary, 204
 definition, 203
 diagram, 206
 empty, 204
 height, 205–207, 210
 ordered, 204
 rebalance, 216, 218–221
 representation, 206, 207
 root, 206
 search, 204
 searching for key, 204
 sibling, 204
treep (AVL) (*see* predicate)
true list, 105
true-listp predicate, 105
truth table, 20, 22–24
Turing, Alan, 113
Turing–Church hypothesis, 114
Turing complete, 113
Turing machine, 5
Turing, Alan, 5
turnstile (⊢), 37
Twelf, 112
two-bit adder circuit, 144
two's-complement
 length, 153
 negation, 152, 154
 negation circuit, 153
 numeral, 150–152, 154
 operator, 151
 sign, 153
 word, 150–152, 154

uncomputable, 5, 113
universal gate, 33, 35, 36
universal quantifier (∀), 51
 empty universe, 52
 negation, 59
universe of discourse, 51

up (increasing order) (*see* predicate)
update, database, 243, 244
URL, 252

value, let∗ name for, 75
variable, 15, 16, 27, 43, 63
 Boolean, 27, 63
 bound, 52, 54
 capture, 58
 free, 52, 54
 let∗, 75
 metavariable, 16
vertical scaling, 251, 261

Web 2.0 company, 244
why inductive?, 92
wiring (*see* circuit, wiring)
word
 computer, 150
 two's-complement, 150, 152, 154
worried?, 16, 37

xnor (logic gate), 34
xor (logic gate), 34
xor-gate (*see* operator)

zeros & ones, 33, 134, 146
zig, zag (AVL rotate) (*see* operator)
zp (*see* predicate)